Systems Research for Real-World Challenges

Frank Stowell
University of Portsmouth, UK

A volume in the Advances in Information Quality and Management (AIQM) Book Series

IGI Global
DISSEMINATOR OF KNOWLEDGE

Published in the United States of America by
IGI Global
Information Science Reference (an imprint of IGI Global)
701 E. Chocolate Avenue
Hershey PA, USA 17033
Tel: 717-533-8845
Fax: 717-533-8661
E-mail: cust@igi-global.com
Web site: http://www.igi-global.com

Library of Congress Cataloging-in-Publication Data

Names: Stowell, Frank A., editor.
Title: Systems research for real-world challenges / Frank Stowell, editor.
Description: Hershey, PA : Information Science Reference, [2019]
Identifiers: LCCN 2017061523| ISBN 9781522559962 (hardcover) | ISBN 9781522559979 (ebook)
Subjects: LCSH: Social sciences--Methodology. | System analysis. | System theory. | Decision making. | Problem solving.
Classification: LCC H61 .S9948 2019 | DDC 300.1/1--dc23 LC record available at https://lccn.loc.gov/2017061523

This book is published in the IGI Global book series Advances in Information Quality and Management (AIQM) (ISSN: 2331-7701; eISSN: 2331-771X)

British Cataloguing in Publication Data
A Cataloguing in Publication record for this book is available from the British Library.

For electronic access to this publication, please contact: eresources@igi-global.com.

Advances in Information Quality and Management (AIQM) Book Series

ISSN:2331-7701
EISSN:2331-771X

Editor-in-Chief: Siddhartha Bhattacharyya, RCC Institute of Information Technology, India

MISSION

Acquiring and managing quality information is essential to an organization's success and profitability. Innovation in information technology provides managers, researchers, and practitioners with the tools and techniques needed to create and adapt new policies, strategies, and solutions for information management.

The **Advances in Information Quality and Management (AIQM) Book Series** provides emerging research principals in knowledge society for the advancement of future technological development. This series aims to increase available research publications and emphasize the global response within the discipline and allow for audiences to benefit from the comprehensive collection of this knowledge.

COVERAGE

- Human and Societal Issue
- Supply Chain Management
- E-Collaboration
- Web services and technologies
- Mobile Commerce
- IT Innovation and Diffusion
- Knowledge Management
- Business Process Management and Modeling
- IT Management in Public Organizations
- Application of IT to Operation

IGI Global is currently accepting manuscripts for publication within this series. To submit a proposal for a volume in this series, please contact our Acquisition Editors at Acquisitions@igi-global.com or visit: http://www.igi-global.com/publish/.

The Advances in Information Quality and Management (AIQM) Book Series (ISSN 2331-7701) is published by IGI Global, 701 E. Chocolate Avenue, Hershey, PA 17033-1240, USA, www.igi-global.com. This series is composed of titles available for purchase individually; each title is edited to be contextually exclusive from any other title within the series. For pricing and ordering information please visit http://www.igi-global.com/book-series/advances-information-quality-management/73809. Postmaster: Send all address changes to above address.

Titles in this Series

For a list of additional titles in this series, please visit:
https://www.igi-global.com/book-series/advances-information-quality-management/73809

Ensuring Research Integrity and the Ethical Management of Data
Cees Th. Smit Sibinga (University of Groningen, The Netherlands)
Information Science Reference • ©2018 • 303pp • H/C (ISBN: 9781522527305) • US $275.00

Handbook of Research on Global Enterprise Operations and Opportunities
Mehdi Khosrow-Pour (Information Resources Management Association, USA)
Business Science Reference • ©2017 • 375pp • H/C (ISBN: 9781522522454) • US $245.00

Examining Information Retrieval and Image Processing Paradigms in Multidisciplinary ...
Joan Lu (University of Huddersfield, UK) and Qiang Xu (University of Huddersfield, UK)
Information Science Reference • ©2017 • 425pp • H/C (ISBN: 9781522518846) • US $245.00

Ontologies and Big Data Considerations for Effective Intelligence
Joan Lu (University of Huddersfield, UK) and Qiang Xu (University of Huddersfield, UK)
Information Science Reference • ©2017 • 632pp • H/C (ISBN: 9781522520580) • US $245.00

Information and Communication Overload in the Digital Age
Rui Pedro Figueiredo Marques (University of Aveiro, Portugal) and Joao Carlos Lopes Batista (University of Aveiro, Portugal)
Information Science Reference • ©2017 • 380pp • H/C (ISBN: 9781522520610) • US $195.00

Handbook of Research on Information Architecture and Management in Modern Organizations
George Leal Jamil (Informações em Rede, Brazil) José Poças Rascão (Polytechnic Institute of Setúbal, Portugal) Fernanda Ribeiro (University of Porto, Portugal) and Armando Malheiro da Silva (Porto University, Portugal)
Information Science Reference • ©2016 • 625pp • H/C (ISBN: 9781466686373) • US $325.00

Inventive Approaches for Technology Integration and Information Resources Management
Mehdi Khosrow-Pour (Information Resources Management Association, USA)
Information Science Reference • ©2014 • 315pp • H/C (ISBN: 9781466662568) • US $205.00

For an entire list of titles in this series, please visit:
https://www.igi-global.com/book-series/advances-information-quality-management/73809

IGI Global
DISSEMINATOR OF KNOWLEDGE

701 East Chocolate Avenue, Hershey, PA 17033, USA
Tel: 717-533-8845 x100 • Fax: 717-533-8661
E-Mail: cust@igi-global.com • www.igi-global.com

I would like to dedicate this text to the UKSS Gold Medallists whose work has inspired so many researchers and practitioners to seek innovative ways of thinking about our world

UKSS Gold Medallists

Professor Russell Ackoff
Professor Stafford Beer
Dr Fritjof Capra
Professor Peter Checkland
Professor C. West Churchman
Professor Humberto Maturana
Professor Ian Mitroff
Sir Geoffrey Vickers

Table of Contents

Detailed Table of Contents

aspects. The contribution and implications of this research are to expand the existing theories of development in order to create the scientific basis for future studies. The study demonstrates the laws and tendencies of development and attempts to define the stage of development, shedding light on a new way to manage systems more effectively, helping to foresee future scenarios, and to prevent the negative circumstances the result from crises. Moreover, the proposed approach may facilitate focusing on the future and on innovation, instead of on vain attempts to preserve an imaginary stability.

Reaching community consensus on water reforms was the motive—and the challenge—for this operational research. While water is comparatively abundant in New Zealand, a pattern of decline in the quality of water resources has persisted. Living up to its "clean, green" image is a significant goal for New Zealand, with high economic value derived from the effects of its globally recognized environmental credentials on key exports like tourism and agriculture. A 2009 government task force suggested that a "business as usual" approach is untenable, and water reform should be a priority. This systems study examines the challenges—and opportunities—facing Kāpiti, a rapidly growing coastal community, with water scarcity and quality constraints that had long prevented them from meeting their sustainable development objectives. The authors used a stakeholder typology to identify system stakeholders, Theory of Constraints (TOC) to examine their perspectives, causal loop diagrams to highlight potential negative outcomes, and TOC to design appropriate interventions.

There are many problems and challenges facing the world. They go beyond problems that can be solved. They present as continual challenges that perpetually shift and transform themselves. The word we use for these situations are issues and they are by nature complex and large scale – they are global and long term. However, action does follow thinking and talking, even if that action is internalized. This chapter explores thinking about complex global issues and taking action.

While Easton's systems theory contributed to political science by demonstrating how the political system meets societal demands with policy outputs, he ignored the state's role in providing security in a hostile international environment. Hence, this chapter builds a sub-systemic governance model, arguing that large input generating groups require sufficient public goods to maintain the domestic political system and the state. Application to the South Korean case demonstrates that public good allocations increased along with the size of the input generating group. A functioning transmission belt, in the form of civic groups and elected local government, facilitated articulation of these demands to political leaders. Disruption of this subsystem in cases with unmet demands from large input generating groups can destabilize the state and its domestic political system. This new sub-systemic model seeks to advance understanding of the operation of the system and open up new areas of research into the persistence of the domestic political system.

Paradoxes and system contradictions in organizations may expose employees to tensions and contradictions, which they can only partially resolve, and from which they may only partially escape. Exposure to double-bind situations leads to stress, anxiety, and other symptoms of behavioral, affective, and cognitive disturbances. When employees are caught between paradoxical injunctions and organizational defensive routines hinder metacommunication, some employees develop dysfunctional coping strategies and end up being burned out. During a phenomenological PhD research on the burnout experience of 14 ICT employees in France and Norway, a stress-inducing pathological communication pattern was discovered and named the "intensification-quality paradox." The research participants had experienced two conflicting demands as a double bind: "do more with less" and "be excellent." The employees with high standards and ideals spent excessive personal energy to do more with less while keeping the quality constant. They got increasingly exhausted, ineffective, and finally burnt out.

This chapter examines issues around sharing subjective knowledge as practice in an organization, exploring what knowledge means to those working with it, how knowledge sharing happens in practice, and what influences its effectiveness. It is suggested that knowledge as practice as internalized by individuals can be difficult to communicate for a range of reasons, many relating to the effects of organizational subcultures on employee self-efficacy. A case study is used to illustrate this, based on fieldwork carried out in a research organization using the appreciative inquiry method to look at influences on the individual practice of knowledge sharing. An environment that promotes cognitive and affective trust amongst an organization's employees may go some way to address these issues.

This chapter reports on an application of systems theory to a complex area of human endeavor, classical dressage. The area is well represented in a rich literature dating back to the time of Xenophon (c.380BC) and has many practitioners worldwide today. The author uses her interpretive systems perspective to explore classical dressage theory and practice and, throughout, uses examples from the classical equitation literature to support and illustrate the analysis presented. The chapter offers a description of classical dressage from considering its component parts and suggests that it concerns not only the "correct" and ethical training and riding of horses but, fundamentally, the personal development of the trainer/rider. The chapter concludes by (1) inspecting the relationship between classical and competition dressage and suggests that they contain important elements that are mutually exclusive and (2) considering the potential contribution of "systems" to the study of the human-horse relationship as complementary to the increasingly popular approach of equitation science.

As described in the GTI, ISIL's transnational tactics in combination with lone actor attacks inspired by the group drove an increase in terrorism to its highest levels ever across Europe and many OECD countries (upwards of a 650 percent increase in

Preface

We cannot solve our problems with the same thinking we used when we created them. - Albert Einstein

INTRODUCTION

Systems and Systems Research for Real-World Problems

I am delighted to be editing this text which is entitled *Systems Research for Real-World Problems*. All the papers that make up the text were first published in previous editions of the *International Journal of Systems and* Society. Invited authors have been given the opportunity to extend the length of their paper into a chapter taking into account developments and any criticisms they might have received since it was first published. By not having a strict word limit imposed upon them has made it possible for the authors to expand upon the ideas expressed in their published paper. There are eleven chapters and each chapter provides the reader with a different example where Systems ideas have been applied. There is an eclectic range of topics including: Environmental concerns; Stress-inducing communications; Sustainability of Socio-economic Systems; Violent Extremism; Complex global issues; Economic Policy and Systemic Persistence; Organisational Power and Classical Dressage.

The reader of this text, especially one new to Systems ideas, might ask what is the shared idea that binds such a wide area of interests? A knee jerk reply might be 'Systems' But this will doubtless encourage the reply '...yes I know that but what ***is*** Systems'. Aye there's the rub, because many, including some of those who consider themselves 'Systems' researchers or practitioners, may give a different answer. Most would begin by referring to the notion of wholeness, which is understandable as it is at the root of all Systems thinking and practice, but this does little to clarify the answer to the question 'what is Systems'? Illustrating the answer with feedback, boundary and influence diagrams does little more than provide examples of ways of modelling *complex systems*. It does not help the inquirer to understand what it is. Although it is not the purpose of this text to provide a history of Systems thinking and practice, as there are adequate texts elsewhere[1], it might be useful to the reader to briefly reflect first of all on why there is a need for and second what is Systems. So for the reader of this text we have attempted to provide some clarification.

Why Systems Thinking?

There is little doubt that the success of the scientific approach to addressing problems has changed our world. Objectivity and the guiding principle of breaking the problem into its smallest component are key features of the scientific method. The scientific approach has come to define the very notion of intelligence. To question the 'scientific way' is, for many, to betray ignorance. The world, for the reductionist scientist, is one that can be defined by mathematical laws where nature is assumed to be something that can be encapsulated by numbers. Our world has become a construct described in terms of a reality that awaits discovery and it is the method of science that provides the key to making sense of the world. But Science, as Patočka, (2016, p117) argues '...simply posits ...reality without investigating the meaning of this reality...' In other words we have conceded that common sense and experience is analogous to ignorance and, in some cases, to mysticism. Cartesian reductionism is the predominant method of investigation into human existence and there are countless examples that illustrate its success. There is little doubt that this method of investigation and discovery has enabled the human race to gain control over its environment. But this control has often come with a social or environmental cost[2].

What is the scientific method? The method of science can be characterised by '…a neutrality of methods/methodologies; objectivity in analysis; reality external to individual; problem situation taken 'as given'; problem solving based upon optimization speed/efficiency/status' (Stowell & Welch, 2012, p135). But when these ideas are applied to complex social situations they are often found to be lacking. Checkland, (1999, p59) explains, '…The crucial problem which *science faces is its ability to cope with complexity*'. Our concern is that the idea of natural science has become the de facto approach to inquiry and we should be aware that its influence has changed the relationship between humans and their environment because it is '…not simply a development, but rather a radical reconstruction of the naïve and natural world of common sense.' (Patočka, 2016, p8). The world, for many, is viewed through the lens of the scientific method. But human and social behaviour is not so easily modelled nor can it be treated as if is something easily objectified. Humans are affected by and in turn affect human experience. We are capable of appreciating value as well as fact, meaning, '…Our situation is ever re-forming with no eternal and universal laws of human behaviour. The analysis of any situation by the actors within will be different because each of our concerns is highly personal. Recognition of the validity of each interpretation implies the recognition of subjectivity which is at variance with the scientific model of objectivity.' (Stowell & Welch, 2012, p135). Systems thinking and practice might offer one way of addressing the kind of complexity and 'messy' situations that are often found in social inquiry.

What Is Systems?

It is not possible to provide a simple answer to the question, 'What is Systems?' in the form of a sound bite, or a tweet. At the risk of being accused of oversimplification, modern Systems thinking can be said to defined by three events[3]. These are evident by key shifts in thinking that inspired the researchers and commentators that followed. These events can be, crudely, divided as follows. The first was the outcome from the Macy conferences[4] in the 1950's, then the development of General Systems Theory c1955 and Von Bertalanffy[5] et al. in 1968 and finally, to Checkland[6] in the 1980's. It should be emphasised that these are by no means the sum total of the important contributions to Systems thinking and practice, but are referenced here to mark the shift in thinking about how we make sense of the world. (This text includes some references for the reader interested in exploring the history of Systems, as there is a vast literature on the subject).

Following the second world war, in the mid 1940's and early 1950's, the Macy foundation[7] sponsored a series of conferences that were devised to bring together a varied interdisciplinary community of scholars and researchers. Participants at these conferences were leading scientists from across a wide area of study. It is true to say that these scholars have had a direct influence upon Systems thinking and practice. The intention was for these scientists to get together and lay the groundwork for the new science of cybernetics. Interestingly the first task was to find a means of communication, which, in its self, is an illustration of the problem that existed then and continues today. Then as now (ironically) this was because they came from different specialist backgrounds and used different terminologies, often when describing the same phenomenon. Despite the seemingly impossible task they did find a form of language that allowed individuals from the different fields to communicate about matters of significance. There were ten conferences followed by meetings in which they attempted to develop a universal theory of regulation and control that would be cross-disciplinary (See Stowell & Welch, 2012 for a summary). The outcome of the Macy conferences heralded what many believe to be the beginning of cybernetics and has inspired a whole generation of practitioners and theoreticians.

For many "Systemists" the beginning of modern systems was in the 1920's when a group of scientists set about a quest for an isomorphism of concepts. Von Bertalanffy, a biologist; Rappaport, a mathematician; Gerard, a physiologist and Boulding, an economist collaborated to seek a commonality between disciplines. They realised that there were principles, models and laws that apply to many areas in a generalised way. Bertalanffy wrote in General System Theory '...Physics is concerned with systems of different levels of generality…nothing prescribes that we have to end with the systems traditionally treated in physics…Rather we can ask for principles applying to systems in general...A consequence of the existence of general systems properties is the appearance of structural similarities or isomorphism's in different fields' (Von Bertalanffy, 1968, p32). This thinking led to what became a General Systems Theory[8] (General System Theory is a general science of 'wholeness'[9]). Earlier Boulding, produced a document announcing a Society for the Advancement of General Systems Theory, in the Journal of Philosophy of Science (1955, p.331). Out of this idea emerged the International Society for General Systems Research, formalised in 1956, followed by a change of title in 1988 to the International Society for the Systems Sciences to reflect its broadening scope.

In many respects the Macy conferences and the meetings of Bertalanffy et al. (1968) shared a desire to find a commonality between disciplines. One outcome could have been the breaking down of barriers between disciplines, but it is not to stretch the point when we say that we are still waiting for that breakthrough! In some respects both these factions' shared similar goals in that they sought general science of 'wholeness'. They shared the same idea that human behaviour can be modelled. Whilst Von Bertalanffy's modelling was based upon observations from the natural world the outcome from Macy was a Cartesian view of the world, a world that can also be modelled; a world where reality is mathematically objectified. Whilst we cannot dispute that models can be built to provide insight into our existence they are not *the* world but an approximation and one developed from a particular frame of reference about the world.

The antithesis to this kind of thinking emerged in the late 20th Century when the prevailing ideas in Systems were challenged. These new ideas materialised from research at the University of Lancaster under the guidance of Peter Checkland. This approach to Systems thinking was no longer based upon a view of the world as a set of 'systems' fixed and waiting to be discovered, described or modelled in some way. These ideas acknowledged that every human situation is unique, challenging the notion that human interactions could be encapsulated easily in a once-and-for-all model. Whilst it is conceded that it might be possible to construct a model of human activity it is, necessarily, an approximation based on evidence gathered after the event; a generalised model rather than one expressive of a unique representation. Although not considered a supporter of the kind of ideas Checkland espoused we can agree with Popper when he remarked '…Models are, from our point of view, nothing but attempts to explain laws in terms of old laws which have already been tested' (Popper[10], 1983, p359).

The very nature of thinking about the world was revisited and new ideas emerged based upon Phenomenology and Interpretivism, marking a shift from ontology to epistemology[11]. These ideas galvanised Systems thinking in the UK and similarly across the Atlantic where C West Churchman and Russ Ackoff were exploring similar ideas. Systems thinking about the world could now be thought of as something each individual experiences in a different way. But critics charged that this kind of thinking was being akin to solipsism. If the world only exists in the mind they said then surely we disregard a common consciousness of things we experience?

It is reasonable to allow that the world could be thought of as something each individual experience in a different way. But to suggest this kind of thinking as being akin to solipsism betrays a misunderstanding of what was being proposed. First, it is debatable that any human being can exist without reference to past, present and an anticipation of their future. Patočka answers the question in this way:

The monadic transcendental subjectivity has no existence properly speaking; it is not given to itself, nor is it the object of any ontic thesis; rather it exists only for the onlooker who, on the basis of the process of reduction, lifts this productive subjectivity out of self-forgetfulness and transposes it into a mode of 'pre-existence' while developing, through ideation, the universal essence of possible constitution in general. (Patočka, 2016, p. 50)

In other words, it is not possible to exist in a self-contained bubble as we are a part of a continuity of existence and as such are subject to its influences. Heidegger provides us with an explanation, which also serves to reject charges of solipsism, he writes:

But this existential 'solipsism' is so far from the displacement of putting an isolated subject-Thing into the innocuous emptiness of worldless occurring, that in an extreme sense what it does is precisely to bring Dasein face to face with its world as world, and thus bring it face to face with itself as Being-in-the-world (Heidegger[12], 1962, p.233).

But it is Thornton[13], who provides us with a robust and succinct dismissal of solipsism; he writes '…As a theory, it is incoherent. What makes it incoherent, above all else, is that the solipsist requires a language (that is a sign-system) to think or to affirm his solipsistic thoughts at all.' in other words the mere fact that language is necessary emphasises its existence and 'of being in the world'.

The research programme at the University of Lancaster went further with its new thinking by differentiating between 'Hard' and 'Soft Systems'. Hard could be described as an approach aimed at achieving an effective outcome against a defined need. Checkland defines it as making use of '...the kind of thinking which is natural to design engineers, whose role is to provide an efficient means of meeting a defined end' (Checkland, 1999, p.139). In many respects hard systems thinking preserves the traditional starting point of the scientific method by taking the problem 'as given'. A practitioner using hard ideas accepts that the world is something, with the right tools, that can be discovered; it is waiting to be discovered. Here the observer, looking from the outside, imagines the world.

The outcome of the Lancaster research programme suggested it was not possible to create a once-and-for-all model, because each situation was unique. This kind of approach was termed soft systems. 'Soft systems' thinking allows for unexpected answers, but requires an approach that is capable of exploring ambiguity. Soft systems is concerned with inquiring into what Ackoff described as a mess[14]. Importantly the practice of soft systems means using a method that will not of itself determine its outcome[15] it is intended to enable an understanding, or an Appreciation, to be gained. Arising from the 'Appreciation' (Vickers, 1983) it may be possible for those (or individuals) to affect the changes thought necessary to enable the System to maintain its relationship with its environment. But it is important to note that, in the case of a method (methodology) used in soft systems, it should be faithful to the phenomenological underpinning of the approach.

It was this kind of Systems thinking that created an apparent division between Hard and Soft Systems. But too much is read into this categorization and it would be wrong to think that this is a schism within the domain. Whilst 'hard' could refer to techniques used in systems engineering Checkland suggests that Hard methodologies are special cases of Soft. He writes that Soft '*...systems thinking provides a structure for debate about change which hopefully will be good quality...*' (Checkland, 1999, p191), providing a sufficiently well defined model that can be used by an appropriate quantitative approach.

Summary of Discussion

In the preceding section I have attempted to summarise the origins of modern systems thinking and practice, but in such a short preface it is not possible to incorporate the wealth of ideas and material that have been published over the past decades. Suffice it to say that lessons have been learnt from the practice and reinvested in the way in which they are applied. For the reader new to Systems I have provided reference to texts that expand on the ideas I have merely touched upon. This is by no means a definitive history of Systems ideas and I acknowledge that other Systems theorists might have a different vision of the development of modern Systems thinking.

But it is clear from the forgoing that to define the notion of *Systems* simply as 'thinking holistically', is naive. Practitioners in other areas of human concern can also claim to take into account the wider issues, say, when designing a building or a bridge. Whilst the basis of Systems thinking is thinking about the whole and not focusing on a part, this is not done in an unbridled way. Ideas such as boundary and what *constitutes* a System[16] are just some of the ideas that a Systems practitioner must appreciate. The above account provides little more than a taster for those who would like to explore the ideas further. The different approaches to trying to gain understanding and explore the human condition are by no means exhausted. In the text we have provided just a few examples of the way in which some of these ideas have been used.

Systems Thinking in Practice

In this text I have included different examples of applying the ideas, or their derivatives, introduced above. I hope that as you read it you will also reflect upon the 'origins' of the ideas and how they have been interpreted. It would not be helpful to come down in favour of one set of ideas or another as each has contributed to the practice and to the further development of the ideas themselves. My own view is that whichever approach is adopted the practitioner should attempt to maintain the logical coherence between the underpinning philosophy and the practice. To mix them can create difficulties. However, there is much to be gained by the cooperation of practitioners applying each of these ideas. Moreover, the inclusion of a Systems researcher within the team of a research project into any area of concern would reflect a positive contribution to interdisciplinary research espoused by various research funding agencies. It is difficult to think of any project where Systems practitioners/researchers cannot play an important, if not vital, role.

INTRODUCING THE CHAPTERS

In this relatively short text we provide eleven chapters in which Systems ideas have been used. The reader will notice the different ways that the ideas have been applied, which hopefully, will be aided by the above discussion. Each of the chapters describes an area of interest where Systems thinking and practice is thought to be useful. The reader will note the range of situations in which each inquiry has been undertaken.

Chapter 1

The first chapter is entitled *Enterprise Complexity Model: Enterprises, Organizational Systems, and Dynamic Capabilities*, written by Professor Raul Espejo. Professor Espejo is a distinguished Professor who has first-hand experience of the work of Stafford Beer and the development of the Viable Systems Model (VSM)[17] that was so influential in the mid and late 20th century. Since that time Raul has continued to develop his ideas and this chapter describes the present incarnation, which he describes as an Enterprise Complexity Model (ECM). This model is visualised as a distributed governance model of an ecology of evolving enterprises. In this case governance is understood beyond directing development; and but as guiding the ecology of a collective of enterprises self-organization.

Distributed governance refers to enterprises at all levels, from local to global. Governance is focused on processes aimed not just at innovation but making things happen now. To be effective, Espejo argues, distributed governance should guide the collective appreciation of the products, services and issues of those within an enterprise. It is this awareness that makes implementation possible because ECM facilitates the alignment of stakeholders' interests and values. Self-organisation is grounded in the relationships and interactions of these stakeholders. It would be a mistake to think of an ECM of as an institutional tool it should be thought of as helping with guiding, enabling and facilitating to increase society's requisite variety to deal with social, ecological and economic challenges. Espejo argues that ECM helps in creativity and in moment-to-moment coordination of actions and alignment of interests. As relationships and interactions evolve a deeper and wider appreciation of issues of concerns will emerge. This is an ongoing process of building ecosystems of enterprises and making their boundaries more meaningful to relevant stakeholders. This ecosystem of enterprises can be related to the evolution of the digital economy in the societies of the 21st century.

Chapter 2

In the second chapter we have a specific class of systems called *Socioeconomic systems,* which is entitled *Sustainable Development and the Sustainability of Socioeconomic Systems.* The chapter is written by Dr. Olga Tikhomirova she suggests that if their functioning and development are based on objective laws of development then, she says, the management of firms should be based on the universal laws of evolution and development too. To this end she offers a framework, with a particular reference to business systems that, she argues, will provide the route to better understanding of the theory of development, sustainability and the sustainable development of socioeconomic systems. In the chapter Dr. Tikhomirova revisits some theoretical approaches to expand the existing theories of development in order to create the scientific basis for future studies. In particular she considers the General Theory of Systems; Synergetics; Social Dynamic and Development. Dr. Tikhomirova's study highlights the laws and tendencies of development from which she suggests a new way to manage 'systems' more effectively. Future scenarios are rehearsed as a means of anticipating possible negative circumstances that are the result of crises. She argues that with such a 'tool' that emphasizes the future and innovation rather than focusing on preserving an imaginary stability. In her paper she reflects on the financial, economic, social and technical elements of the enterprise as parts of a connected and holistic system

Chapter 3

Chapter three is written by Robyn Moore and Victoria Mabin and is entitled, *Reaching Community Consensus on Reforms for More Sustainable Urban Water Management Systems: The Case of Kāpiti, New Zealand.* This chapter is an example of another important area in which Systems thinking has proved beneficial to a community. In this project the authors employed Causal Loops as a means of modelling the situation of interest. Causal loop diagrams[18] (CLD) are the foundation of Systems Dynamics (SD), which can be modelled further using a software package to run on a desktop computer. SD provides the inquirer with the means to explore 'what if' questions. This helps in making informed decisions before implementing change. In their chapter the authors say that in 2009 a government task force suggested that water reform should be a priority. It was this government initiative that was the motive for Moore and Mabins' work and this chapter describes their attempt to reach a community consensus on water reforms. They point out that whilst there is no water shortage in New Zealand its 'Green' credentials are important. New Zealand benefits economically on key exports such as tourism, agriculture, forestry and fisheries derived from its globally recognised environmental credentials. The Chapter provides an interesting account of a community-focused systems study where there were concerns about the declining quality of its water. The challenges and opportunities facing the community is described and is an example of a situation that cannot be 'solved' simply. Here we see the importance of collaboration if a successful outcome is to be achieved. The authors applied the Theory of Constraints (TOC) and a stakeholder typology that was used to identify system stakeholders and examine their perspectives. Causal Loop Diagrams were constructed to explore and anticipate any negative outcomes. This study in a community resource management setting tests the effectiveness of the combined problem-structuring framework.

Returning to the study almost a decade later shows the approach to be resilient. During this time the local authority had reached a level of consensus with its community on a collection of pre-emptive responses to the pressures and impacts of population and a warming climate. Here is evidence for the benefits of consensus building, which is supported by the community satisfaction surveys and the residents' responses to some significant policy changes. The results of the study show encouraging signs and insights gained from applying framework might help to inform local authorities and others charged with reaching consensus. The lessons learnt from the study have encouraged the researchers in using the approach in other areas of concern such as housing and conservation management.

Chapter 4

Ian Roderick is the MD of the Schumacher Institute in Bristol and is the author of this chapter, which is entitled *An Exploration of Thinking About Complex Global Issues and Taking Action.* Ian has some first-hand knowledge of the challenges facing us all. As Ian points out many of these problems cannot be solved and to add to the complexity they continually transform themselves thereby creating new challenges. They are by nature complex and large scale; in short they are global problems.

Roderick argues that we have to learn to accept that such issues do not go away, they mutate and spawn. Ian says that we often enter into a situation that he refers to as, schismogenesis[19], which is the escalation of behaviour caused by the failure to ensure equity He cites the 'arms race' as an example that once entered can lead to the worst kind of system change; a system collapse. Complex global long-term issues impinge on our lives today and we formulate policies and plans to deal with them, as they are perceived to affect us- piecemeal fashion.

In this chapter Ian revisits traditional, rational policy-making and suggests that a systems approach might to a better understanding of the complex processes taking place. The idea he advocates is to explore complex issues to a higher level of adaptation such that escalating process can be controlled and perpetually transforming challenges are held within safe boundaries. The knowledge gained from such exercises may create a feasible system change that will work before in adapting to complex global issues.

Roderick's chapter raises our awareness of some complex global issues, what they are, their nature and the patterns they present. He considers what it means to do something about them acknowledging the almost impossibly of doing a great deal at the global level. But even without Nations acting in concert, he argues we can all do more a smaller scale. A good starting point he says is that we can all take action in our own neighbourhood, through personal involvement we will want to affect others and to effect a 'system' change. Such small beginnings create a change that might contribute to improving the global condition.

Chapter 5

In this chapter, *Revising Sub-Systems Models to Analyze Economic Policy and the Question of Systemic Persistence* James Schopf explores and criticises Easton's systems theory[20]. Easton's behavioural approach greatly contributed to the field of political science by providing a holistic framework to demonstrate how the political system might function to meet demands through its policies. For Easton the political system was organic that comprised various functional parts, which can be summed up as system, input, output, demand, support and feedback (not dissimilar to the idea behind black box). Dr. Schopf criticises Easton because he says that Easton's only interest was in the political system's persistence. All Easton's model required was the existence of community. He says argues that if a community is to survive in a hostile international environment they require state-provided security. To address this problem Dr. Schopf offers what he calls a sub-systemic governance model, which is capable of explaining domestic political system and state persistence. For the long-term maintenance of the domestic political system he says that his model predicts the need for large input generating groups to provide sufficient allocation of public goods.

In his chapter he describes an application of the model that he says demonstrated that the share of public goods increased along with the size of the input-generating group. The model acted as a transmission belt, in the form of civic groups and elected local government, from which the articulation of public good demands could be represented to political leaders. An important lesson learnt from this exercise was that, the increased pressure from unmet societal demands in the domestic political system could be destabilised by large input groups. Schopf argues that his idea for a sub-systemic model will advance understanding of the operation of the system and in turn open up new areas of research into the persistence of the domestic political system.

Chapter 6

In this chapter, entitled *The "Intensification-Quality Paradox" in ICT Organizations*, Solveig Evenstad explore the paradoxes and system contradictions that exist in organizations subjecting employees to tensions, contradictions and oppositional tendencies. To add to the tension they feel relatively powerless as they can only partially resolve them and from which they feel that they cannot escape. Dr. Evenstad writes that exposure to, what she refers to as, double bind situations negatively affects attitudes and behavior, leading to stress, anxiety and other symptoms of behavioral, affective and cognitive disturbances.

The paradoxical injunctions presented in this chapter are "do more with less" and "be excellent" which together constitute a double bind. The frustrating nature of this paradox was expressed quite precisely by an employee: "*Doing more with less, doing more with less. And, I'm thinking, how can I do that? Doing more will have to mean that I have to do what I'm doing less well*". It is frustrating, especially to employees who are highly engaged and want to deliver excellent results. These employees commonly respond by increasing their efforts, despite feeling overworked and overburdened. When employees are caught between paradoxical injunctions and the organizational defensive routines hinder metacommunication, some employees develop dysfunctional coping strategies and end up being 'burned out'.

The chapter recounts a PhD research project on the burnout experience of 14 ICT employees in France and Norway. The double bind that is identified in this study was called 'intensification-quality paradox' and classified as a *stress-inducing pathological communication pattern*. She says that the injunction of 'do more with less' implied work acceleration and intensification for the research participants. At the same time, they had to deliver excellent results as they always did. The excellence ideal is both a personal standard and an organizational standard. The employees with high standards and ideals spent excessive personal energy to do more with less without lowering the quality. They got increasingly exhausted, ineffective and finally burnt out. It is the intensification-quality paradox, she says, that is a threat to organizational sustainability.

Chapter 7

In this chapter Frank Stowell expands upon the discussion of organizational power and how it might be reflected upon. The chapter is entitled *Organizational Inquiry, Power, and the Metaphor Commodity*. In this chapter Stowell argues that organizational interventions based on models of organization are unsatisfactory because they are based upon a certain level of predictability. This, he argues, is because each organization is unique and cannot be encapsulated by a single ontological model. Organizations exist in a turbulent environment and must change to survive, but change creates tensions that threaten its stability. Individuals and groups then use their 'soft' power in an attempt to shape the outcome into one that they feel is less threatening. The way that this power is used will have a significant effect upon decisions and outcomes. As a means of trying to gain insight into the way that 'soft' power is used the notion of 'Commodity' as a metaphor for power is introduced as the means of bringing out hidden and imagined manifestations of power. The intention is that by thinking in terms of 'Commodity' the participants gain insight into the way in which it is being used and the effect it might be having upon discussions. In this way the participants provided an opportunity to embrace the concerns of others and factor this into their discussions.

Chapter 8

Enterprise wide integrated systems (ES) are a common feature of large organizations but research has yet to provide a socio-technically informed approach to facilitate implementation. There is a need, the authors say, to reflect upon the impact of technology on professional working practices at the time of procurement. In their chapter entitled *Enterprise Systems, Power, and Improvisation: Equipping Universities for Mass Production?* David Wainwright and Teresa Waring explore the socio-technical approach to power and improvisation in ES's. In this chapter they take a critically informed review and synthesis of the literature. The outcome of their literature research led them to develop an analytical framework to explore the concepts further. They did this through a longitudinal, ethnographic[21] study of an ES within a UK university. This led to a framework that they call combined 'circuits of power-improvisation' (CPI). This framework will, they argue, facilitate a better understanding of ES implementation, socio-technical theory and practice. The lessons they learnt from the practice can contribute to a better understanding of some of the problems that emerge from the rapid introduction of digital technologies. The study highlights the importance of the power and improvisation that new digital technologies often bring to an organisation, some intended and some unintended.

Chapter 9

Penny Hart's chapter, *Factors Influencing Knowledge Sharing in the Knowledge Workplace,* is based upon some field research carried out at a Government department whose 'product' could be described as providing expertise to a Government ministry concerned with equipment provision and maintenance. The department manages customer work through programmes, using external suppliers where possible. However, the career pattern for many members of staff means that many move on to other areas after a cycle of three years or so. This pattern of change meant that sharing knowledge was important to the efficacy of the group. Dr. Hart's chapter examines some of the issues emerging from sharing knowledge as practice She explores what knowledge means, how knowledge sharing happens and what influences its transfer. Hart raises the issue that in practice knowledge, which is internalised, is difficult to communicate. There are many reasons for this including the influence of organisational culture and personal status. Which she discusses within the chapter. The fieldwork was undertaken using the Appreciative Inquiry Method[22] that was used as a means of gaining insight into the influences upon knowledge sharing.

Chapter 10

Daune West's chapter is an example of how Systems thinking can be applied across a diverse range of human concerns. In this chapter she describes how she applied systems theory to complex area of human endeavour, namely Classical Dressage. Dressage she writes has a rich history of literature dating back to the time of Xenophon (c.380BC) and has many practitioners across the globe today. In her paper Dr. West gives us a description of Classical Dressage theory and practice that she presents through a number of systems concepts, which she illustrates by means of systems tools. Dr. West's background in Systems relates to the notion of interpretive systems and through that lens she describes how Classical Dressage can be seen as being concerned not just with the 'correct' training and riding of horses, but also about the personal development of the trainer/rider.

The paper concludes by presenting a description of the component parts of a 'classical' or 'academic' approach to equitation. Throughout, examples from the classical equitation literature are provided to illustrate the analysis presented.

Chapter 11

Anthony Masys chapter is another 'unusual' situation of where Systems thinking has been applied. In this case he has applied these ideas to try to gain understanding of areas of extreme human conflict. He cites the terrorist attack in Paris (2016), Belgium (2016), USA (2016), Manchester (2017) and London (2017), to underscore his concerns. Global Terrorism Index, (GTI), 2016, describes the transnational tactics used by ISIL used in combination with lone actor attacks drove an increase in terrorism to its highest levels ever across Europe and many OECD countries (upwards of a 650 percent increase in deaths since 2014). The attacks by ISIL in Paris, Brussels and in Turkey's capital Ankara, were amongst the most devastating in the history of these countries and reflect a disturbing return of the transnational group-based terrorism (GTI, 2016). Masys argues for the need of a greater understanding about the radicalization and recruitment of terrorists. He describes the use of Actor Network Theory (ANT)[23] as a systems lens to open the 'blackbox[24]' of terrorism. ANT facilitates his use of this method of Systems practice in this investigation and argues that the process of ANT translation enabled him to highlight how dynamic networked actors shape radicalization. In his chapter Masys applies Functional Resonance Accident Model (FRAM)[25] methodology (e.g. Hollnagel, 2012). FRAM is based on the notion of describing outcomes arising from the variability of everyday performance. It is a method for modelling complex socio-technical systems and is used in this chapter to analyse how radicalization activities (as described through

ANT) take place and where and how intervention strategies can be designed to interfere with the radicalization process

The Chapters

Each of the invited authors has contributed a chapter that I felt illustrated different ways in which Systems ideas have been interpreted and applied. The topics are varied and serve to provide the reader with an illustration of the wide range of areas of interest in which Systems can play a part. These examples are, of course, only a small sample of the areas in which Systems ideas are used. The chapters are grouped according to my interpretation of the different areas of Systems thinking but I have avoided stating what I believe these to be. The order of chapters coupled with the brief summary of modern systems outlined in the preface will, I hope, inspire the reader to think about the origins of the methods described. I do not claim that my interpretation of the thinking behind each paper, implied by the order in which they appear, to be without debate. If a healthy discussion emerges from the text because of this it will have provided the readership with an added bonus.

I hope that you enjoy this text,

Frank Stowell
University of Portsmouth, UK

ENDNOTES

1 For example: Baecker, D. (2016). *Schlüsselwerke der Systemtheorie, 2nd ed.*, Wiesbaden: VS Springer; Beer, S (1972) *The Brain of the Firm*, Wiley Chichester; Capra, F., & Luisi, P.L., (2014), the Systems View of Life: A Unifying Vision. Cambridge University Press, Cambridge; Checkland, P.B. (1999) *Systems Thinking, Systems Practice, a Thirty Year Retrospective.* Chichester: Wiley; Churchman, C.W.(1979). *The Systems Approach and its Enemies.* Basic Books, New York; Vickers, G. (1983). *Human Systems are Different.* London: Harper and Rowe;

2 e.g. Knoblauch, J.A. (July 2, 2009). Plastic Not-So-Fantastic: How the Versatile Material Harms the Environment and Human Health. *Scientific American.* Retrieved from https://www.scientificamerican.com/article/plastic-not-so-fantastic/

3 For a valuable account of the development of ideas, the method of science and Systems see Checkland, P.B. (1999). *Systems Thinking, Systems Practice, a Thirty Year Retrospective*. Chichester: Wiley, chapter 2.

4 See Stowell, F.A., & Welch, C. (2012) *The Mangers Guide to Systems Practice: Making Sense of Complex Problems*. John Wiley & Sons; Macy Conferences. (n.d.) Retrieved from https://en.wikipedia.org/wiki/Macy_conferences; sblachfellner. (March 7, 2012). Looking Back in History: The Macy Conferences. Retrieved from http://emcsr.net/looking-back-in-history-the-macy-conferences/

5 See BCSSS. (2018). The Center. Retrieved from www.bcsss.org; von Bertalanffy, L. (1968). General Systems Theory: Foundations, Development, Applications. New York: GeorgeBraziller.

6 Checkland P.B. (1981, 1999). Systems Thinking, Systems Practice. A Thirty Year Retrospective. Chichester, UK: John Wiley

7 Useful references can be found in Pias, C. (ed.). (2003). *Cybernetics – Kybernetik: The Macy Conferences 1946–1953.* Zürich/Berlin: Diaphanes

8 See Bertalanffy et al. c1940, 1950,1956 ; Sadovsky, 1992; as cited in Stowell, F.A., & Welch, C. (2012) *The Mangers Guide to Systems Practice: Making Sense of Complex Problems*. John Wiley & Sons;

9 General Systems theory- General System Theory, therefore, is a general science of 'wholeness'. 1) There is a general tendency towards integration in the various sciences, natural and social. 2) Such integration seems to be centred in a general theory of systems. 3) Such theory may be an important means of aiming at exact theory in the nonphysical fields of science. 4) Developing unifying principles running 'vertically' through the universe of the individual sciences, this theory brings us nearer to the goal of the unity of science. 5) his can lead to a much-needed integration in scientific education. Taken from Ludwig von Bertalanffy, as cited in Panarchy. (2000-2017). Ludwig von Bertalanffy- passages from General System Theory, 1968. Retrieved from https://www.panarchy.org/vonbertalanffy/systems.1968.html

10 Popper discusses the strengths and weaknesses of subjectivity in Popper, K.R. (1979). Objective Knowledge-An Evolutionary Approach. Great Carendon Street: Oxford University Press. (pp41-44)

11 Phenomenology – see Husserl, E. (2002). *The Phenomenology Reader,* Dermot Moran and Timothy Mooney (Eds.) London: Routledge, pp 124-133; For Interpretivism, see Schutz, A. (1943). The problem of rationality in the world." *Economica, 10* (38): 130–149; *Sociological Theory And Philosophical Analysis*. (1970) Dorothy Emmet and Alasdair MacIntyre (Eds.). London: Macmillan; Ontology -the nature of being; Epistemology; is the study of the nature and scope of knowledge and justified belief.

12 Heidegger, M. (1962), Being and Time. Translated by Macquarrie, J., & Robinson, E. Oxford, UK: Blackwell Publishing

13 Thornton, S.P. (n.d.) Solipsis. Retrieved from http://www.iep.utm.edu/solipsis/

14 "Every problem interacts with other problems and is therefore part of a set of interrelated problems, a system of problems I choose to call such a system a mess" from *Redesigning the future*, 1974

15 See Churchman, C. W. (1971), The Design of Inquiring Systems: Basic Concepts of Systems and Organization. New York: Basic Books, Inc., p249

16 See Checkland, P.B. (1999) Systems Thinking, Systems Practice, a Thirty Year Retrospective. Chichester: Wiley for a valuable discussion; Stowell, F.A., & Welch, C. (2012) The Mangers Guide to Systems Practice: Making Sense of Complex Problems. John Wiley & Sons, pp82-93

17 Beer, S. (1972) *The Brain of The Firm*, London: Allen Lane the Penguin Press.

18 Causal loop diagrams can be developed through the identification of key variables in the system of interest and then identifying the causal relationships between them. By linking together several loops, a model of a particular problem or issue can be created. e.g. See Stowell, F.A., & Welch, C. (2012) *The Mangers Guide to Systems Practice: Making Sense of Complex Problems*. John Wiley & Sons, pp62-64.

19 Bateson, G. (1958). *Naven*, Second Edition. Sanford: California: Stanford University Press.

20 Eastons Systems Theory is a conceptual framework for analysing politics and can be used in understanding actual forces operating in a political system.

21 Ethnography is the branch of anthropology that involves trying to understand how people live their lives. In this approach the researcher observes society from the point of view of the subject of the study.

22 e.g. AIM – see Stowell, F.A., & Welch, C. (2012) The Mangers Guide to Systems Practice: Making Sense of Complex Problems. John Wiley & Sons. (pp49-57); Cooray (2010), End User Driven Development of Information Systems-Revisiting Vickers notion ofAppreciation, Thesis (unpublished). Portsmouth, UK: University of Portsmouth; Stowell, F.A.. & West, D. (1994). Client Led Design. Maidenhead. London, UK: McGraw Hill.

23 Attributed to Callon, M. (1991). *A sociology of monsters: essays on power, technology and domination.* John Law (Ed.) London: Routledge; and Latour, B. (1992). Shaping technology/building society: Studies in Sociotechnical change. Wiebe E. Bijker and John Law (Eds.) Cambridge, Massachusetts: MIT Press, pp. 225-228.; their actor-network theory includes people, objects and organizations.

24 There are many references to the notion of Black box, e.g. Ashby, R. (1956). An Introduction to Cybernetics. London: Chapman & Hall Ltd.; Norbert Wiener N. (1961). *Cybernetics: Or Control and Communication in the Animal and the Machine. 2nd edition.* Cambridge, Massachusetts: MIT Press; Glanville, R. (2009). The black boox: Vol. III, 39 Steps. Echoraum, Vienna; Examples contained in Stowell, F.A., & Welch, C. (2012) *The Mangers Guide to Systems Practice: Making Sense of Complex Problems*. John Wiley & Sons, pp27-31

25 Hollnagel, E. (2004). *Barriers and accident prevention*. Aldershot, UK: Ashgate.

Chapter 1
An Enterprise Complexity Model:
Enterprises, Organizational Systems, and Dynamic Capabilities

Raul Espejo
Syncho Research, UK & World Organisation of Systems and Cybernetics, UK

ABSTRACT

In this chapter, the author proposes an enterprise complexity model (ECM), which is visualized as a methodology to achieve the distributed governance of an ecology of evolving enterprises. Governance is understood as guiding the enterprises self-organization towards policies creating, regulating, and producing products and services for society. Self-organization is grounded in the communications and interactions of stakeholders. The purpose of an ECM model is not institutional development but guiding, enabling and facilitating interactions of all kinds with the support of current and disruptive technologies to increase society's requisite variety to deal with social, ecological, and economic challenges. An enabling context helps the branching of the enterprises' creativity into all kinds of innovations, forms of coordination, and operational alignment of their interests. Quality of life, fairness, and social justice are the values driving this ecology of enterprises towards a deeper and wider appreciation of issues of social concern.

DOI: 10.4018/978-1-5225-5996-2.ch001

INTRODUCTION

This paper is a development of one published in IJSS, Vol 2, No1 pp.1-22 (Espejo, 2015a). This new version is the outcome of feedback coming from multiple talks and lectures delivered at conferences and courses in several parts of the world. In these interactions participants and students have made conceptual and methodological contributions to its ideas and have also illuminated instances of their application. Further research has followed for a couple of years. The new version offers a significant development of the Viplan Methodology (Espejo, 1993) and clarifies its relevance to the governance of ecologies of enterprises.

The core of this paper is methodological; it aims at exploring how to produce an ecology of enterprises in the context of a guiding enterprise. The integrating framework is an Enterprise Complexity Model (ECM), which is an extension of earlier work on the Viplan Method (Espejo, 1989b), Viplan Methodology, (Espejo, 1993), and is grounded in Beer's Viable System Model (Beer, 1979, 1981, 1985). An ECM goes beyond individual enterprises and is used in the sense of an innovative undertaking in society; it offers a model of collaborative enterprises more than any particular institutional form. It is a model of an ecology of enterprises, guided by a leading one, dealing with the complexity of multiple environmental agents. It is in the nature of this enterprise to operate in a context of challenges and opportunities, which requires of its ingenuity and capabilities, since its complexity is significantly smaller than that of its environment. In other words, this environment's complexity is much larger than the enterprise's response variety. The challenge for the enterprise is to find ingenious strategies to bridge this complexity gap[1]. Today's digital economy and technology can play an increasingly important role in making this bridging more effective and socially acceptable. I offer responses to these challenges from the perspective of systems thinking and cybernetics.

How is it possible for an enterprise to perform effectively under increasing social, ecological and economic demands? How does it achieve viability in these challenging circumstances? These are questions for which Beer's Viable System Model (Beer, 1979, 1981, 1985) gives powerful answers. In this contribution I offer a methodological extension of this model in the network economy as enabled by the digital society. The digital society shares with the Viable System Model (VSM) its focus on complexity as understood in this paper, that is, on the huge number of possible states, or variety, of any situation[2]. The digital society, as we are witnessing it today, is enabling networking and is grounded in technologies with large capacity to create as well as map all kinds of situational states. Today enterprises can recognise and manage people's relational variety at the most disaggregated levels; algorithms, artificial intelligence, 3D printing, engineering services and so forth are making this apparent all the time. Rather than dealing with aggregations and averages, enterprises can focus, through algorithms, on individual transactions; services can be tailored to specific and individual needs. Distributed responses to people's widespread needs can be managed not only by service providers but by themselves through enabling technologies, adding flexibility and convenience. Computer networks can increase in orders of magnitude an enterprise's research capabilities. An enterprise can increase much more its performance through inclusion of others in a network of autonomous units, rather than remaining as an independent unit.

The VSM provides a powerful heuristic for the practical accounting of the huge number of possible states constituting an enterprise and its environment. *As is discussed later, this accounting can help to develop its capabilities to respond effectively to challenging environments.* The Enterprise Complexity Model (ECM), as offered in this paper, is a methodological extension of this model. It is a guiding model of the enterprise's need to manage the ever-increasing complexity of its environment. Examples of this management and its connections with the digital economy are given throughout this paper.

The opportunities and threats that enterprises are experiencing in their surroundings suggest that they need to invent and develop ingenious organisational forms, often beyond their formal boundaries, to match these challenges. The ECM offers a methodology for the on-going re-invention of an enterprise's organisational forms to bridge environmental demands. I call this methodology the *Viplan Methodology*, which uses ECMs as ingenious forms to bridge the enterprise-environment complexity gap. ECMs are helpful to visualise and develop the enterprise's dynamic capabilities[3], and it offers a heuristic to create, maintain and develop innovative organisational forms through the use of technology.

The VSM offers a template to map an enterprise's resources and their relationships according to their systemic purposes. In the context of the ECM I understand an enterprise as a system striving for, and maintaining, by itself or in collaboration with other enterprises, its viability in a complex environment. Whether by itself or in collaboration with others, it needs *developing capabilities to create, regulate and produce products and services* for stakeholders, customers and other environmental agents, which are part of its value chains and externalities. The VSM postulates that for these purposes, the enterprise by itself -if it operates as an independent unit- or as a network -if it operates together with others- constitutes five systemic functions; policy, intelligence, cohesion, coordination and implementation (Espejo 2003). All the resources producing this enterprise have one or more of these systemic functions, or in other words, all resources can be mapped onto one or more of these systemic functions. *Together the five interrelated systemic functions constitute an organisational system* (Espejo & Reyes, 2011). An enterprise achieving viability on its own is an organisational system by itself. However, if it is striving for viability in collaboration with other enterprises, it is the collective that constitutes the organisational system.

The Viplan Methodology uses the *Viplan Method* to *diagnose* and *design* an enterprise's structure. In its diagnostic mode, Viplan is a method to assess whether this structure is adequate or not for its ascribed purposes, or in other words, whether it is an organisational system or not. Through diagnosis it is possible to work out archetypical organisational shortcomings (Espejo, 2008a) that offer guidance for improvements. In its design mode it works out, from the ascribed purposes, the configuration of its own and/or collective resources and relationships that are necessary to produce an effective structure. In this paper I use this latter mode to model the dynamic adaptation of an existing enterprise to a changing environment. The main construct for this purpose is the enterprise complexity model (ECM) within the framework of the Viplan Methodology.

This paper makes first the distinction between enterprises, institutions and organisational systems. These distinctions provide its systemic foundation. Secondly, the key constructs of this paper are complexity and requisite variety (Ashby 1964). Thirdly, I develop the methodological support to study and develop ECMs. This is a learning methodology supported by two loops, the cybernetics loop underpinned by Beer's Viable System Model and the developing loop (Espejo, 1993; Espejo and Reyes, 2011). The fourth and final section of this paper explores the contribution of the ECM to our understanding of an enterprise's performance in the network economy, with a focus on governance.

ENTERPRISE, INSTITUTIONS AND ORGANISATIONAL SYSTEMS

In the network economy the formal enterprise usually is different to an organisational system. This is an important distinction. In the network economy more and more enterprises *name* as their core transformation, transformations of inputs into products that go beyond their inner capabilities. In this case, to produce these transformations they require the contribution of other enterprises; they are naming an *organisational system* that is larger than the enterprise in focus. Making viable this organisational system is a means of making viable an extended enterprise. *From Viplan's perspective this organisational system is the system-in-focus to which the VSM is applied.* The enterprise's success will depend on its ability to form a viable network of resources and enterprises. The participants in this network of course can and do change over time.

Public Sector Organizations

In the public sector these networks, as already said, often go beyond a particular institution. These institutions often do not create, regulate and produce their services by themselves. For instance in education, the ministry responsible for a country's education is unlikely to deliver educational services by itself. The ministry, as an institution, depends on the collaboration of thousands of schools and millions of families to deliver its services. Unless it succeeds in creating this larger organisational system, the ministry will fail in delivering the country's educational policy (Espejo & Reyes, 2011). The system in focus is the educational system and this is the "enterprise" that needs diagnosing and development and not the ministry on its own. The ministry is a *public institution* for which the total educational system is the *organisational system*[4]. The institution alone, in this case, is not an organisational system. Of course a study may focus its attention on the institution, but it will need to recognise that the institution's performance is highly dependent on the quality of its relations with the other contributors to the organisational system. Otherwise the study may reinforce the customary fragmentation of public services. Of course if individual *public institutions* create, regulate and produce their services, such as could be the case of autonomous public and private schools, they are organisational systems. In recent times it is common to find public-private partnerships (PPP). This type of organisation is becoming a form of delivering services to society. These public institutions are leaders of *service delivery networks* in collaboration with the private sector. The focus of the Viplan Methodology is the viability of these PPP networks rather than of the public institutions on their own. Thus, the system-in-focus is larger than the public enterprise and its constitution can be more flexible and susceptible to on-going changes. Instilling flexibility in enterprises and institutions is a dynamic process that may require considering an enterprise complexity model with boundaries beyond the public institution (West et al., 2014).

Lead Enterprise

Globalisation is being exploited by many enterprises through the creation of networks. For instance a company's manifest transformation may be producing computers; however at closer scrutiny it becomes apparent that it is a *lead enterprise* to produce computers, with the support of many other enterprises. In this case the system in focus is the organisational system that creates, regulates and produces the computers, which is a network of enterprises that creates, regulates and produces the products agreed by the network with the lead enterprise. While each of the networked enterprises may well be an organisational system in its own right, the lead enterprise depends on the network to achieve viability. This kind of structure requires a quick reconfiguration of resources. The core competencies of the lead enterprise support the clustering of a wide range of enterprises; these enterprises can vary according to environmental demands and local circumstances. This flexibility allows the lead enterprise to develop its own dynamic capabilities in collaboration with the network.

The Viplan Methodology, together with the Viplan Method, offer a heuristic to develop and assess an enterprise's ECM over time. This heuristic, as will be elaborated below, is grounded in the management of complexity. A core competency for enterprises today is managing relationships and interactions that are continuously changing as new technologies and environmental conditions make possible changing configuration of resources and therefore asserting different strategies to manage complexity. This is a process of creating new structures based on the existing ones. In this sense Viplan is about supporting significant structural adjustments, even changing identities, as new conditions emerge.

Another instance of a lead enterprise is that of scientific parks, which have been set up near universities around the world for decades. Scientific parks provide common services to enable the viability of high technology enterprises. However, in this case, each enterprise maintains its own concerns and identity. In this case the involved enterprises may be weakly aligned in terms of transformations, but they share the enabling context of a lead enterprise. The transformation of the Scientific Park as an enterprise is providing services to enterprises producing high tech products. Thus the transformation of the lead enterprise is *orthogonal* to those of the peer enterprises under its umbrella. Making the Scientific Park, as an umbrella enterprise, viable does not imply that it is producing sophisticated software applications or transducers for industrial applications or many similar possibilities, which are the concerns of its embedded enterprises; it implies providing common services to a network of enterprises. However, in the longer run the viability of a scientific park will depends on successful tenants and the collective organisation they form together; some form of collaborative complexity management is taking place.

In all cases, even if its transformation is not aligned with the transformations of some or all of the networked enterprises, the lead enterprise strives for its own viability and therefore develops an orthogonal transformation to those of the network. This lead enterprise is providing regulatory rather than steering capabilities to the enterprises using its services; it possesses its own primary activities, such as providing technological infrastructure, meeting facilities and printing facilities, and possibly several others, to the embedded enterprises.

Peer Enterprises

An alternative organisational form is that of enterprises aligning their interests to constitute a larger enterprise. Each has an independent transformation of its own, but they agree to produce together a larger, encompassing, transformation. For the Viplan Methodology the system in focus is the larger organisational system formed by the member enterprises. Provided they achieve cohesion they are forming a larger organisational system. This is the case, for instance, of a cluster of aligned regional enterprises. Regional policies may enable producing a collaborative cluster. What is particular to this case is the alignment of their interests. One enterprise may operate as a catalyst, as primus inter pares, but it has its own transformation; all enterprises depend on external resources supporting the cluster and steered by regional or sectorial policy and management bodies. This idea can be extended to develop virtual relationships with similar and complementary enterprises in different regions and with larger complementary enterprises of a global nature, forming what O'Callaghan calls extended dynamic clustering (Nachira et al., 2007). What is of particular relevance today is that these forms of collaboration are supported by ICTs that few years back where unavailable (Avril & Zumello, 2013; Johnson, 2012; Christensen. et al., 2009; Tapscott, 2009; Teece, 2008). Standards for collaboration are evolving constantly, making possible routine exchanges and communications between enterprises as if they were parts of the same enterprise.

In summary, methodologically, Viplan makes the distinction between enterprises, institutions and organisational systems. The system in focus implied by an organisational system can be much larger, but also smaller, than either an enterprise or an institution. It is the organisational system in focus that is considered the response to the opportunities and challenges posed by a relevant problematic environment. Naturally, if an enterprise or institution creates, regulates and produces its own products/services, and is the focus of the study, then it can be the system in focus, whose structure and identity are studied by the Viplan Method. However, if the focus is one institution that does not create, regulate and produce its products/services but only does one or two aspects of this triplet, then the organisational system will be larger or smaller than the institution, as was the instance of the educational system explained above.

ECM'S: COMPLEXITY AND REQUISITE VARIETY

Enterprises operate in a world that constantly surprises them with new possibilities; in the language of complexity this means that agents in the environment are generating unexpected states for which these enterprises so far neither have interest nor response capacity. In these circumstances enterprises either create these responses or risk underperforming and failing. In other words, in a dynamic and increasingly technologically sophisticated world, the number of possible relevant states that they are dealing with is increasing all the time, possibly beyond their immediate performative capabilities, and unless they develop commensurate response capacity they will underperform or fail altogether (Ashby, 1964).

The variety, even of simple situations, is exceedingly large (Beer, 1979). Thus, observers in general, pay more attention to collective behaviours than to moment-to-moment individual interactions and in doing so they choose the particular aspects that they want to observe. To avoid being swamped by variety observers, including participants within, ascribe purpose to the collectives. Indeed, this is also necessary to the collectives themselves if they are to maintain their identity and make meaningful their interactions. This implies that, to some degree or another, they need to observe their interactions holistically rather than piecemeal. In a cognitive sense their situation is defined as a black box, with inputs and outputs. Attenuation of a situational variety is unavoidable; this is a reflection of our limited cognitive capacities. A myriad of processes of co-evolution of human interactions and technology are responsible for transformations in our world. These are learning cycles that emerge from perceived threats and opportunities.

Dealing with threats such as terrorism or global warming or dealing with opportunities such as introducing new services in the market or empowering communities to make their own decisions, requires socially far more than trial and error; situations are too complex and time is too short. We need ingenuity, inventiveness, creativity and resources to produce desirable outcomes and avoid undesirable ones. These are all problem-solving situations that share a common structure; the need to attenuate (reduce) situational complexity and amplify response complexity. Performance depends on the balance achieved between attenuation and amplification. Quite naturally these are learning processes that help improving an enterprise's complexity model.

This is fundamentally a relational construct; it is about interactions between regulators and regulands, that is, situations relevant to them (Ashby, 1964). Operating in any context poses challenges –it is a learning experience- making apparent that one way or the other enterprises –as long as they remain in operation- are always matching situational varieties. Enterprises, as regulators, are constantly striving to regulate a challenging surrounding. In this practice, through interaction, they develop variety operators, that is, particular strategies and practices to deal with the variety of their environment (cf. Espejo & Reyes, 2011). What is of significance is that for any situation unwittingly a wide range of variety operators co-evolve, thus defining the communications between the enterprise and its situational environment. The quality of these operators defines the enterprise's performance. Not any variety management is adequate or even viable. An imbalance between amplifiers and attenuators implies a waste of resources and often poor performance. This may be happening precisely in challenging situations where the enterprise should be striving for more and more ingenious forms of problem solving. Thus, from the perspective of complexity, the question is how to develop more ingenious and effective forms of variety management. Indeed, not only these forms are influenced by available technologies, but also, in challenging situations and uncharted grounds, the search for these forms might be the trigger of new technologies and scientific discoveries. ECMs are proposed as forms of variety management for specific problematic situations.

Situational challenges for an enterprise emerge from imbalance between the attenuation of the situation's variety and the amplification of its responses' variety. When, in any relevant situation for an enterprise, new developments beyond its capabilities emerge, such as labour unrest, or a new scientific discovery or a technological breakthrough, they may trigger a loss of competitiveness or inadequate performance in comparison with those producing similar services. The implication is that its ECM is proving ineffective and its regulatory cycle with the environment -encompassing its attenuators and amplifiers in use- is requiring learning; possibly the new developments imply redressing the balance of these variety operators. Indeed, often these new scientific and technological developments affect this balance, making attenuators or amplifiers or both stronger or weaker. Not surprisingly new global communication demands are challenging enterprises' response capacity everywhere. Proliferation of variety that is not accompanied by a commensurate attenuation of environmental variety, traps enterprises in non-learning situations; they fail to understand the consequences of the evolving new states. They risk loss of control, reduced performance and damage to people and organisations. Enterprises need a holistic appreciation of the situation. This holism is at the core of the enterprise's complexity model. This view implies that to learn and improve performance enterprises have to improve in tandem their capabilities to observe and articulate environmental distinctions and design and produce responses.

A consequence of the new technologies is increased connectivity –at all structural levels, from the local to the global- connectivity that enterprises have not experienced before; less and less they can operate in isolated niches and more and more they need to appreciate the consequences of their actions beyond their local environments. It would appear that the new technologies are increasing people's holistic (i.e. systemic) appreciation of their organisational world. In this sense technologies are changing the nature that enterprises are confronting in their quests for effectiveness. Dealing with these challenges requires not only cajoling other enterprises and people to accept an increased number of products and services but listening to their views and concerns through effective communications. These are essential learning loops to maintain a dynamic stability. An enterprise's performance is experienced as communications that produce desirable outcomes for stakeholders; it depends on communication channels that amplify and attenuate complexity and also on transducers that relate the enterprise to environmental agents. These experiences are underpinned by knowledge, resources and technologies that change and reconfigure the enterprise's dynamic capabilities.

Enterprises in the network economy are constantly reconstructing their relationships with other enterprises and institutions. The speed of this reconstruction is increased by the new technologies, which allow reconfiguring resources, developing new dynamic capabilities and creating new organisational forms and cognitive capabilities. The invention and deployment of new technologies play a key role in this evolution. Enterprise complexity models offer a way to understand how to use new technologies[5] in the constitution of effective economic and social processes.

To sum up, enterprises are constantly confronted with challenges that exceed their response capacity. To maintain viability they require ingenuity, inventiveness and stamina. ECMs offer a systemic view of the opportunities that an enterprise needs to develop viability in a complex world. This view emphasises the co-evolution of an enterprise in its self-constructed environment and the entrepreneurial efforts that are necessary when this enterprise is confronted to a challenging and demanding environment. In these circumstances ingenuity in the use of technologies is essential. The key conceptual constructs for these models are variety and more specifically the Law of Requisite Variety (Ashby, 1964).

THE VIPLAN METHODOLOGY: BRIDGING AN ENTERPRISE´S COMPLEXITY GAP WITH ITS ENVIRONMENTAL COMPLEXITY

Enterprises today are evolving in environments that are significantly different to those of the industrial economy of the 20th century. We are now talking about the 4th Industrial Revolution, with big data, supercomputers in our pockets, bitcoins and blockchains, internet of things, 3D printing and many more (Schwab, 2017), as enabling technologies for enterprises. New enterprises are driven by this collaborative and enabling technologies that make possible self-organising networks and new organisational forms that potentially can be less hierarchical and more participative and inclusive.

The Viplan Methodology was first published under the name of the 'Cybernetic Methodology' in the paper 'Management of Complexity in Problem Solving' (Espejo 1993). Since then it has had a considerable evolution and in the context of this paper it highlights innovation and new enterprise models (see Figure 1). It is a learning platform that bootstraps the development of new dynamic capabilities to the current structure of the enterprise. It is the outcome of two learning loops. As can be seen in Figure 1, the inner (diagnostic) loop is concerned with improving the enterprise's current structure, which among other aspects, aims at improving its adaptation mechanism and therefore its capacity to assess capabilities and potentialities (productivity and latencies). This mechanism gives a solid platform to observe the enterprise's problematic environment and to decide whether to develop new capabilities aligned with the enterprise's evolving purposes and values. This is the purpose of the second outer (development) loop. In this loop the ECM is proposed as a model of the strategies both to develop the enterprise's ecology of enterprises and to interact with environmental agents. Implementing these strategies requires a well-structured organisation, which is provided by the implementation of the diagnostic loop.

The Inner Loop: The Viplan Method and Diagnosing and Improving the Cybernetics of the Enterprise

The Viplan Method (Espejo, 1989 b) has mainly been used for diagnosing organisational structures of firms, companies and governmental institutions. Beer's book 'Diagnosing the System for Organizations' (Beer, 1985) helped highlighting their structural shortcomings, but the VSM was used to a lesser extent to support

Figure 1. The Viplan methodology and the ECM

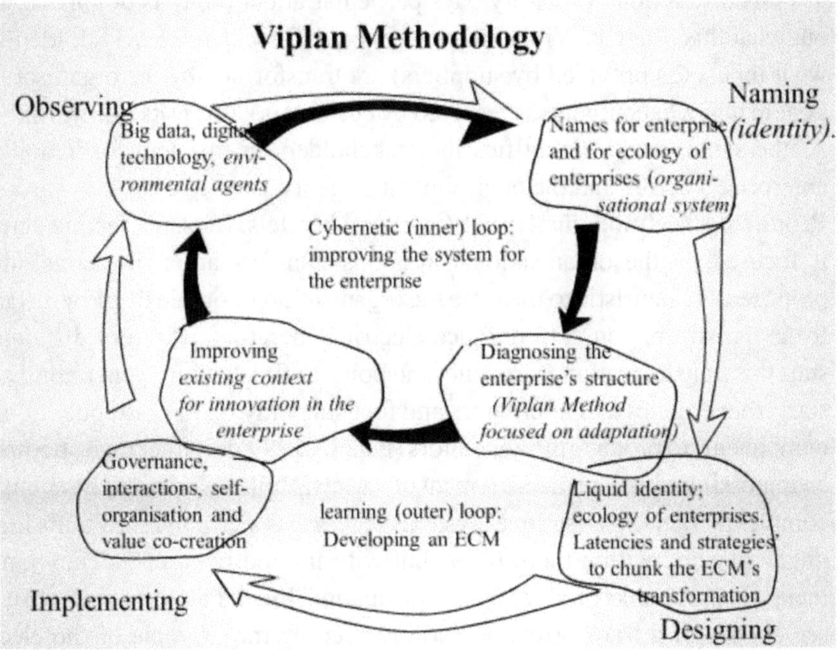

new organisational transformations, which is the focus of this discussion and my purpose is making the method more in line with the technological developments of the digital economy.

The five activities of the Viplan Method (Espejo et al., 1999) are driven by the need to measure an enterprise´s complexity, or in Ashby´s terms, by measuring the number of its possible states. Since counting the number of possible states of a situation is practically impossible (Beer, 1979), the trick for this measurement if matching the varieties between parts that need to be in interaction to achieve a desirable outcome; in Beer's terms for each "brick you need a brick layer" (Beer, 1979, p 31). The VSM is a model of the variety underpinning the necessary interactions of an organisational system to achieve viability in a complex environment. As discussed above, if the organisational system fails to match the relevant variety of its environment, then we may expect that it will not be viable. Viplan offers a method to work out the necessary parts and interactions to do this measurement.

Briefly, the five activities of the Viplan Method are:

1. **The Black-Box Identity:** To support measuring a system´s complexity, that is, an enterprise's complexity, we use a black-box description. In this mode the enterprise is observed from the outside and its identity is defined by what observers see it does. This doing is the outcome of the system's multiple interactions that define the *organisational system's boundaries*. In Viplan, following Beer's dictum that 'an organisation is what it does' (Beer, 1985, Pp 99) this discussion of identity pays particular attention to its doing. To work out what this doing is, Viplan names systems (Checkland, 1981). It identifies what inputs (as provided by suppliers) are transformed by the organisational system into what outputs (as received by customers); this is its *transformation*. At the same time, it identifies the stakeholders in this transformation; the enterprise's actors and the environment's agents[6].
2. **Producing Technological and Structural Models:** Viplan's second activity is focused on the organisational system's transformation; its modelling is proposed as a heuristic to measure the organisation's complexity. For instance, if the transformation is to produce electric generators, it is very different to start this transformation from 'nuts and bolts' to finish with a generator than to start from engines and alternators and focusing only on the assembly of these components to produce the generators (Espejo, 1989a). Rather than attempting an impossible absolute measurement of variety, Viplan works out the chunks of complexity that organisational systems *manage* as self-contained units highly interconnected within themselves, but with limited interdependency among them[7]. These chunks produce its transformation. In a diagnostic mode, observes see chunks of a transformation as illustrated by the example of the electric

generators. This is the enterprise's selected strategy to contain the variety of its self-defined transformation -producing generators. For each chunk the enterprise develops an organisational unit to match relevant environmental complexity. We can assume that through a learning process the enterprise has reached a strategy to match the variety of the environment with autonomous chunks of organisational complexity, implying that these chunks, as set up in the enterprise, are an empirical measure of its relevant environmental complexity. The situation is different when the purpose is not the diagnosis but the design of an organisational system. This is the situation when an ECM is relevant. In this case we only have partially the enterprise's learning as a reference; a new chunking of the current transformation or a new transformation altogether may be necessary. This is the case of an innovation or a new business model. For this transformation it is necessary to create a new complexity management strategy, or Enterprise Complexity Model, which may require partnerships and agreements with other enterprises. The challenge is designing a collaborative enterprise, which may make use of competitive advantages of distributing the management of environmental complexity along various complexity drivers; what we refer as innovative technological and structural models.

3. **The Organisational System's Unfolding of Complexity:** It is only when the enterprise's technological and structural models are embodied in *organisational units* that we have a template *for* an organisation's *enterprise complexity model*. In Viplan those self-contained organisational units producing the enterprise's transformation at different structural levels, define its *primary activities*[8]. and its *unfolding of complexity* (Espejo, 2003). This is the organisational system's strategy to manage the complexity producing its transformation. Each of these chunks needs autonomy to produce something cohesive in a complex environment. Whether there is a one-to-one matching between the chunks identified by structural and technological models and primary activities is precisely a point that, as we will see later, is clarified by the *enterprise complexity model*.
4. **Mapping the Organisational System's Resources Onto the Recursion-Functions Table:** The Table Recursion/Functions (figure 2) maps regulatory functions (functional resources like pricing, staff training, financial reporting and others to recursion (Figure 2 shows three levels of recursion; at the most general level COMLIS and next a number of primary activities at each of the two levels below). The table offers the configuration of the enterprise's resources creating and regulating the enterprise's primary activities. These resources are structured as regulatory functions that may be centralised or decentralised according to the enterprise's strategy to deploy resources and the available *technologies*. In particular the table makes apparent the enterprise's strategy to

distribute complexity. For instance, particular technologies may permit more centralisation without bureaucratisation. Other technologies may allow more decentralisation without losing cohesion. As new technologies emerge, different configurations of resources are possible. With new technologies the distribution of complexity can change not only within the enterprise, but perhaps, more of the enterprise's transformation, and related chunks, can be done by external agents, allowing a slimmer and perhaps more effective structure. As alternative distributions of complexity become necessary or technologically feasible, the enterprise's black-box identity -its technological and structural models and its unfolding of complexity- may change. Anticipating and discussing these changes is a purpose of the ECMs.

5. **The Organisation's Operational Identity: The Enterprise's Relationships for Cohesion and Adaptation Using the VSM:** The enterprise's performance is the acid test for the adequacy of its strategies to manage complexity. This performance is the outcome of the quality of the enterprise's structural relationships, which are mapped by the VSM (see Figure 3). In that figure the numbers in brackets are the same as those in the following list;

Figure 2. Table Recursion/Function

RECURSION \ FUNCTION			Systems Support	Assessment of Information Requirements	Intelligence on Sources	Resources Audit	Choice of Information to Acquire	Negotiation with Information Suppliers	Contract Administration	Response to Enquiries	Sales and Marketing	Liaison With Other Bodies	Pricing	Staff Training	Recruitment	Personnel	Financial Reporting	Capital Expenditure	Budgetary Control	Cost Control	Research & Development
COMLIS Ltd.			●	●		●			●		●	●	●	●	●	●	●	●	●	●	●
	Library Services		●	●		●					●	●	●	●					●	●	●
		Collection Acquisition	●	●	●	●	●	●													
		Collection Organisation	●			●															
		Information Provision	●			●				●											
	Specialised Reports		●	●	●	●					●	●	●	●					●	●	●
		Problem Definition	●	●		●				●											
		Information Services	●	●	●	●	●	●													
		Report Preparation	●			●															
	Online Services		●	●		●					●	●	●	●					●	●	●
		Data Capture	●	●	●	●	●	●		●											
		Information Processing	●			●				●											

Figure 3. Viable system model: reference model

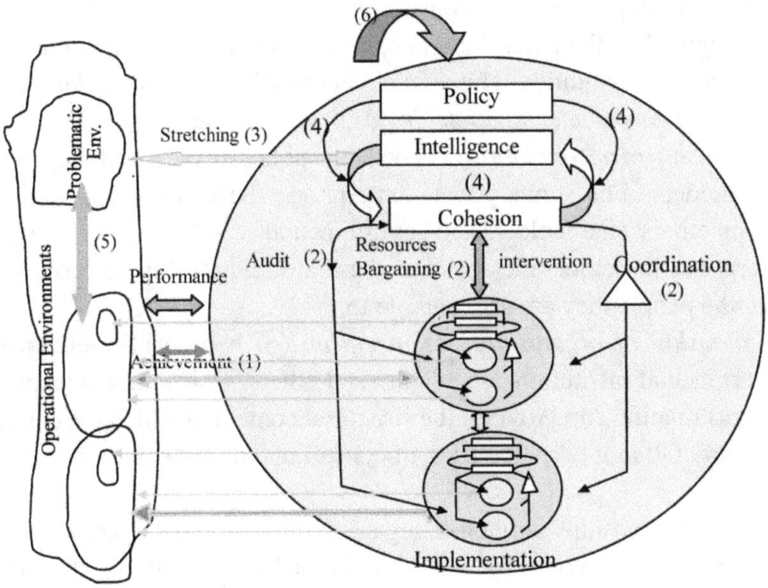

The focus is now on relationships and not on transformations:

- *Relationship of achievement* (1) between the primary activities implementing the enterprise's policies and those environmental agents affected by, and affecting, its products, services and externalities.
- *Relationship of cohesion* (2) between all the *actors* constituting the enterprise's *cohesion relationship* (e.g. entrepreneurs, managers, administrators, operatives and so forth); in the extremes these relationships can be governed by the authoritarian tendencies of some or by the mutual respect among all actors, enabling autonomy within a cohesive enterprise. This mechanism is focused on the 'inside and now'.
- *Relationship of stretching* (3) between those concerned within the enterprise with its 'outside and then' and agents such as competitors, regulators, manufacturers, scientists and others in its *problematic environment*. This relationship may be influenced by dynamic markets, changing technologies and other socio political factors. This *stretching* puts pressure on entrepreneurs to invent new responses and change the *status quo*
- *Relationship of policy-making* (4) between policy-makers and related actors and resources for cohesion and stretching, contributing to policy-making in the enterprise. Policy-makers need to receive a balanced influence of those

focused on the 'inside and now' and those focused on the 'outside and then'. If the former dominate the relationship policies are likely to be conservative and short sighted. If the latter dominate, policies are likely to be unrelated to the enterprise's capabilities. This is the enterprise's *mechanisms for adaptation (related to environmental stretching as mentioned before).*

- *Relationship of accountability* (5) between policy-makers and the enterprise's stakeholders. For some private enterprises these are relationships with, among others, shareholders and environmental stakeholders. For government enterprises in democratic societies these are relationships of policy makers with the people they are accountable to.
- *Relationship of organisational citizenship* (6) between the enterprise and governmental institutions, regulators and others, which are contextually part of. This relationship provides the structural context in which the enterprise is evolving. Often it takes the form of constraints to this evolution.

These six relationships define the organisational system's operational identity and together with the previous four activities of the Viplan Method help to diagnose the enterprise's organisational identity and structure (Espejo, 2008a) and support the final activity of the inner loop of the Viplan Methodology; improving the existing context for innovation and development in the enterprise. This is a platform to bootstrap the learning (outer) loop of the methodology in the process of developing the ECM. This is elaborated in the next section of this paper.

The Outer Loop: Transformation and ECMs

An ECM is a model of the strategy of an ecology of enterprises to manage the complexity of their interactions with agents in their environment. It should be an ingenious form of deploying resources to bridge the complexity gap between an enterprise (as an ecology of enterprises) and agents in its environment in the quest of developing its dynamic capabilities and improving performance.

As already anticipated, this complexity management entails the creation, regulation and implementation of an enterprise's policies. The key aspect to appreciate is that the ECM is a methodology focused on problem solving rather than on the viability of a particular enterprise. The VSM is a model of effective ways to manage the complexity of a self-constructed organisation taking place in a highly complex environment. The ECM's key concern is working out its boundaries, finding ways of understanding and operationally measuring the complexity of its chunks, reconfiguring its resources taking into account technological developments and the complexity of the environment and building a cohesive and adaptive structure. In order to reconfigure resources it is necessary to understand their systemic contributions to the extended model of the enterprise, that is, to its ECM. Are they supporting aspects of its policy, intelligence, cohesion, coordination or implementation, at different recursive levels of the ECM?

The starting point for all this is the entrepreneurial concern of producing a desirable transformation to bridge a complexity gap and improve a situation. Producing this transformation is what the policy is about. Additionally, this transformation helps clarifying the *systemic purposes* of the enterprise's resources. The enterprise can be a business, a charity or any other organisational type. Thus, methodologically it is necessary to agree the enterprise's transformation. From this transformation it follows the discussion of extending or modifying the boundaries of the enterprise, followed by possible structural and technological models, unfoldings of complexity and alternative distributions of resources and regulatory discretion for the ECM.

Transformation: Technological and Structural Models for the ECM

The first step to model and enterprise's complexity is defining its problem-solving space. The question is the mismatch between its current "ECM" and environmental challenges. This step starts clarifying the enterprise's new transformation.

This transformation (in the outer loop) is bootstrapped to the diagnosis of the cybernetic (inner) loop in figure 1. This diagnosis, as explained above, uses the five activities of the Viplan Method. Can resources be better deployed, and can relationships be more effective? However, as the diagnosis is done, it is natural to work out changing environmental conditions and whether or not a new ECM is necessary. The enterprise is learning from its environment, which may require adaptation to produce more effective responses. This is a matter of ingenuity; the enterprise may need new organisational forms to bridge the variety gap with its self-defined surroundings (Espejo, 2017).

Reconsidering what is going on, naming a new transformation, or finding better ways to do the current one, is at the core of extending the viability of any organisation. This is a trigger to develop an ECM with the support of the outer loop of the Viplan Methodology. In the network economy new forms of transforming inputs into outputs are emerging constantly. Information and communication technologies (ICTs) are making possible networks and alternative forms of association for enterprises. Producing an enterprise's services can take place in different parts of the world and with different technologies. Discussing the enterprise's boundary is essential to reinvent the strategy to carry out its revised purposes. More and more these boundaries entail several, if not many, enterprises. The organisational system that it implies -the system in focus- is likely to be larger than the current enterprise; the ECM needs to consider resources and relationships beyond the enterprise. The named transformation implies tasks that are performed by other enterprises. This fact implies developing relationships of cohesion for this new network of enterprises, which requires coordination and integration systems to overcome, among other aspects, distance and time differences. Product and procedures standardisation between so far different enterprises become essential; enterprise data interchange, communication protocols (e.g. blockchains and operational coordination) of B2B interactions. Software and many other coordination technologies are necessary to build up trustful relationships. These technologies are enabling increasingly reliable communications throughout the organisational system, regardless of distance and time differences. These developments are making possible new and more flexible organisational forms.

Defining the problem-solving space helps appreciating the complexity gap between the enterprise and its self-defined environment. This is the role of the new ECM. In this case, an aspect of problem solving and therefore of the ECM, is visualising a new network of enterprises -a new business model- to produce the new problem-solving system. The boundaries of this network are defined by the technological and structural models for the *new transformation* (outer loop in Figure 1).

The outer learning loop of the Viplan Methodology is bootstrapped to the current enterprise's technological and structural models, that is, a new ECM, with a new problem-solving space, is bootstrapped to the current one. This is proposed as a heuristic to assess the organisation's strategy to manage complexity. Methodologically, Viplan requires that technological and structural models are produced to visualise a new cluster of networked enterprises producing the organisational system.

Producing new products and services, as enabled by new technologies, transform inputs into products and services for the stakeholders emerging from the revised problematic environment. Producing this new complex transformation entails different interrelated chunks of complexity or tasks, which can be deployed in different forms. This new technological model is a preliminary map of how to produce these products and services. In practice these maps co-evolve from the interactions between chunks of complexity emerging from the technologies at hand and the agents producing environmental pressures.

As implied above, breaking a transformation into manageable, self-contained tasks, is an important strategy to manage its complexity (Simon, 1981). A small consultancy task may be within the scope of one person. However, manufacturing an airplane requires many, largely self-contained, tasks (i.e. chunks of complexity). These tasks need to be properly coordinated and integrated through an effective organisational structure; thus decisions about this chunking are of key importance for effective organisation. If the chunking leaves large aspects of significantly interdependent activities distributed between different chunks, then the organisational system will require additional coordination efforts between chunks that otherwise would be largely autonomous. In any enterprise there are teams for each task or chunk of complexity; they absorb in full the contingencies that may take the task off course. The workflow starting from supplies and finishing with products is composed of several or multiple chunks that define the enterprise's technology to produce them. When these chunks are too many it is natural to cluster them in larger chunks that define the enterprise's strategy to manage its complexity. This chunking is the outcome of the enterprise's experience managing the complexity of a transformation and as such it could be the source of practices for learning purposes. In the case of successful enterprises these models may well be good practices that are used to learn and improve performance. Good practices from other enterprises support the development of this enterprise (cf. Cisco, 2007). It is natural to expect that multiple practices or technologies emerge to produce the same transformation suggesting the possibility of multiple technological models.

In the language of the Viplan Methodology these are alternative forms to manage the complexity of a transformation; these are learning strategies. Perhaps, the simplest way to start this learning process is to model the current way in which the enterprise is producing its products. However, staying with that model and not searching for new practices would be limiting the enterprise's learning. This search for a new technological model, in line with new technologies, is at the core of ECMs.

Furthermore, the complexity of an enterprise is more than the interrelated chunks producing its products. Were that the case the enterprise would be limiting its value adding to managing the production processes with no concern for environmental externalities. In fact the complexity of an enterprise extends beyond producing particular services or products; though these products and services are produced with particular characteristics, for agents or customers located in different places, operating at different time zones and so forth, the environmental externalities are increasingly more relevant. These are all required strategies to manage complexity. The complexity that an enterprise manages emerges from key *complexity drivers or variety operators* closely related to an enterprise's technology, but increasingly they need to account for new environmental pressures. These models refine the technological models to account for product types, geography, market segments and time in different environments. And, it is only when the system in focus is the *extended enterprise*, that is, when the enterprise is the new ECM, bridging the complexity gap emerging from its problematic environments, that the enterprise is measuring its relevant environmental complexity.

If an enterprise has product design capacity it may subcontract the manufacturing of some or all of its parts leaving to others their implementation. This is a typical case where the organisational system, as an enterprise complexity model, is different to the enterprise (e.g. see Nike's example below). In fact an enterprise's strategies to produce a transformation can be many and new technologies are offering opportunities for new, environmentally sensitive, forms of its chunking and their connecting. The technological and structural models in use are co-evolving with new technologies. At any time, current models can be superseded by an ingenious new form of producing the transformation. Each time this is the case the enterprise may take different decisions about the structure and relationships to produce it, perhaps creating a different, innovative, *organisational form*. In addition to choosing the technology for a transformation -as explained before- the enterprise may decide the geographic sourcing of their suppliers, the distribution processes, markets to serve, approaches towards time and strategic externalities.

From the perspective of complexity these decisions imply the selection of *variety operators* for the enterprise (Espejo, 2015 b). The design and selection of these operators is particularly relevant in today's globalised, environmentally sensitive, societies. These operators define an enterprise's new strategies to produce and make available their products and services, intertwined with the technologies-in-use. Each of these decisions generates complexity that the organisation needs to contain structurally in ways that enhance its capacity to respond to environmental pressures. As the VSM shows, structurally this implies autonomous units, within autonomous units (i.e. *a recursive structure*). For a *lead enterprise* variety operators are both other enterprises and the communication and management systems that are necessary to produce a cohesive and adaptive organisation. This is a heuristic in the generation of ECMs. As new technologies suggest alternative structural and technological models, most likely their structural mapping will require an alternative unfoldings of complexity and new considerations about the distribution of autonomy and discretion. These are platforms to work out necessary structural changes using the VSM and the Viplan Method. In the extreme these new technologies may show that the current transformation of an enterprise is unviable, suggesting that before reconfiguring its resources, it needs questioning its identity and purposes altogether.

Returning to Nike, assuming that it defines its transformation as providing sportswear and apparel to worldwide markets, in practice it means doing this transformation through other enterprises distributed globally, and taking into account more regulatory demands. The chunking of its transformation includes tasks that are responsibility of distant enterprises. Nike is the lead enterprise, and together with its contractors constitutes the organisational system. These manufacturers are primary activities amplifying Nike's variety in relation to its markets, but need more functional capacity to respond to local, national, global regulatory demands as is discussed in the next section.

RECONFIGURING RESOURCES AND RELATIONSHIPS FOR THE ECM: DISTRIBUTION OF DISCRETION IN THE ORGANIZATION

A complementary strategy to close the complexity gap between and enterprise and its environment is reconfiguring resources and relationships. This strategy is closely related to the Viplan Method, in particular the activities 3 and 4; the organisational system's unfolding of complexity and the mapping the organisational system's resources onto the Recursion/Function table (figure 2). Perhaps one of the most powerful strategies to amplify an enterprise's complexity is devolving problem solving and decision capacity to small teams in the enterprise. In the language of the VSM this implies unfolding its complexity into primary activities as far as possible. In complexity terms this is the case where resources and functions are possible to decentralise.

Some products are more likely than others to make possible this straight unfolding of complexity. For instance, in technological terms the production of materials for plastics is likely to be less flexible than the provision, for instance, of consultancy services. In plastics it is unavoidable to use large machines. On the other hand it is natural to have small management consultancy teams. This would suggest that a flexible structural amplification of complexity is more restricted in a plastics plant than in a consultancy firm. These are precisely the challenges for ECMs. Of course, technology and ingenuity can overcome –to a degree- these restrictions. Automation can reduce the number of people working for a plastics machine to small largely autonomous teams. Ingenuity can create for them multifunctional teams with capabilities to create, regulate and produce product lines for particular markets, all using the same large equipment. Even, and perhaps this is less easy to see, mixed strategies to manage complexity can produce significant increase in variety amplification. With current technology, it is possible to have resource centralisation, using scarce specialised resources, and functional decentralisation for service delivery. This is a case where technology permits centralising product creation and regulation, thus gaining the advantage of scarce expert resources, at the same time of decentralising less specialised resources for service delivery (Christensen et al., 2009 illustrates this situation for health services). What makes this situation consistent with the heuristic of pushing down complexity unfolding is that structurally and technologically it is possible to link specialised central resources with service deliverers by bringing them together into *virtual autonomous teams* (Espejo & Reyes, 2011, Chapter 8).

These new ECMs offer different strategies to manage the enterprise's complexity. The new technologies are changing the nature of these enterprises. Complexity considerations and new technologies offer opportunities for innovative ECMs for enterprises producing large products such as aircrafts. In these cases, key business functions such as finance, sales and marketing are likely to be centralised reducing the scope for entrepreneurship at lower structural levels. In this example, at a first glance, the unfolding of complexity is likely to be skewed towards the top. In other words, the scope for local autonomy appears to be restricted by the product and its technology. However, this needs not be so. An enterprise like Airbus is the primary (lead) enterprise of a much larger organisational system, which is managing the cohesion of a wide range of enterprises in different European countries. Each of these enterprises needs to be viable in its own right and therefore it is in Airbus's benefit to enable their viability. Similarly, these local enterprises might well be networks of more specialised enterprises, whose viability is equally necessary. And so forth. This ECM is very different to that of a large all-embracing enterprise, which in the extreme conflates enterprise and organisational system.

These examples highlight the amplification of an enterprise's complexity. Equally important is to consider attenuators in the ECMs. In Nike's case, in addition to the amplifiers already discussed it needs attenuators of environmental complexity. Without effective means to reduce undesirable variations between distributed manufacturing operations, that is, without attenuators of those enterprise's varieties, Nike may find itself in difficulty to maintain its reputation and success. This latter aspect shows that developing mechanisms for organisational cohesion, which respect the autonomy of their suppliers, are necessary for Nike's viability. Furthermore, it needs to manage relationships between subcontractors, customers and a range of other agents. Developing an ICT platform, such as blockchains (Tapscott & Tapscott, 2016), to manage an extended value chain is a powerful attenuator of environmental variety, which not only standardises interactions but makes them more trustful. The more Nike enables direct interactions between local suppliers and between these and customers, the more of its environment's relevant variety will be absorbed in the environment itself, reducing the residual variety that it needs to manage directly. Similarly, there are a range of variety operators that Nike needs designing in order to have an effective ECM. Also, it needs making viable its own enterprise transformation; that is, of its orthogonal transformation to those transformations of the enterprises constituting the organisational system it leads. Say, it needs capacity to produce designs and business networks (i.e. its own primary activities) to produce sports related products. These transformations; the design, developing and maintenance of relationships are platforms for its own learning. These transformations are its strength and also its Achilles heel. It offers the strength of the great flexibility to reconfigure resources and develop new capabilities should the circumstances so require, but it has the challenge of building up relationships with distributed enterprises which use different standards and make, among other aspects, more difficult complying with safety requirements.

ICTs are changing the nature of organisational relationships. Relationships dominated by high cost technologies, only accessible to a few are likely to reinforce centralised relationships. On the other hand, distributed technologies, such as 3D printers, are becoming accessible to many and support distributed, but demanding, relationships. Information technology enterprises have implemented changes in relationships between customers and suppliers over time. A company like CISCO has implemented, supported by ICTs, an innovative business model, which has enabled effective relationships between suppliers and customers of its products. Developing and maintaining these digital platforms appear to be CISCO's primary activities, which add significant value to customers and define their core competencies.

Equally, in the late 1990s and early 2000s Dell developed an imaginative enterprise complexity model, making an increased use of the network economy to sell computers directly to individual and corporate customers. One of their core competencies was, and to some degree still appears to be, developing and maintaining web applications and logistic networks to relate customers to suppliers bypassing distributors. The challenge for them, in addition to making competitive whatever manufacturing activities they have, is developing and maintaining web and logistic competencies to support a large number of enterprises contributing to the delivery of its products. Dell to make possible direct services to individual customers has managed relationships between customers, suppliers and logistic enterprises. A more traditional business model would have required a large number of intermediaries and distributors. Indeed, Dell, enabled by ICTs, invented a new model to manage the complexity of its environment some time back. However, its more recent apparent loss of competitiveness suggests that ECMs need to be reinvented continuously and, for instance, blockchains may be offering this opportunity.

ECMs in the financial sector have experienced much change in the past 30 years, perhaps not always for the good (Haldene, 2010). In the UK starting with the Big Bang in the late 80s- functions such as registering share transactions, which were in the traders' back office, are now enterprises in their own right. Many more transactions are now happening in significantly shorter periods of times. Managing these transactions became a parallel transformation to the more glamorous share dealings of the London Stoke Exchange. The chunking of financial services and their derivatives changed the ECM of financial services often at the expense of economic activities, as the 2008 crises made apparent.

As said before, a focus on relationships among producers is an ECM for regional and local economies. Clear local policies and flexible technologies make possible the development of enterprises as clusters. Today's ICTs, more than ever before, allow the clustering of several enterprises in one organisational system. This has been the goal of the EU project Digital Business Ecosystems reported above. This project made clear that open source software allowing Business to Business commerce (B2B), Enterprise Data Interchange, (EDI), internet integrated Computer Aided Design (CAD) and so forth were infrastructures or enabling technologies for new economic models, less reliant on unfair free markets. In this case government institutions provide the leadership for viable clusters. These efforts to enable clusters require that the participating enterprises develop relationships of organisational citizenship (relationship (6) in figure 3) by aligning their interests within the cluster as well as with the relevant institutions in the region or locality.

An important methodological tool for modelling the distribution of functional resources in an organisational system is Viplan's Recursion/Functions Table (figure 2). From the perspective of the ECM this table is an instrument to test whether better management and organisational practices, including new technologies, permit a more effective distribution of resources. Through the VSM this is a way to appreciate the organisational consequences of their reconfiguration. Though the table by itself permits design and diagnosis, its refined use requires assessing the performance of the system in focus (i.e. the organisational system in focus) in its problematic environment. It is through this review that the systemic consequences of any proposed reconfiguration of resources can be assessed.

This is the concern of the next section.

ECM and Performance

The enterprise's performance is the acid test for the adequacy of its strategies to manage complexity. An ECM, as a tool for governance, is a tool to adjust/change the enterprise's structure to develop its performance. These changes are driven by many factors, but most importantly, for the purpose of an ECM, by changes in policies and by communication and information technologies. Thus, as an ECM unfolds, it is necessary to observe how technologies alter communications between environmental agents and organisational system's actors. Is the enterprise developing adequate capabilities to build up its intended ECM in an uncertain environment? Maintaining an adequate performance in a dynamic environment requires constant structural adjustments. The ECM, as a model of the structural strategies to manage complexity, helps working out improvements, small and large, in the enterprise by developing communications and reconfiguring the distribution of resources.

The ECM is a dynamic model that has as a starting point the current situation of an enterprise, whose response to its total and problematic environments is reflecting its current performance. The enterprise's performance is measured by its current achievements and latencies in its communications with agents in the environment.

They are depicted in figure 3 by relationship (1) -with the total environment- and by relationship (3) -with the problematic environment. Key performance indicators are used to measure this performance. The current enterprise has actualities -what it does today-, capabilities -what could do with a good organisation-, and potentialities -what its latencies are in the environment (Beer, 1981). Also for the ECM, the enterprise's performance uses actualities, capabilities and potentialities, but now the focus is on the complexity of communications rather than on products or service. In this context a*ctuality* is measured by the relational complexity of the current communications of the enterprise with its agents, *capability* is measured by the complexity of the communications that the enterprise could match in the environment, with existing technologies and competencies, and potentiality is defined by the latent complexity of its communications with agents as emerging from technologies, in the problematic environments. This is the complexity implied by the ECM, which reflects the new boundaries of the enterprise and its "potentialities". To deal with this complexity the enterprise needs investing in new technologies and developing new structures. *Achievement* is the ratio between actuality and capability, *latency* is the ratio between capability and potentiality and *performance* is the ratio between actuality and potentiality. Achievement structurally relates to the enterprise's *achievement relationship* (relationship (1) in figure 3); is the enterprise matching the communication demands of the intended network? Latency structurally relates to the *stretching relationship* (relationship (3) in figure 3); is the enterprise aware of its communication latencies, that is, how much more it ought to develop new communications? Is it exploiting the opportunities offered by the problematic environment? Is it developing its capabilities? And the *performance relationship* relates to the ECM bridging the organisational system´s complexity gap with its environment.

The structural aspects of these communications can be appreciated in figure 4, which guides the practical development of the ECM. Improving an enterprise's performance is the purpose of the cybernetic (inner) loop in figure 1. It requires bridging its self-defined complexity gap -or potentiality- with environmental agents. For this it needs capacity to assess current capabilities and the resources necessary to develop new capabilities. The ECM models strategies to manage the complexity of environmental agents' communications starting from the current *organisational structure.* The better is this structure the better is likely to be its appreciation of the problematic environment and the clearer are the opportunities to develop the enterprise towards a selected ECMs. The challenge is dealing with the communications implied by new capabilities. The *inner loop* of the Viplan Methodology offers a heuristic to diagnose and improve the current enterprise's structure. The modelling supporting this loop, and the selection of key performance indicators, is done with the Viplan Method (Espejo and Reyes, 2011). Additionally, developing the enterprise towards its ECM requires seeing the enterprise as part of a *development loop (i.e. outer loop* in Figure 1*)* defining and creating new communication capabilities. This is a learning loop whose purpose is appreciating in conversations with environmental agents alternative problematic environments and visualising latencies for developing new communication capabilities. These are potentialities for the enterprise and point at the resource requirements and their configuration to develop the enterprise implied by its desired ECM. Developing these capabilities imply investment programmes and structural changes. The *development loop* uses an ECM of the interactions between the enterprise and its environmental agents to work out new capabilities (see Figure 4). This model proposes strategies to manage environmental complexity based on ICTs and other technologies. It helps bootstrapping new policies to an existing organisational structure that makes possible the development of new capabilities.

From figure 4, considering that the complexity of communications is its variety, managing the relational variety of the enterprise with its environment, requires:

1. Reducing the environmental variety without destroying it. This strategy has the potential of increasing the environmental variety that the enterprise is matching by fostering policies that increase the self-organisation and self-regulation of the targeted environmental agents. Self-organisation and self-regulation deal with most of the variety within the environment itself an decrease the residual variety that needs the enterprise's attention. This is a way of doing more with less. Indeed, the very same communication technologies can help to do this.

Figure 4. Template to model the interactions between the Enterprise and the Environment of the ECM

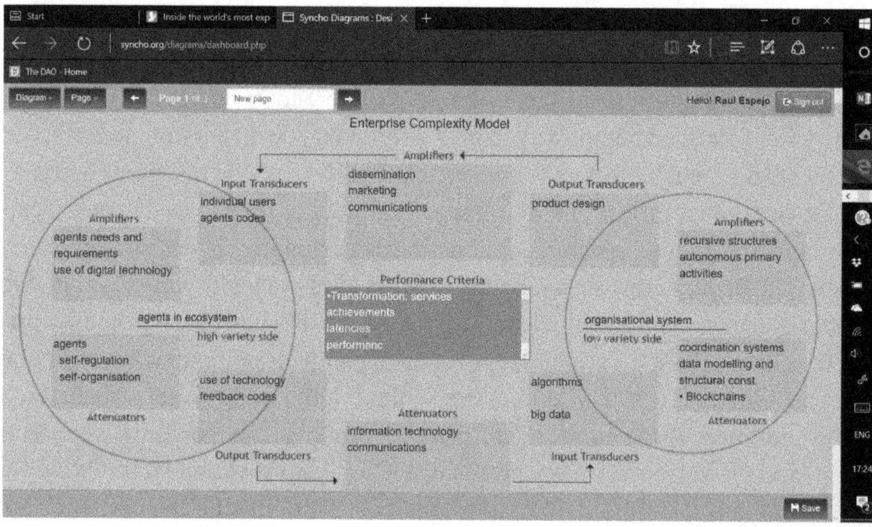

2. Increasing the enterprise's variety through collaborative enterprises building the ECM; these collaborative enterprises increase the enterprise's variety as they share with the enterprise its doing transformation. They constitute the evolving innovative ECM; they communicate the enterprise's variety to the environment, provide distribution networks and so forth.
3. Managing the residual variety left unattended by environmental agents before it reaches the enterprise, by using variety attenuators, such as information filters, communication devices and so forth. This is the *variety engineering* relevant to the enterprise. Thus, an enterprise's ECM is constituted by the organisational system's structure as it evolves enacting ECM interactions with environmental agents. A purpose of the ECM is offering ingenious variety engineering strategies for the enterprise to manage environmental complexity and to develops new capabilities. To check whether the variety engineering model is adequate it is necessary to assess whether the complexity management strategies are balanced regarding amplification and attenuation of variety. For instance, an enterprise's communication attempts to develop a network may increase other enterprises' awareness of its products and services. However, if this amplification triggers a demand for communications that the enterprise is unable to respond well to, then the development of new communications may not succeed. The enterprise may develop tools to improve the other enterprises' willingness to produce part of the services by themselves in the environment. But, if after these changes the enterprise still does not match satisfactorily the building up of communication requirements, it may experience a backlash; the enterprise did not pay enough attention to balancing amplification and attenuation. It fails to develop new capabilities. Figure 4 is one of multiple templates to study these strategies to manage complexity. These templates are learning models and therefore need to be in constant evolution supported by new technologies and possibly social networks to allow the enterprise to monitor its achievements, latencies and performance.

From the recursive nature of the VSM it may be expected that multiple ECMs will be necessary as the primary activities of the organisational system become particularly significant in the management of environmental complexity. The same ideas as developed above apply to all these primary activities.

Coda

The strategies that an enterprise use to create, regulate and implement its transformation depend on social, cultural, economic, technological and other factors, whose relevance emerges as it co-evolves with relevant environmental agents. *When environmental pressures mount an enterprise can vary significantly the configuration of its resources and this can happen in a relatively short term.* It is apparent that the days when a particular configuration could remain unchanged for long periods are over and flexibility is critical for viability beyond survival (Espejo, 2008a).

Enterprises are constantly experiencing significant changes in their environments. Ingenuity in managing the complexity of their relationships is fundamental to success. New digital technologies offer possibilities for multiple complexity management strategies, which are changing the nature of organisations. The industrial economy of the 20th century is being superseded by the networked economy of this century (Benkler, 2006; Schwad, 2017). This change is transforming the strategies to manage complexity. Creative forms to match the requirements of an enterprise are emerging today. From the rather centralised design, production and distribution of products of the industrial economy of the past century, the networked economy is allowing decentralised production models and distributed production processes in collaboration with new enterprises.

Dynamic enterprises are interested in networks to reconfigure structures and response times; they make possible the production of sophisticated products, quicker and closer to customers. They are less interested in production processes from 'nuts and bolts' to finished products. Globalisation is making this ineffective. This new kind of enterprise may decide that it is more challenging and effective to manage the *communications* that are necessary for an effective value chain than to manage in full the production and distribution activities of this chain. For one *transformation, as well as for alternative innovative transformations* it is possible to produce a wide range of enterprise complexity models, which may support reconfiguring resources in the organisation and from there the emergence of different capabilities. The enterprise may explore alternative primary activities to those producing this transformation today. An ECM transforms the resources, communications and relationships of the existing enterprise. In the Viplan Method, technological and structural models help hypothesising new unfoldings of complexity and distribution of resources. The Viplan Methodology recognises that these models are platforms for discovering latencies and for innovative implementation (i.e. for bridging the enterprise's complexity gap with its environment).

They provide the platform for developing communications and eventually for hypothesising new unfoldings of complexity. For instance, for an enterprise in a dynamic environment, its primary activities are less likely to be those producing its products, which can be subcontracted, and more likely to be those enabling communications and relationships between them and suppliers, customers and other agents. In the latter case the enterprise is servicing its customers with the products named in its identity but 'its doing' is managing the communications between producers, customers, suppliers and other agents. Primary activities for this enterprise, in this case, may be internet based ICT platforms to enable the communications entailed by the value chain. For public sector institutions similar considerations may apply; rather than providing traditional services directly, public institutions may manage the communications of public-private partnerships (PPPs) to provide them. What is particular to these situations is the ingenious use of digital technology to increase performance and competitiveness. The potentials of digitalisation in a wide range of industries such as publishing, transportation, retailing, marketing and so forth are indeed enormous (the 4th Industrial revolution). In all these industries enterprises are developing innovative *communications* with other enterprises and suppliers and customers. This is implied by the strategy of blockchains (Talcott & Talcott, 2016). These are strategies to develop *many to many interactions* between enterprises through distributed self-organising interactions, to make them more trustful and reliable. They need models to make easier and more cost-effective the *matching of complexities* in these interactions; new technologies are opening the space for *these matchings, and offer opportunities for ECMs.*

The enterprise's ECM, that is, its strategy to manage relevant complexity in highly dynamic niches/markets may require an on-going reconfiguration of resources, communications and relationships to produce a more responsive organisation. The problem-solving heuristic of the Viplan Methodology emphasises the capabilities of the *network economy.* The enterprise models of the industrial economy need to be replaced by network models underpinned by digital platforms. The challenge for enterprises is transforming their old industrial economy based models into network economy models that make use of new opportunities. Clustering the activities of small enterprise or individuals in different configurations can produce much more fluid organisation structures (Bauman, 2012). The cybernetics of the network economy supports decentralisation and fluidity in ways that the industrial economy did not. In the XVI century (Martin Luther's Church reformation in 1517) the invention of printing weakened the power that the Church had over information and knowledge production in Europe, and enabled, particularly in the protestant north, the emergence of early modern capitalist enterprises in the form of print shops. Equally ICTs in the digital economy have the potential of weakening the power of the well connected to the benefit of the more talented people. Our increased autonomy and in particular our capacity to do more by ourselves, makes it possible to increase the use of self-organisation and self-regulation in the environment to the benefit of social, environmental and economic enterprises. These facts, emerging from the networked information economy, can influence enterprise models to drive today's activities.

The Viplan Methodology offers a learning platform to develop an enterprise's latencies towards dynamic capabilities. The Viplan Method offers directions to design effective structures. The variety engineering model is used to model communications and interactions of all kinds, but in particular the communications and interactions between enterprises of all kinds.

This paper has proposed an Enterprise Complexity Model (ECM), which is visualised as a distributed governance model of an ecology of evolving enterprises. Governance is understood far beyond directing development; it is understood as guiding the ecology of the enterprises' collective self-organization. The challenge is steering their self-organization towards values and policies creating, regulating and producing products and services that are deemed as necessary for society in general and for the economy in particular. This view of distributed governance includes enterprises at all levels, from the most local to the most global. Governance is focused on processes aimed at innovation as well as making things happen today.

REFERENCES

Ashby, R. W. (1964). *An Introduction to Cybernetics*. London: Methuen.

Avril, E., & Zumello, C. (Eds.). (2013). *New technology, organizational change and governance*. Macmillan. doi:10.1057/9781137264237

Beer, S. (1979). *The Heart of Enterprise*. Chichester, UK: John Wiley.

Beer, S. (1981) Brain of the Firm (2nd ed.). John Wiley.

Beer, S. (1985). *Diagnosing the System for Organizations*. Chichester, UK: John Wiley.

Benkler, Y. (2006). *The Wealth of Networks: how social production transforms markets and freedom*. New Haven, CT: Yale University Press.

Checkland, P. (1981). *Systems Thinking, Systems Practice*. Chichester, UK: John Wiley.

Christensen, C. M., Grossman, J. H., & Hwang, J. (2009). *'The Innovator's Prescription: A disruptive Solution for Health Care*. McGraw Hill.

Cisco Systems. (2007). *Benchmarking a Global Business: Semiconductor Manufacturer Infineon Leverages Cisco's Experience*. Retrieved from http://www.cisco.com/web/about/ac79/docs/success/Infineon_Final.pdf

Eisenhardt, K. M., & Martin, J. A. (2000). Dynamic capabilities: What are they? *Strategic Management Journal, 21*(10-11), 1105–1121. doi:

Espejo, R. (1989a). PM Manufacturers: The VSM as a diagnostic tool. In R. Espejo & R. Harnden (Eds.), The Viable System Model: Interpretations and Applications of Stafford Beer's VSM. John Wiley.

Espejo, R. (1989b). Cybernetic Method to Study Organisations. In R. Espejo & R. Harnden (Eds.), *The Viable System Model: Interpretations and Applications of Stafford Beer's VSM*. Chichester, UK: John Wiley.

Espejo, R. (1993). Managing Complexity in Problem Solving. In R. Espejo & M. Schwaninger (Eds.), *Organizational Fitness, Corporate Effectiveness Through Management Cybernetics*. Frankfurt: Campus Verlag.

Espejo, R. (2003). *The Viable System Model; A Briefing About Organisational Structure*. Syncho Ltd. Retrieved from www.syncho.com

Espejo, R. (2008a). Observing organisations: The use of identity and structural archetypes, *Int. J. Applied Systemic Studies*, *2*(1/2), 6–24. doi:10.1504/IJASS.2008.022791

Espejo, R. (2015a, January-June). An Enterprise Complexity Model: Variety Engineering and Dynamic Capabilities. *International Journal of Systems and Society*, *2*(1), 1–22. doi:10.4018/ijss.2015010101

Espejo, R. (2015b). Performance for viability: Complexity and variety management. *Kybernetes*, *44*(6/7), 1020–1029. doi:10.1108/K-02-2015-0044

Espejo, R., Bowling, D., & Hoverstadt, P. (1999). The Viable System Model and the Viplan Software. *Kybernetes*, *28*(6-7), 661-678.

Espejo, R., & Reyes, A. (2011). *Organizational Systems: Managing Complexity with the Viable System Model*. Heidelberg, Germany: Springer. doi:10.1007/978-3-642-19109-1

Homer-Dixon, T. (2001). *The Ingenuity Gap*. London: Vintage.

Johnson, S. (2012). *Future perfect*. Allen Lane.

Nachira, F., Nicolai, A., Dini, P., Le Louarn, M., & Rivera Leon, L. (2007). Digital Business Ecosystems. *European Commission, Information Society and Media*. Retrieved from www.dogital-ecosystems

Schwab, B (2017). *The Fourth Industrial Revolution*. Penguin.

Simon, H. A. (1981). *The Sciences of the Artificial* (2nd ed.). Cambridge, MA: The MIT Press.

Tapscott & Tapscott. (2016). *Blockchain Revolution: How the Technology Behind Bitcoin Is Changing Money, Business and the World*. Penguin.

Tapscott, D. (2009). *Grown up digital*. New York: Mc Graw Hill.

Teece, J. D. (2008). *Technological Know-How, Organizational Capabilities, and Strategic Management: Business Strategy and Enterprise Development in Competitive Environments*. London: World Scientific Publishing Co.; doi:10.1142/6147

West, J., Salter, A., Vanhaverbeke, W., & Chesbrough, H. (2014). Open innovation: The next decade. *Research Policy*, *43*(5), 805–811. doi:10.1016/j.respol.2014.03.001

ENDNOTES

1 This is my interpretation of Thomas Homer-Dixon's Ingenuity Gap (Homer-Dixon 2001).

2 Ross Ashby proposed variety, or the number of possible states of a situation, as a measure of the complexity of that situation (Ashby, 1964).

3 Eisenhardt and Martin (2000) define dynamic capabilities as 'the firm's processes that use resources – specifically the processes to integrate, reconfigure, gain and release resources – to match and even create market change'.

4 Public institution, in this context, is a formal body in a society created by law to deliver a public service (Espejo & Reyes 2011).

5 "Technology ("science of craft", from Greek τέχνη, techne, "art, skill, cunning of hand"; and -λογία, -logia[2]) is the collection of techniques, skills, methods, and processes used in the production of goods or services or in the accomplishment of objectives, such as scientific investigation. Technology can be the knowledge of techniques, processes, and the like, or it can be embedded in machines to allow for operation without detailed knowledge of their workings" (Technology in Wikipedia, 2017).

6 In activity 5 of the Viplan Method we discuss an operational description of the system's identity, in which observers are within the system; its identity is then defined by relationships rather than by transformations.

7 This is a means to reduce the residual variety relevant to management. Most of the variety is self-contained in the autonomous units managing the chunks.

8 Primary activities are those directly producing the enterprise's products and services.

Chapter 2
Sustainable Development and the Sustainability of Socioeconomic Systems

Tikhomirova Olga
Saint Petersburg National Research University of Information Technologies, Mechanics, and Optics, Russia

ABSTRACT

The management of socioeconomic systems should be based on the universal laws of evolution and development. We need to discuss and create a framework for better understanding the theory of development, sustainability, and the sustainable development. To solve this problem, it is necessary to study several theoretical approaches from which we alter the terminology and some relevant methodological aspects. The contribution and implications of this research are to expand the existing theories of development in order to create the scientific basis for future studies. The study demonstrates the laws and tendencies of development and attempts to define the stage of development, shedding light on a new way to manage systems more effectively, helping to foresee future scenarios, and to prevent the negative circumstances the result from crises. Moreover, the proposed approach may facilitate focusing on the future and on innovation, instead of on vain attempts to preserve an imaginary stability.

DOI: 10.4018/978-1-5225-5996-2.ch002

BACKGROUND

This paper was originally published as a result of years studies of economic crises and behavior of socioeconomic, especially business systems. Studies showed, that socioeconomic systems are nonlinear dynamic systems and their behavior are fully compliant with general laws of systems functioning and development. This Chapter provides opportunity to expand our understanding of development, sustainability, sustainable development. Here we develop the idea that crises, nonequilibrium, chaos are the natural stage of evolution, and we need to learn these lessons of history to create a new win-win future.

INTRODUCTION

This chapter shows how the management of firms can be based on the universal laws of evolution and development. It aims to discuss and create a framework for better understanding the theory of development, sustainability, and the sustainable development of socioeconomic system, with a particular reference to business systems.

To this end, in this chapter we will:

1. Study the main theories of development.
2. Discuss the essence of evolution and the development of the socioeconomic system.
3. Describe the main structural transformations over the last 100 years, considering the evolutionary change of socioeconomic systems on the global scale (highly complex nonlinear systems), structural transformations, and their implications in the global context.
4. Construct a holistic model of evolutionary dynamics of socioeconomic systems, allowing a better understanding of the systems as a whole, in interaction and interdependence with their elements over time. This model will be based on the coherent interaction and coevolution of the system's elements and subsystems.
5. Show how holism (systems thinking) in management and in corporate government can be an effective approach to sustainable development.

This chapter refers to several theoretical approaches from which we alter the terminology and some relevant methodological aspects: General Theory of System (Bertalanffy, 1968); Synergetics (Buckminster Fuller, 1975), Schumpeter's entrepreneurship theory (1949); the theory of social dynamics (Sorokin, 1937); the theory of social action (Parsons, 1949); Durkheim's theories of sociology and labor (1983 and 1900); Spengler's theory; the theories of economic cycles by Kuznets,

Mensch and Elliott (1971–1973); theories of development of nonlinear system by Haken (2000) and Prigogine (1989–1994); and chaos theory as popularized by Gleick (1987). Theory of economic and financial crises of Kindleberger and Aliber (2005). Furthermore, this chapter is based on a postulate concerning the transition of socioeconomic systems from unorganized to self-organizing systems. We can find a similar direction and approach to management in the works of Beck & Cowan(2006) and of Warnecke (1993).

The chapter provides an analytical overview of the financial and economic crises from 1900 to present time. The data was analyzed using the system-synergetic approach.

The contribution and implications of this research are to expand the existing theories of development in order to create the scientific basis for future studies. The study demonstrates the laws and tendencies of development and attempts to define the stage of development, shedding light on a new way to manage systems more effectively, helping to foresee future scenarios, and to prevent the negative circumstances the result from crises. Moreover, the proposed approach may facilitate focusing on the future and on innovation, instead of on vain attempts to preserve an imaginary stability. To this end, we consider the financial, economic, social and technical elements of the enterprise as a connected holistic system.

The chapter consists of the following section:

1. On the understanding of the universal laws of evolution (historical travel and study of theories of development and evolution).
2. Sustainability and sustainable development of the business system: concept and nature
3. Development and transformation of the global socioeconomic system (the study of statistics and some data).
4. A holistic model of corporate systemogenesis.
5. A new approach to economic evolution and crisis: crisis and innovation.

ON THE UNDERSTANDING OF THE UNIVERSAL LAWS OF EVOLUTION

First, we consider the dialectical theories of the classical German philosophers (Hegel, 1807; Marx, 1894; Fichte, 1792–1810; Kant, 1770–1780s). Friedrich Engels systematized the three laws of dialectics: the transformation of quantity into quality; the negation of the negation (thesis, antithesis, synthesis); and the unity of opposites. According to the first law, within any period of development of society, changes gradually accumulate until further change cannot be accepted, triggering the leap of revolutionary transformation. According to the second law, within any stage of development, mutually antagonistic forces interact as a system. The third law states that any stage of development is the result of the negation and synthesis of the previous stage; development thus has the character of a spiral, and could be characterized as a three-step process.

Auguste Comte's "social physics" divides society into two components: social statics and social dynamics. The social system requires a more "intimate harmony of the component elements and more complete subordination to the external world" (Comte, 1856, p. 353). According to Comte, "Humanity cannot herself act otherwise than by her separable members; but the efficiency of these members depends upon their working in co-operation" (Comte, 1856, pp. 354–355). In other words, social progress (that is, development) depends on the cooperation of the separate system elements (individuals). Social progress and development are the quintessence of positive knowledge. Comte states that "systematic study of the Past is necessary in order to determine the future" (Comte, 1856 p. 357).

Herbert Spenser's theory postulates the unity of the decomposition and creation of the system: "evolution is the transition from an incoherent undefined state to determined organized state; complex systems are formed from a combination of simple ones" (Spenser, 1852, p. 237).

Evolution is the transition from an unconnected homogeneous number of elements to interconnected heterogeneous elements (the system). So, the process of social evolution is the process of social differentiation and the emergence of social strata: "at that time as parts of mass are integrated, they are differentiated" (Spencer, 1897, p.277).

The social evolution of Émile Durkheim depicts the evolution of societies from the mechanical solidarity of preindustrial society to the organic solidarity of industrial society. In simpler societies, people are associated by personal charisma and traditions, but in complex industrial societies, they are united due to interdependence, diversification, more specialized tasks, and the division of labor. The organic society is considered democratic, because it is based on freedom of choice and diversification of elements. On the contrary, the mechanical society tends towards totalitarianism (Durkheim, 1893, 1900).

Pitirim Sorokin's "Social dynamic" defines the dynamics of social systems as "any kind of motion, modification, conversion, alteration or 'evolution', in short, any change this logical entity in time, whether it changes its place in space, or are we talking about the modification of its quantitative or qualitative aspects" (Sorokin, 1937–41, p. 98). The most important element of the development is the "unit", which undergoes changes while concurrently maintaining its identity. The system may "change itself without losing its identity" (Sorokin, 1937–41, p. 99).

Sorokin believed that the main causes of social change are immanent factors: "systems' changes are the result of their own resources and properties". External factors may have an influence on the system, up to the destruction of the system, but they do not have any impact on the preset path of development. The system has a built-in mechanism of development: "socio-cultural systems choose the way of existence by itself" (Sorokin, 1937–41, p. 807).

The structural functionalism of Talcott Parsons (along with his Structure of Social Action)and Robert Merton states that the social system is made up of the actions of individuals (Parsons, Shills, 1976). In the process of interaction, individuals are faced with a variety of choices about how they might act, given that their choices are limited by a set of physical and social constraints.

Oswald Spengler (1918) considered evolution as a changing of cultural–historical types. He distinguished the concepts of culture and civilization: culture is focused inward and on development, while civilization is focused outward and on mere expansion. Yet civilization is the destiny of every culture. Whereas culture is "things-becoming", civilization is the "thing-become"—that is, the transition from culture to civilization is the transition from dynamics (development) to stagnation (Spengler, 1918, 1922).

Another important theory of evolution in terms of economic dynamics was proposed by Joseph Schumpeter. The basis of Schumpeter's theory is "creative destruction", which indicated that development ("economic mutation") destroys the old and creates the new. "This process of 'creative destruction' is the very essence of capitalism" (Schumpeter, 1942). Thus, we must destroy our present to gain the future, and ordinary things to gain innovation; we must create chaos from order, in order to develop.

Other relevant theories of development in this period are the theory of long waves of Kondratiev (1935), the wave theory of Elliott (1938), and, more recently, the theory of spiral dynamics of Beck and Cowan (2006). These theories are based on the cyclical dynamic—on repeatability, the return to the past—where cycles can be determined and predicted.

These traditional paradigms are carried out by the need to avoid the nonequilibrium state and instability. The positivist sciences are based on laws that are limited in their possibility of explaining and describing reality: for example, it is impossible to find an appropriate geometrical function to describe mountains, rivers, trees.

In the 1970s, scientists began to examine the causes and patterns behind the chaotic behavior of various systems, finding that the apparent disorder has a logic and organization—that this "organization" differs from the traditional positivist understanding of order.

Questions of nonlinear dynamics and the development of nonlinear systems were investigated by various new theories, such as synergetics, chaos theory, and dissipative systems (Haken, 1996; Prigogine & Nicolis, 1989; Prigogine & Stengers, 1994; Gleick, 1987).

A relevant contribution was given by synergetics, which holds that:

- Evolution is a process of formation of order from chaos;
- Development occurs through self-organization;
- Instability—the nonequilibrium state of the system—is the basic, initial state of every system;
- The state of equilibrium and orderliness is the stage of evolution;
- The development of the system is carried out through the sequence of nonequilibrium and
- Equilibrium states (chaos and order);
- Chaos is the creative force, not the destructive force;
- A complex system has a number of possible alternative ways of development;
- The creation of a whole by simple addition of initial structures is impossible.

Thus, according to the synergetics theory, the development process requires an increased complexity in space (e.g., geographic expansion, regional expansion, acquisition of material assets) and an expanding of areas of influence and interests. At the same time, development requires the differentiation of elements and parts of system and transitions from a state of disorder and homogeneity to a state of order and diversity. Development could follow two paths: the smooth evolutionary path or the revolutionary leap. Therefore, development can be considered as movement: through time (the transition from the past to the future); through space (the creation of new structures); and in terms of qualitative changes (innovations).

Considering all these theories of development, we can conceptually classify them into two groups:

1. **The cyclic-wave theories, where cyclical development is reversible:**
    a. The theory of economic cycles (Kondratiev waves; the Grand Supercycle theory of Ralph Elliott (Elliott, 1938, 1946; The "Kuznets swing" of Simon Kuznets, 1930; and others);
    b. The theory of cycles of innovation (e.g., of Joseph Schumpeter (Schumpeter, 1942), and spiral dynamics (Beck, Cowan, 2006)).
2. **The theories of the dialectical evolution of the system and self-organization, where dialectical development is irreversible:** These theories are based on the well-defined dialectics of development, competition, and competitive evolution:

a. Hegelian dialectic; Karl Popper (the competitive development of scientific knowledge);
b. Comte's "social physics";
c. Ilya Prigogine (competition of the trajectories of the development ("attractors") and the elements themselves (the process of self-organization); the systems approach and others.

SUSTAINABILITY AND THE SUSTAINABLE DEVELOPMENT OF THE BUSINESS SYSTEM: CONCEPT AND NATURE

In this section, we attempt to explain the concepts of sustainable development and sustainability, according to general systems theory and synergetic theory.

According to general systems theory, sustainability is the ability of the system to keep its movement on the given trajectory, or the ability of the systems to support the given mode of functioning, in spite of external and internal disturbances (Gigch, 1978). The main types of sustainability are equilibrium, homeostasis, and the stationary mode.

Equilibrium is the condition of a system in which all competing forces are balanced. The balance of a socioeconomic system is not always the optimum from the social justice point of view.

The equilibrium should be considered in two aspects: the static (point of equilibrium, such as equilibrium prices, or the equilibrium of supply and demand) and the dynamic, or growth, which means the balanced process of development.

Homeostasis is the stable state of a system in its interaction with the external environment.

The stationary mode is the mode of functioning of the system in which the basic system parameters remain unchanged. The opposite form of functioning of the system is the period of transformation. We can see the transition from one state of the system to a qualitatively new state. In the transition period, transformational effects are observed—structural shifts and transition recession.

The generalized scheme of the concept of sustainability is showed in Figure 1.

What, so, is sustainable development and sustainability of a system? The question is whether we will be able to achieve sustainability of the system, increasing its effectiveness and sustainability.

The UN's concept of sustainable development involves a balancing of economic growth, environmental protection, and inexhaustible resources. In the Report of the International Commission for the United Nations Environment and Development Program, sustainable development is considered as development that meets the needs of the present generation, but without threatening the rights of future generations

Figure 1. The generalized scheme of the concept of sustainability of business systems (Source: Author's work)

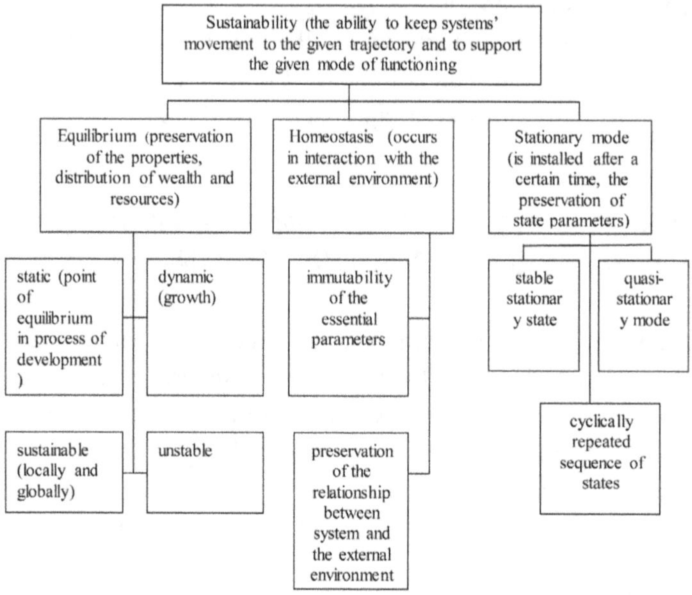

to such opportunities. Sustainable development means building an effective One-Planet Economy, a new understanding of the "good life" (a higher quality of life), and environment protection, and all this requires cultural changes. Much of the progress in sustainable development is being achieved by the business sector and local governments (Adams W.M., Jeanrenaud S.J., 2008).

Generally, all concepts of sustainable development are based on three hazy terms: economic growth, environmental protection, and inexhaustible resources.

But what do we have to do? How should we manage our businesses for sustainable development?

Sustainable development and sustainability are different concepts: sustainability refers to the state of the system, whereas sustainable development is a dynamic concept referring to movement, change, evolution, and innovative development. Thus, it is not easy to achieve sustainability and development at the same time.

However, in this case, it raises the question of whether the nature of sustainable development and sustainability is compatible with the UN declaration. Sustainability is achievable on the local level (financial, economic, social stability). But on the global level, the social systems could only achieve a state of dynamic equilibrium, which is not sustainability in a long term.

However, we must also understand that sustainable development and economic growth are different. Growth is just one of the stages of development; hence, we need to restrain growth and seek development. Because the main goal of economic growth is increase of effectiveness and income, it does not consider the natural environment. In contrast, the main goal of development is to create innovations to cope with the natural system's integrity and to save energy. Economic growth is therefore just one stage of the development towards the more advanced concept of sustainable development. For sustainable development, we have to learn how to manage all the stages of development, including the stages of growth, stability (dynamic equilibrium), transition, recession, and crisis. If we try to safeguard a single stage of development, we will not be able to move on to the next stage.

In this framework, crises are inevitable, and they give an opportunity to update and to produce further development. This is connected with the nature of development: development is carried out by transformation, but transformation means unsustainability. Sustainability is possible within some stages, but not at every stage of development. Sustainable development is the understanding the whole process of development. We must achieve development, but not every aspect (economic, ecological, social) of it separately. Only by acting in this way can the consequences of the crises be creative, rather than destructive; crises will then provide growth, environmental protection, and human well-being.

DEVELOPMENT AND TRANSFORMATION OF THE GLOBAL SOCIOECONOMIC SYSTEM: EXPIERENCE AND DATA

What happens in the modern socioeconomic systems? What global transformations are implemented in the last 100 years? What structural and organizational changes can we observe?

In this section, we analyze the transformation and structural changes in business systems from the 1900s to the present.

Before we formulate our theory of evolution of socioeconomic systems, we should study the historical data which help us to find out some structural and organizational transformation of businesses. It is well known that a universal scheme of system evolution involves a succession of periods of organic growth (that does not lead structural changes) and irreversible structural transformations. The development of socioeconomic systems inevitably leads to their integration and growth because of objective universal regularities in the evolutionary process. The main goal of this process of consolidation is ensuring the survival of the system and its sustainability. As a result of these processes we can see creation of new organizational forms such as corporations, conglomerates, multinational corporations, holding companies etc.

vertically integrated structures, venture capital funds, clusters, etc. Consolidation could be vertical (the majority) and horizontal (for example, General Electric or Du Pont). The mergers and acquisitions (M&A) is still the main method of ensuring the survival of business organizations, their enlargement and a leading market position in the long term. However, when a direct merger of assets is not possible (e.g., because of antitrust laws), businesses create strategic partnerships, alliances, associations, clusters, etc.

The main factors contributing to the consolidation of business structures and creation of new organizational forms are the following.

The first is *the macro-level factors*, namely strengthening and accelerating of the economic growth. Economic growth in the countries-customers can increase profitability and strengthens the position of companies in international expansion. Economic growth in the countries-sellers can attract foreign direct investment, including the mergers and acquisitions, construction of new space from scratch (so called greenfield investment), etc.

The second are *industry factors*, namely:

- The growth of global competition (for example, oil and gas, automotive, pharmaceutical, telecommunications, financial industry);
- Increasing pressure of the market because of falling prices (e.g. the oil industry);
- Overcapacity in key markets (e.g., automotive industry);
- High speed of technological change or deregulation of the market (banking and financial sector).

In the 1980s the most multinational transactions for the acquisition of assets, mergers and acquisitions were carried out with the aim of capturing new markets, undervalued businesses and unrelated activities. But in the 1990s the largest transactions occur within a single industry, and the main goal was restructuring: the reorganization of key processes, the reorientation of the competitiveness at the international level. The growth of global competition and increasing pressure of markets are forcing modern socioeconomic system to focus on core business. Activation of international deals of mergers and acquisitions is largely caused by the growth of sales of non-core assets and acquisition of enterprises operating within the same technology and production process.

Third is the *group of micro-level factors*, namely:

First: The acquisition or enhancement of intangible assets to enhance competitiveness.

To ensure the effective functioning and development of socioeconomic systems among other things we need to have certain specific benefits in non-material assets: brand name, research and development, intellectual assets, knowledge management and marketing capabilities, etc. These non-material assets are universal across the whole economy system and can be used repeatedly regardless of the geographic location of some company or socioeconomic system. While the growth of a such non-material assets and accumulation of corporate knowledge and information, socioeconomic system (firm, company, enterprise, conglomerate and any others) receives tremendous opportunities for geographic expansion and internationalization, that means size increase and enlargement.

In this case mergers and acquisitions are more preferred to enhance the organizational and managerial knowledge and skills, while direct “gfeenfield” investments are more preferred if we mean the aim of enhancing technological and production skills. The lack of non-material assets for successful competition may be offset by borrowing them from the acquiree. Direct investment in the construction of new facilities may lead to increased competition and falling profitability of business entities. Also, often assets are acquired in those industries, where businesses have a strong competitive advantage.

Second: The allocation of risks (“risks’ spread”). Acquiring assets in various countries, socioeconomic systems assume that the covariance of the industry in several economies will be less than in a single country.

Third: Increasing income of socioeconomic systems by acquiring assets in a rising market. This is one of the most simple and common methods of increasing the efficiency of companies throughout the history of entrepreneurship: capital (money, finance assets) flows into those sectors and regions where it ensured the most rapid growth.

Fourth: The acquisition of additional intellectual, human and other resources. Mergers and acquisitions as a form of consolidation are most preferred in compare with direct investing in zero cycle for business entities in some industry segments since the acquisition of local firms provides the necessary intangible assets, ready untwisted brands, distribution network and delivery network, a certain society of partners, working system of supplies, raw materials. In addition, the acquirer of the assets receives a pool of knowledge of the specifics of the local market and consumers.

Fifth: Reducing level of duplication of function, staff and departments through reorganization and restructuring of the combined company.

Sixth: The growth of capitalization of companies.

Seventh: Increasing of the resistance and sustainability against of hostile takeover. Takeover of larger company is more complicated and more expensive than smaller one.

Eighth: Expectations of synergy growth: it is assumed that the growth and development of the combined company will be higher than each of them separately.

Next, the fourth group of factors of structural and organizational transformation of socioeconomic systems is the *technological factors.*

The necessary of reducing communication and transportation costs lead to the desire of businesses systems for international expansion and concentration of its competitive capabilities. In addition, the need for constant innovation and Research & Development forces the enterprises to interact with the aim of funding researches and accelerate the bringing to market of a new product. Technological changes lead to a reduction of the product life cycle and force the business systems to search and introduction of innovation, which in turn changes the conditions of competition and the market structure.

Also technological changes create a new markets and areas of business activity (e.g. information technology and telecommunications). Mergers and acquisitions allow componies to react immediately to technological changes and to have access to a new sectors or markets, to strengthen its competitive capabilities. A wave of mergers and alliances in the late 1990s – 2000s clearly illustrated this postulate. The most of deals of mergers and acquisitions were in the field of telecommunications, media industry and information technology, that demonstrated the desire of business systems to capture new markets created by the development of information technology, especially the Internet, and to expand its presence through the provision of a broader integrated range of services on a global level.

In turn, todays' concentration of business and the creation of largest corporations due to new technological possibilities which are represented by information technology in the twenty-first century. The development of information technologies leads to the possibility of increasing the size of the business empires and extent of management control that enables you to effectively manage a larger business structures. The fifth and last group of factors are the government factors. Consolidation of businesses and socioeconomic systems contributes to the liberalization of markets and the creation of favorable conditions for investments at the state level, the opening up of markets, regional economic and political integration. In addition, the larger the company the less risk in countries with high political risks.

Based on the previously described theories of the principles and laws of evolution, we should explore global structural and organizational changes in the socioeconomic systems from the five points of view:

One: The market of Mergers and Acquisitions, Foreign Direct Investment as a main illustration and characteristic of the global structural transformational shifts.
Two: Structural characteristics of the socioeconomic systems in the different period.
Three: Priority of the energy resource for the activities of the businesses.
Four: Priority of the industry sectors from the perspectives of the profitability.
Five: The main factors which caused the crises as a one stage of development and evolution.

The study of some statistics data showed that the activity of mergers and acquisitions (M&A) is clearly cyclical in nature and have a c\very strong connection with economic / financial instability and crises. We now consider the statistical data on investment and on mergers and acquisitions for the largest deals of the second half of the twentieth century. We being with FDI and M&A (see Figures 2 and 3).

We can see that, as FDI as M&A are steadily increasing, the result of these transactions is a high concentration of assets and the creation of monopolies and oligopolies. We can also see that the activity of M&A increases just before and immediately after the years of a crisis.

Figure 2. FDI worldwide, 1970-2016 (Source: authors work adapted from OECD Factbook Statistics, (2017); UNCTAC World Investment Report (2016))

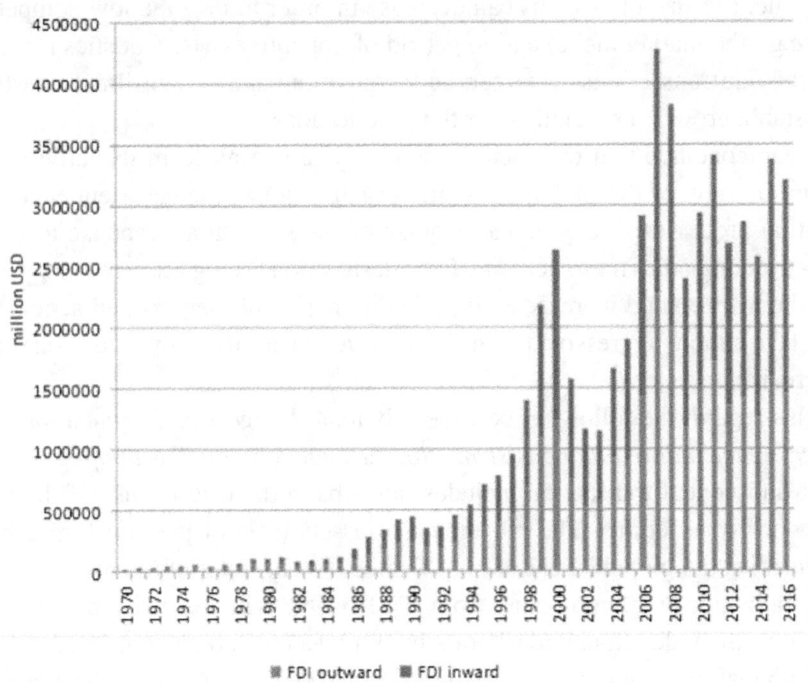

Figure 3. M&A worldwide, 1985-2016 (Source: authors work adapted from OECD Factbook Statistics, (2017); UNCTAC World Investment Report (2016))

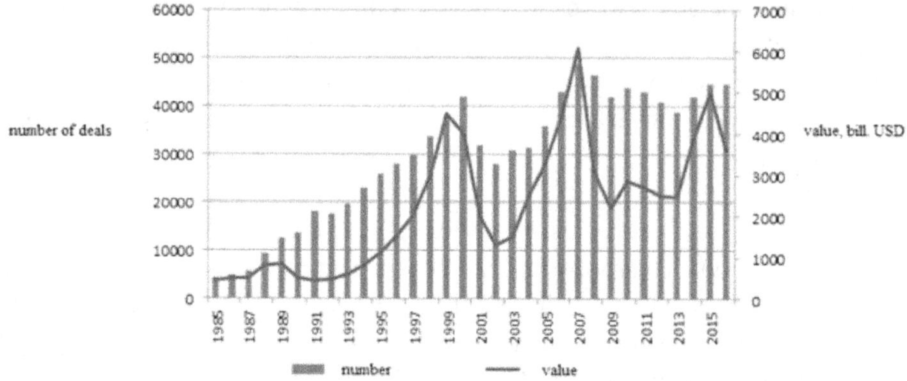

Hence, economic and financial crises show drastic fluctuations of parameters and violations of the equilibrium state. If we compare the development of businesses with the evolution of biological systems, we find that, in periods of crisis, most systems increase their size (this recalls the giants of biological evolution). In business, these processes are equivalent to mergers and acquisitions—business consolidation as an attempt to survive the crisis situation. However, we find that the opposite process is also true: the size of systems can decrease in order to take the low competitive niche (e.g., the market niche) and to get rid of noncore assets. Statistics regarding the worldwide transactions of mergers and acquisitions and foreign direct investment shows stable growth, especially over the past decades.

We can conclude that the socioeconomic systems transform its activities and structure in view of the global structural-organizational changes, tendencies and innovation processes. Mergers and acquisitions are a natural response to regime shifts – the companies try to get away from destruction (being acqueree or squeezed out of the market) and increase activity in the market of mergers and acquisitions so that take a more aggressive position and have the ability to introduce disruptive innovation.

In this regard, the following becomes obvious. *Mergers and acquisitions have an interference nature that consistent with the cumulative build-up of assets*. That is, each subsequent transaction includes (absorbs in the truest sense of the word) all previous transactions. The movement of assets (sale of previously purchased companies) at the global level does not change the character of the processes.

Capitalization of the companies from 1900 to 1998 is increased by more than 100 times: thus, the index Standard&Poor's in 1900 was 6.2 points, but in 1998 it was 1085,5 points; in 2009 it was 756,07 points, in 2007 – 2306,23. At the same time index becomes global: if the S&P500 is for the US market, today we see S&P1200 – for the global market; the index values are steadily increasing: 29.12.2008 S&P1200 =996,33; 30.12.2009 S&P1200=1318,61; 31.12.2010 S&P1200=1438,40; on 23.12.2011 S&P1200 =1325,98; 24.12.1012 S&P1200=1052,96. The value of the S&P500 at the end of 2008 amounted to 869,42; at the end of 2009 – 1128,78; at the end of 2010 - 1256,77; at the end of 2011 – 1257,60; at the end of 2012 amounted to 1402,43. The maximum value of the Dow-Jones Industrial in 1900 was 70.7 points, in 1920 - 109.9 points, and in 1998 - 9457,9 points; it reached maximum values in the pre-crisis 2007 – 13264,82 points, in 2011 – 12151,41; 2012 -1387,66. Herewith, the base of index calculation is changed from 12 included companies to 30.

Dynamics of the index Dow-Jones Industrial for different years is shown in Figure 4.

Visualization of the processes at the above figures allows us to suggest that in the process of industrial development of society over the last century there is a steady cumulative process of capital accumulation. This fact, in turn, means the interference nature of the evolution of socio-economic systems. The volume of transactions in value terms clearly steadily increasing, that can be explained primarily by the fact that the larger company is more expensive on the market than smoller one. In addition a longer period of analysis allows us to identify the long wave which is now in the process of growth in comparison with the beginning wave of development in the 1970-1980s. The sharp decline that we have seen in the post-crisis period can be explained by the desire of businesses to accumulate capital and increase the liquidity of assets. A distinctive characteristic of transactions in the second decade

Figure 4. Dynamix of the index Dow-Jones Industrial (Source: authors work adapted from www.djaverages.com (2017))

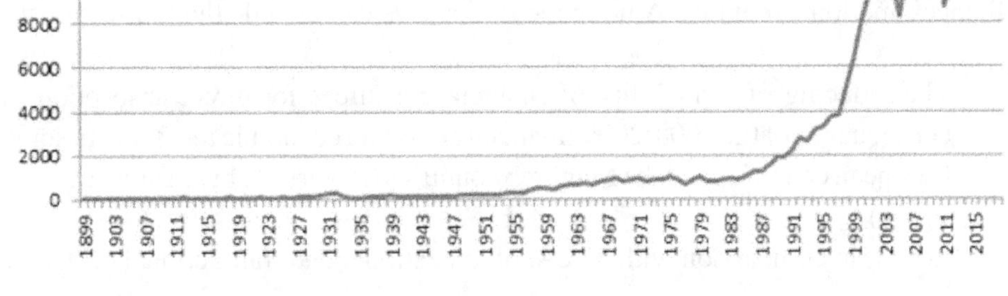

of 21 century (2010s) is the focus on a well-considered and strategically justifiable purchases, which gives a synergy effect in the production and/or sales. Among M&A transactions in 2010s we can clearly identifies two categories of financial transactions: the first type is aimed at simple profit and financial sustainability, and the second type is strategic transactions aimed at sustainable strategic development and ensuring global sustainability in the long term.

If we compare the nature of mergers and acquisitions from the point of view of its influence on the global structural and organizational changes and the on the development of socioeconomic systems, it is possible to identify the following.

In the early 2000s investments in mergers and acquisitions had a character which we could name "organic growth", that is, M&A transactions consisted of small acquisitions that do not lead to major changes in the structure and functioning of companies. While the transactions of the second decade of the 21st century (2010s) cuold be characterized as "inorganic growth", that means that the transformation and structural changes in companies, as well as strategic development lays in tendency with current global trends: energy efficiency and conservation, use of renewable and clean energy, growing needs of a global infrastructure increasing interdependence and interconnectedness of the network society in the field of business and in the social sphere.

So, if we compare the amount of M&A deals (valued from 50 million USD) that has strategic and financial aims, we can see the following. In 2010 the volume of transactions with a financial nature amounted to about 24.4%, the volume of transaction with a strategic nature – 75,6%; in 2011 the volume of transactions with a financial nature was 75.8% and 24.2% respectively. For transactions worth more than 1 billion USD, the ratio is the following: in 2010 - 66.7% a strategic nature and 33.3% financial; in 2011 – 60% and 40% respectively (Source: PWC, special report, Assembling value, 2012, http//www.pwc).

In addition, during the period of 2010s the number of M&A deals with small and undisclosed value increased, that could be caused by pessimistic expectations of investors, by the recession in European countries, by the problems and the difficulty of access to credit, political uncertainty in several countries of the world.

Another characteristic of the organizational and structural changes of the second decade of 21 century is increase of activity in the development of strategic partnerships, alliances and joint ventures. Among the reasons we can include the following:

1. The reducing of availability of financial resources for direct absorption in consequence of the 2008-2009 financial crisis and economic recession in some European countries and the Maghreb countries (Algeria, Libya, Tunisia, Syria, Egypt)
2. Lower, in comparison with M&A, organizational and transactional costs.

3. Lower in comparison with M&A, the riskiness of the transaction.
4. Joint ventures, strategic alliances, etc. more flexible and adaptable.
5. Antitrust laws and barriers that limit the possibility of direct merger.
6. The growth and dissemination of knowledge about the production of goods and services between organizations, and between countries.
7. Increased international division of labor.
8. The concentration of companies on strategic objectives.

According to PWC, among the areas of business development in 2012 the most preferred was the establishment of strategic alliances, the creation of joint enterprises, and only in third place – the merger and acquisition of other business [PWC Trendsetter Barometer: Business outlook report, and Manufacturing Barometer: Business outlook report]. Among the companies which involved in industrial production the relationship of these tendencies of development are less obvious, but the main aims of such deals are becoming more similar – it is strategic investments that provide long-term sustainable business development and synergies from the cooperation in production, sales activities, research and development, access to technologies and results of intellectual activities of yesterday's competitors. If we consider the regional activity, the highest number of M&A transactions realizis in Asia and other emerging markets. And this is the hallmark of this period of organizational and structural transformations of socioeconomic systems. Merger, consolidation and the size growth of companies, creation of strategic alliances and joint ventures lead a number of structural and organizational changes in the companies-participants, in particular:

- Changing of corporate management in accordance with the newly created organization, even while maintaining a high level of autonomy of participants.
- Changing the organizational structure.
- Changing the environment of functioning of assets (production, sales, human, etc.), which must be adapted to the joint organization.
- Reorganization of business processes.

By *structural transformation*, we refer to the various actions of businesses that lead to changes in size, ownership, assets, activities, etc. It includes investments, mergers and acquisitions, and diversification of business. Table 1 summarizes the transformations of business regimes.

Table 1 shows that the development of businesses and of entrepreneurship involved:

One: Only two human-invented production technologies: the steam engine and the electrical engine.

Table 1. Summary of the organizational and structural transformation of business regimes

Period	Main characteristic of the business context	Priority of industry sectors	Priority of energy resource	Main factors which caused the crisis
End of 18th century – 1830s	Separate enterprises in developed countries were scattered	Textile industry	Water (the water mill and engines)	1790s, the beginning of the 19th century: terror in France, the Napoleonic wars, the boom in canal company shares, prices of raw materials in the US and Europe; the rapid development of construction and railways
Second half of the 18th century			Coal for melting cast iron (coke)	The rapid development in the construction of canals and railways, which became objects of speculation
19th century	The spread of the steam engine	Railways, the production of steam locomotives, shipping, the production of steam ships. steam engineering and technology	Coal begun to be used in transport	Heavy industry became an object of speculation
End of the 19th to the beginning of the 20th century	Horizontal mergers, the emergence of monopolies, cartels, trusts, and syndicates. The production of material-intensive products and services	Heavy industry, steel, chemistry, transport (railway, water), concentration of financial markets,	Coal, oil	1856–1857 crisis: end of the Crimean campaign, speculation in heavy industries, agricultural economic society, railway construction
1848		The gradual shift to the electric power industry and electrical technology	The first production of oil from a drilling hole	
1920-1930 (1950)	Vertical integration, businesses united within a single production process	Heavy industry, steel, chemistry, transport, electric power industry and electrical technology	Coal, oil. The role of coal was reduced gradually	1929-1933: overproduction and financial crisis. Crises of the 1950s and 1960s: the currency and financial crises.
1960-1970	The creation of conglomerates	Oil and oil products, oil refining, communications, transport	Oil	Energy crisis 1974-1975.
1980s	Sectoral concentration of the businesses, the reshuffling of assets			Crisis of 1979–1982: oil prices increase, speculation on real estate and agricultural lands in the USA
1990s	Horizontal mega-mergers, creation of transnational corporations and global companies	Oil	Oil, the search for alternative energy sources. The share of coal in the energy sector was about 20%	Crisis 1997-1998: a wave of bankruptcies of the world's largest companies, speculation on the capital market
2000s – present	Megamergers, the involvement of the developing markets in M&A. Production of innovation.	Microelectronics, computer science, biotechnology, genetic engineering, additive technologies	Oil, the intensification of search for new environmentally friendly energy (natural, bio-fuels, etc.) and shale gas.	2008-2009: the global economic recession, financial crisis
The Future	Business as innovation networks is based on information and knowledge. Clusters – the most effective form of business because of synergy.	Biotechnology, robotics nanotechnology, the global informational network	Ecofriendly, renewable sources of energy	?

Source: official, business and historical information; analysis and calculations by the author

Two: Only three types of energy source: water, coal, and oil (natural gas).
Three: Innovations that changed the world include steam engines, electric motors, aircraft, computers and microelectronics.

We can thus point out that evolutionary processes are irreversible. Indeed, we believe that the increasing consolidation of businesses, and the interdependence and reliance on industrial production, will inevitably lead to a deep crisis. This is because the process will persist so long as businesses do not change to a new business model of functioning and development. We judge that, to avoid this future crisis, firms should consider a new business model based on innovative energy saving or the adoption of alternative energy technologies; the international division of labor; and the cooperation and synergy of knowledge, not based only on production synergies.

A HOLISTIC MODEL OF CORPORATE SYSTEMOGENESIS

The theoretical laws of development, empirical data and statistics that were presented above allow us to formulate a theory of the evolution of the socioeconomic systems, which we call holistic model of corporate systemogenesis.

So, in this section we define our idea of a universal algorithm for the development of socioeconomic systems: the *algorithm of corporate systemogenesis*.

Organizational and structural changes in socioeconomic systems occur as a result of development and evolution. These changes can be in the next forms: the creation of diversified conglomerates with a lack of interaction between components of structural elements: the creation of clusters and alliances, that are based on production and innovation synergies and internal division of labor (see figure 5.)

We should pay attention, that the fifth period of development (figure 5) completes the cycle and provides new trend towards horizontal mergers at higher level of Mega-mergers. Therefore, in the VI period, that ongoing today, we can observe the search for new developing markets with high growth potential in the long term. The statistical data presented above confirm this statement by the fact that the much more of investments (in global scale) are carried out in developing markets.

It is obvious that in a view of further development it is necessary to create new markets and conditions of functioning. So, if in the coming decades will not happen rejection of outdated organizational forms, management methods and trends of development, we should expect that the evolution movement will be cyclical and the cycle from the first to the sixth period will be repeated. It is possible that seventh period will be characterized by a high degree of interaction and cooperation of all structural units constituting the socio-economic system. Activity of such elements will be based not on simple cooperation in production, outsourcing, but on the

Figure 5. Periods of organizational and structural transformations of socio-economic systems (Source: author work)

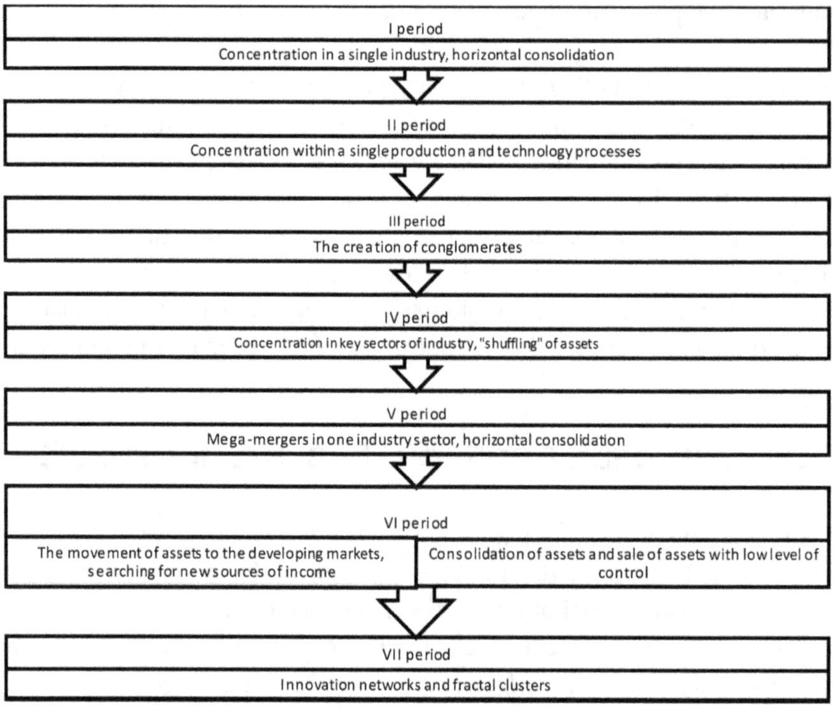

synthesis of innovation and the symbiosis of knowledge. Competition in the sphere of material production and technology as a phenomenon of the industrial epoch will be replaced by the competition of ideas. This is due to the fact that the price competition, product differentiation are the methods of the industrial competition of the twentieth century, which are based on the consumer preference of a product. But strongly interacting business entities compete in the field of innovation, knowledge, ideas. They are obsessed by the idea to be the first; however, they collaborate and help each other because they are united by a single strategic goal.

As a result of generalization and analysis of all available information and statistics we could form a matrix that reflects the interdependence of organizational forms of business systems (BS), the level of competition and interaction, the value of financial and innovation capital (see Table 2), and how they change in the process of evolution.

Let's explain the matrix. At high level of competition with high value capital (importance) there is creation of a vertically integrated structure with a low degree of interaction. The lower the value of competition as a fight for the market, the higher the possibility of creating an effective innovation network is based on the interaction and cooperation of the elements of a business system and synergy of knowledge.

Table 2. Matrix of organizational and structural transformations of socioeconomic systems in the process of evolution (Source: author work).

Innovation capital, knowledge	*A fundamentally new*	A lot of innovative solutions, competition between different formats and technologies, capturing market, increasing the volume of production, cooperation with neighbouring industries (e.g. VHS (JVS) vs Betamax (Sony) in 1970s; DVD-R/RW (Pioneer) vs DVD+R/RW (Philips) at the end of 1990s)	The monopoly on knowledge. Licensing. The winning technology takes the leading monopoly position. M&A, industry consolidation	Venture capital investments, strategic alliances (e.g. in the field of electric vehicles). Internal R&D	Innovation networks, innovation synergy
	traditional	Companies unite in unions and alliances on industry sector or market bases	Licensing, outsourcing and specialization on certain market niches	Internal R&D, acquisitions to obtain access to technology	Strategic alliances, JV with the purpose of production synergy and technological synergy, the using of "cloud" technologies
Financial capital	*Low value*	The creation of joint ventures with industry leaders (or market), niche specialization SME's	The creation of public company to finance development by market leaders. Franchising, consolidations of industries and markets	Venture capital investments, acquisitions of startups	Production synergy, technological synergy, joint ventures and strategic alliances
	High value	The creation of the enterprise, company, etc. M&A, vertical integration, creation of holdings. Increase market share, growth in production volumes, size growth of business systems	Financial-industrial group, the acquisition of market leaders, vertical and horizontal integration, market consolidation	Industry specialization of socioeconomic systems. Shuffling of assets (the exchange and sale of non-core assets and ill-managed assets). Niche specialization of SME's	Integration on a territorial basis, creation of clusters. The creation of small niche enterprises
		high	low	*low*	*high*
		competition		**interaction**	

The involvement of a growing number of markets in mergers and acquisitions, the establishment of various unions and alliances leads to global restructuring and the changing of organizational forms of socioeconomic systems in the world.

Next. Empirical analysis presented in the previous sections showed that mergers and acquisitions of the last decades lead to the creation of an industry leader (market), that is monopoly. This fact confirms the steady process of consolidation and concentration of capital and business. This process forms the centripetal processes of evolutionary dynamics and provokes the increasing isolation of socio-economic systems at the macro level, which in turn may entail an increase in social entropy (that is chaos, or crises (economic, political, finance, etc.))

Now, let's formulate the theory of evolution of socioeconomic systems. Algorithm for corporate systemogenesis presented in Figure 6.

By *systemogenesis* we mean the sustainable development of socioeconomic systems as nonlinear systems in accordance with evolutionary laws.

Figure 6. The proposed algorithm for corporate systemogenesis (Source: Author's work)

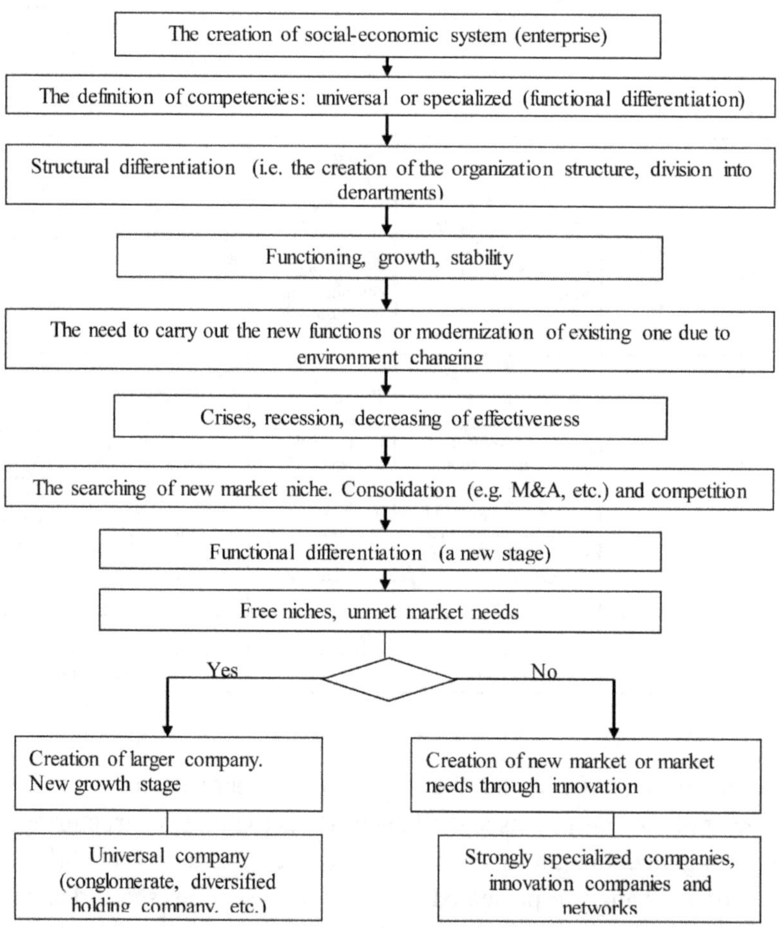

We consider this development as an irreversible wave, where the complex system grows by itself and absorbs its own properties, which already existed on a lower level. The whole process of evolution is illustrated in Figure 7.

The development of socioeconomic systems is accompanied by the growth of interactions between its elements (people or departments). Interaction takes shape in various forms of cooperation, such as different unions, strategic alliances and partnerships, joint ventures, mergers and acquisitions, and clusters.

Evolutionary processes are accompanied by synergistic effects (Baker & Sinkula, 1999).

This is because it is not possible to create the integrity of the system by simple additions to the original elements, which would inevitably lead to interference of their localization and functions.

All of the above mentioned stages of development can coexist and coevolve. In other words, we can simultaneously observe microenterprises at the unorganized spontaneous level of development, mesoenterprises with higher level of organization, and self-organizing business systems. We consider all the levels of development not to be mutually exclusive, but to be complementary; the development is a transition from one stage—micro, meso, or macro—to the next. Even if we can provide effectiveness at each stage of development, we consider the real essence of development to be the sequential change of states of the system.

But what happens within socioeconomic systems? How does the process of development occur?

Figure 7. The scheme of the model of corporate systemogenesis (Source: Author's work)

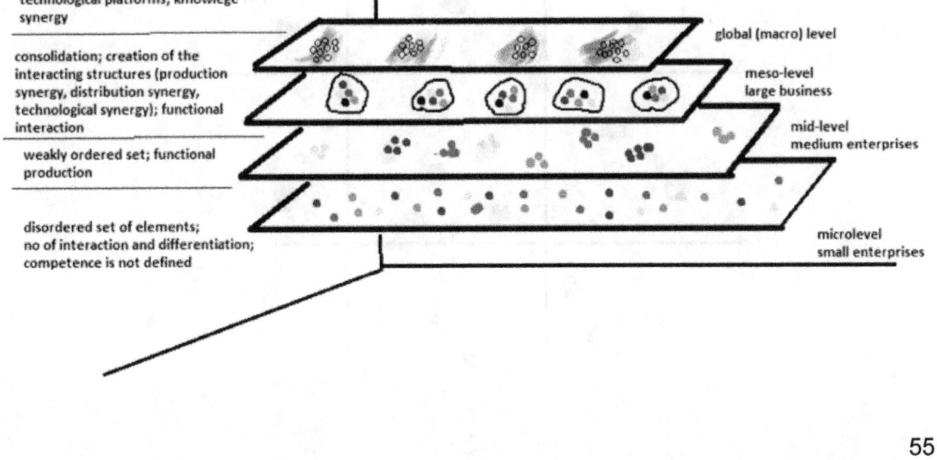

The complex behavior of the socioeconomic system is formed by the interaction of a number of simple behaviors between its main elements—such as its people and workers. These elements of social groups operate independently right up to point of instability and crisis. At that moment, they begin to interact and cooperate (a similar behavior is observed in physical systems by Prigogine & Nicolis 1989, pp. 15–18). The transition from one state to another and the leap forward to the new level of development are the results of the competition between two processes of interaction and random action.

These processes have different directions of impact: interaction between the elements makes the system ordered, while the chaotic random actions of the elements lead to disorder. Figure 8 shows an example of the elements of a system in a certain spatial localization.

The elements are uniformly distributed in space, with no interaction among them. The state of the system is characterized by complete disorder, with each element behaving randomly and chaotically. Examples of such a system include small enterprises that do not interact. Figure 9 shows a different situation.

In Figure 6, we can see the process of clustering. The elements of the system are spatially closer to other similar elements (e.g., the typical industry clusters). We can observe interactions between the elements within each cluster, while there is no interaction among the clusters.

Changes in the external environment cause two different reactions: if the elements interact, this can lead to the creation of clusters, or to a contrary reaction where they move away from each other, leading to a broader geographical distribution, as in the search for new market niches, or the expansion throughout regions.

Figure 8. Spatial localization of a system's elements (Source: Author's work)

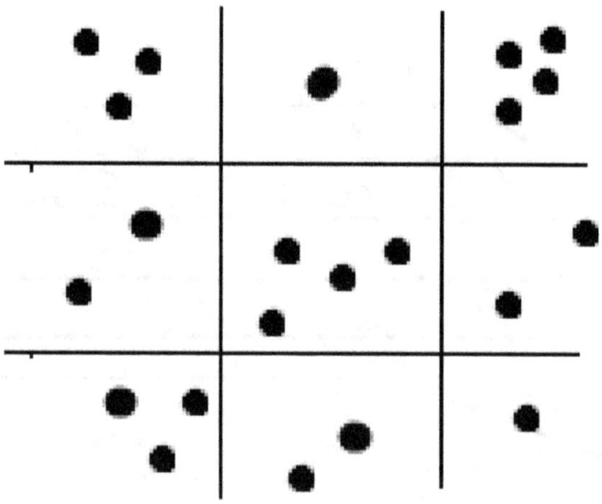

Figure 9. The well-ordered totality of the system's elements (Source: Author's work)

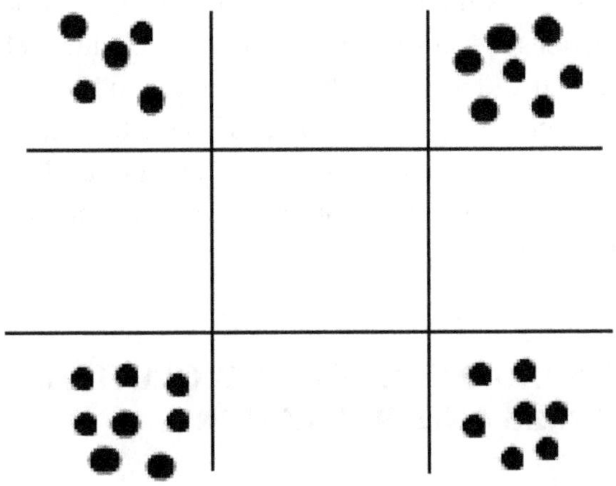

Furthermore, the formation of clusters requires a change in the external environment, with favorable conditions that force businesses to unite and cooperate.

The development of socioeconomic systems is initiated by the actual or expected global regime shifts (systemic transformations)—namely, by functional and structural changes in the global system. These regime shifts include abrupt and irreversible transformations (e.g., changes in state regulation of an industry, fundamental technological changes, financial, economic, and sectorial crises, the development of new high-tech products, and the development of new sectors of the economy or business diversification). Thus, the impetus for development may counter balance a crisis, forming a protection from its consequences, and even the expectation of increased profits in some industries.

The path towards development depends on the location of businesses, both in industry and in the global business system. If similar companies prevail in the industry, they will try to preserve their integrity, independence, and efficiency growth. A problem can be represented by managers trying to maintain the independence of the managed company, in order to protect their salaries. They are then not interested in cooperating or coordinating with other enterprises or competitors to create of unions, alliances, joint ventures, mergers, acquisitions, or other forms of interaction, as this may cost them their working place and salaries, or reduce their level of management control. For this reason, managers will be interested in increasing the value and size of their firms in order to protect it, following the logic that the larger and more expensive the firm is, the less attractive it will look for acquisitions or other forms of external attack. However, the effect of such managerial behavior could be the opposite: buying larger companies leads to a more profitable synergy effect, and this makes the enterprise more attractive for acquisition. Another conclusion we can make from the study is the following. In the dynamics of organizational and structural transformations of socioeconomic systems we could see obvious long wave: according to statistics, the FDI – 1970-2011; according to M&A – 1985-2011. The peak of both waves occurs in the pre-crisis 2007. This fact indicates a significant role of the crisis in structural changes and transformations. This fact leads us to the need to explore the structural transformation of socio-economic systems in the context of the global financial and economic crises.

A NEW APPROACH TO ECONOMIC EVOLUTION AND CRISIS: CRISIS AND INNOVATION

In this section, we analyze the relationship between development and economic crises. As we stated above, development means movement and change, while sustainability means the state of the system. We thus need unsustainability for development. What is unsustainability in social and economic systems? We consider unsustainability to be crisis. Table 3 shows the most important economic crises and the impact they had on the development of business.

As Table 2 points out, during the second half of the nineteenth century, and during the whole twentieth century, the global socioeconomic system experienced several major crises (1847–1848; 1857; in the 1950s and 1960s; 1974–1975; 1979–1980; 1997–1998; 2008–2009). Each of these had its own causes and its own consequences. However, each crisis led to a significant regime shift on the global level: the transition to new technologies (the steam engine was replaced by the electrical engine), or to new energy (water, coal, oil), etc. So, we can assume that the crises are not only an objective reality, but also constitute the need for a search for new opportunities.

Crisis (or chaos) is a constructive beginning that provides new opportunities and conditions for innovation and sustainable development. Therefore, we can assume that crises are not only an objective reality, but precisely the need to search for new opportunities.

How exactly does a crisis offer new opportunities for business?

During a crisis, we have to find the most effective methods for working, to optimize the business processes, to reduce the production costs, and to search for new market niches and opportunities.

Production needs to become more flexible and effective. In the period of stability, managers try to save the results that have been achieved and to ensure the growth of profitability, while in the period of crises, they are forced to invest in development and innovation. Stability often becomes stagnation, and stability must be disrupted if progress is to be achieved.

Table 3. The most important economic crises and their impact on the development of business

Period	The changes led to the development of the business systems
1900s	Increasing of concentration of businesses, mergers and acquisitions, monopolization and consolidation of companies
1929-1933 (The Great Depression)	A gradual abandonment of the concept of a free market as a self-regulating system. Transition to a partial state intervention in the market relations (USA, UK, France, Sweden, Denmark, Norway); as well as full centralization and the direct intervention of the state up to the nationalization; the creation of close relationships between state and private large corporations (Germany, Italy, Spain)
Crisis after II World War	A third scientific and technical revolution, radical changes in technology and production in developed countries. Development and intensification of the international division of labour, international expansion. The growth of international interdependence and the concentration of business. The consumption of energy resources increased; the transition from coal to oil could be observed in developed countries. State support of the modernization of production capacity and investment in the technical development of businesses. Rapid development and growth of production and economy in Japan, Germany, and Italy. Obtaining of independence of former colonies and "decolonization" of economy.
Economic crisis 1974-1975	A deep economic crisis in all sectors of the U.S. economy, which spread to all developed countries. The growth of energy dependence and, as a result, the intensification of the search for alternative energy sources. The share of own production and processing in developed countries increased, intensifying scientific research into energy and resource saving approaches. Development and introduction of innovative technologies and equipment. Liberalization of the market, the opening of the customs borders in the EU countries In the post-crisis period; intensifying consolidation of the businesses. Business activity was restored at the expense of stimulation of demand. Communication technologies and computers were developed. Globalization and the active expansion of the businesses from developed to developing countries.
1979-1982	The creation of favorable conditions for the development of production (the reduction of taxes, the increased availability of credit resources). The focus was shifted from the demand to the supply. The activation and the rapid development of information technologies, telecommunications and communication technologies. Globalization spreads around the world.
1990s	A new industrialization and the restructuring of the economy accompanied by the growth of the financial sector in most developed countries (e.g. economic policy of Margaret Thatcher in the UK). The growing consolidation of the western European countries leads to a restructuring of the industry and strengthening of the needs in a single currency in the European area.
2005	The rapid development of China and the transfer of production to developing markets (Russia, China, India, etc.) lead to deindustrialization of the USA and western Europe.
Crisis 2008-2009	Transfer to innovative processes and technologies. The search for alternatives to oil and traditional energy resources. Real changes in the economy and business do not occur, but there is a global understanding of the need for a change in management and development.
2010-2017	A process of reindustrialization, restructuring and diversification of the economy from non-production, financial sectors to real production industry with high added value and knowledge (aircraft industry, pharmaceutics, machine building, power engineering). There is a process of deglobalization, crisis of euro and the focus on the domestic market. We can observe increasing problems of European identity, unity and economic interdependence between developed and developing countries.

Source: Author's work

If we look at the most important innovations of the second half of the nineteenth century and of the whole 20th century, it becomes evident that the most important inventions and advances in science and technology were the driving forces for economic and social development only if they were accepted and propagated in a global environment (cf. the "institutionalization" of J.Schumpeter). But how do innovations become an engine of economic progress?

As an example, the steam engine was invented in 1780s. The global industrialization and boom in the railroad, metallurgical, and construction sectors led to the crisis of 1847–1848 in the industrialized countries of Europe and the United States. In 1834, the electric engine was invented, and this innovation started to work up the market. In 1903, the first airplane laid the foundations of a new era. The increased demand for oil as a fuel source caused price increases.

And in 1970s and 1980s we faced an oil crisis.

The invention of atomic weapons during the Second World War, and the distribution of nuclear energy, including the environmental disaster in Japan in 2011, set us the task of reducing our energy dependence and energy use and searching for alternatives to oil. Next, from the invention of the computer to collapse of the so-called dot-com bubble, only about 70 years passed. But social networks have seized the initiative, so that the boom of the computer company has not yet fully come.

Hence, we can assert that many years pass from an idea to its commercialization and wide distribution; the boom takes place cyclically—on average, every 70 years.

The last crisis of 2008–2009 did not lead to any structural or ideological changes in management, development or functioning of businesses. It did not become the cause of innovation and did not force us to search for radically new solutions. Therefore, we believe that a new wave of deterioration is inevitable in the near future, until the coming of a quality change and the development of business and the world economy as a whole. The crisis is an opportunity of which no one has yet taken advantage.

In addition, an invention itself does not guarantee development—for example, the steam engine itself did not have such a large-scale impact on the markets as did the construction of railways and steam vehicles. Innovations, therefore, accumulate: each innovation absorbs previous ones and is based on them.

We believe that the basis of development of socioeconomic systems is the sequential change of states from chaos to order, from instability to sustainability. Evolution is like a snowball: it starts at an original point (an idea or invention), and as the evolution progresses, it covers more and more systems: companies, industries, markets and countries.

The essence of the crisis as the natural state of the system – identification or the creation of new innovative opportunities in technology, methods of production, factors of effectiveness, approaches to competition and others. Higher density of crises in the 20 century and the early 21 century compared to previous periods due to increased interdependence, increased international competition, and the absence of a transition to a qualitatively new level of development.

SOLUTIONS AND RECOMMENDATIONS

The development of the socio-economic system is realized on the base of the general laws of development of systems. Socio-economic systems are nonlinear dynamic systems, so the study of their behavior should be based on theories of nonlinear dynamics. Economic and financial, political and social crises are inevitable in the course of system evolution. In this chapter we studied the major theories of evolution. We defined the general concept of sustainability and sustainable development. We founded that these concepts are not identical. So, we formed the theory of corporate systemogenesis, that means the evolution of the socioeconomic system that is based on sequential change of states of order and chaos. In order to ensure a prosperous future for the human community we need an understanding that sustainable development is possible only at the global level. We can't create Paradise in a particular state and even continent. Systems laws, which are based on the tendency of the system to establish equilibrium, can lead to the destruction of civilization. So we have to create conditions for functioning and development of socioeconomic systems to sustainable development and growing prosperity are possible for all elements of the system - that is, all human society.

FUTURE RESEARCH DIRECTIONS

The behavior and development of socioeconomic systems is still complex and challenging task from the point of view of modeling. it is obvious that periods of crisis and economic recession are natural and inevitable. Today we can observe the creation and introduction of new technologies and inventions in the production process and in our everyday life. This may indicate the beginning of a new stage of development through political and economic crises, that we can see in set of European countries, Africa and Asia: both in developed and developing countries. Further studies need to be associated with the study of new production technologies, based on the digitalization, customization, energy efficiency, virtual and additive manufacturing. New industrialization and new forms of organization of production, new level of life and welfare of the population, social and economic problems must be investigated from the standpoint of nonlinear dynamics and the General laws of development of systems.

Further studies will be devoted to the modeling of general laws of development of socio-economic systems and creation of models that can predict the behavior of the world economic system in the future in order to reduce the negative effects of crises and the formation of the win-win scenario of sustainable development of the world community in a global strategic perspective.

CONCLUSION

The analysis of some theories of development allowed us to describe our view of the universal laws of development, such the increase in complexity, the differentiation, the transition from instability to sustainability, the accumulation of changes and the transition from quality to quantity, and abrupt character of development.

The analysis of some theories of development allowed us to describe our view of the universal laws of development, such the increase in complexity, the differentiation, the transition from instability to sustainability, the accumulation of changes and the transition from quality to quantity, and abrupt character of development.

The twentieth century socioeconomic systems experienced several abrupt regime shifts. Each period of transformation was connected with global changes, innovations, and economic crises. We could see that inventions gradually spread, becoming object of speculation that cause other crisis. The threat of crisis forces businesses to interact and to unite with each other, in the belief that a larger company is more sustainable. Our analysis of the statistical data on FDI and M&A showed the dynamics of the processes of concentration of assets and consolidation of businesses.

On these, assumptions we constructed a holistic model of the development of socioeconomic systems, based on holism and the systems approach. The algorithm of development is based on the gradual transition from an unorganized set of small and medium enterprises to self-organized clusters and unions.

Through the analysis of the theories of sustainability and sustainable development, we pointed out how sustainability and sustainable development are different concepts.

The synergetic methodology helped us to draw the conclusion that the crisis has a constructive essence and that development is possible through refusal from past, transition from stagnation to innovation.

Systems and holistic management thinking are necessary if the concept of competition is to be changed. The purpose of competition should not be to a win a war with competitors, but to cocreate a unique product.

We should change our attitude to crises, considering them as creative instead of destructive. Life without crises is no utopia, as crises are necessary to focus on constant innovation as new opportunities for the firms.

The right kind of development on the global level is possible only by creating a win–win scenario.

In further research, we will make use of Herbert Spenser's theory, Sorokin's "social dynamic", the theory of Schumpeter, the three laws of the dialectic, and synergetic methodology.

REFERENCES

Adams, W. M., & Jeanrenaud, S. J. (2008). *Transition to Sustainability: Towards a Humane and Diverse World*. International Union for Conservation and Nature, Publications Services.

Baker, V. E., & Sinkula, J. M. (1999). The Synergistic Effect of Market Orientation and Learning Orientation on Organizational Performance. *Journal of the Academy of Marketing Science*, *27*(4), 411–427. doi:10.1177/0092070399274002

Beck, D., & Cowan, C. (2006). *Spiral Dynamics. Mastering values, leadership and change*. Moscow: Exmo.

Bertalanffy von, L. (1968). General System theory: Foundations, Development, Applications. New York: George Braziller.

Bloomberg Businessweek. (2010). *Dow Jones Industrial Average*. Retrieved from: https://www.bloomberg.com/quote/INDU:IND/members

Bresnahan, T., & Trajtenberg, M. (1995). General Purpose Technologies "Engines of Growth?". Stanford University.

Bresnahan, T., Trajtenberg, M., & Helpman, E. (Eds.). (1998). General Purpose Technologies and Economic Growth. The MIT Press.

Buckminster Fuller, R. (1975). *SYNERGETICS. Explorations in the Geometry of Thinking*. Macmillan Publishing.

Durkheim, E. (1893). The division of Labour in Society. New York: Free Press.

Durkheim, E. (1900). *Sociology and its Scientific Domain*. Moscow: INFRA-M.

Elliott, R. (1938). The Wave Principle. Gainesville, GA: New Classics Library.

Feldman, R. A., & Watson, C. M. (n.d.). *Central Europe: From Transition to EU Membership*. Retrieved from http://www.imf.org/external/pubs/ft/fandd/2000/09/feldman.htm

Gigch, J. P. (1978). *Applied General Systems Theory*. New York: Harper&Row, Publishers.

Gliek, J. (1987). *Chaos: Making a New Science*. National Book Foundation.

Gorton, G., Kahl, M., & Rosen, R. (2005). *Eat or Be Eaten: A Theory of Mergers and Waves*. Cambridge, MA: National Bureau of Economic Research. doi:10.3386/w11364

Grove, N. (1974, June). Oil, the Dwindling Treasure. *National Geographic*.

Haken, H. (2000). Information and Self-Organization. Macroscopic Approach to complex System. Springer-Verlag.

Kindleberger, C.P., & Aliber, R.Z. (2005). *Manias, panics and crashes. A history of financial crisis*. Academic Press.

Kitchin, J. (1923). Cycles and Trends in Economic Factors. *The Review of Economics and Statistics*, *5*(1), 10–16. doi:10.2307/1927031

Kondratiev, N. (1935). The Major Economic Cycles. *The Review of Economics and Statistics*, *18*, 103–115.

Kuznets, S. (1971). *Economic Growth of Nations: Total Output and Production Structure*. Cambridge.

Kuznets, S. (1973). *Population, Capital and Growth*. Norton.

Lipsey, R. G., Carlaw, K. I., & Bekar, C. T. (2005). *Economic Transformations*. General Purpose Technologies and Long-Term Economic Growth.

Mensch, G. (1979). *Stalemate in Technology: Innovations Overcome the Depression*. Ballinger Pub. Co.

Merton, R. K. (1949). *Social Theory and Social Structure*. New York: Free Press.

Nsouli, S. M. (1999). *A Decade of Transition An Overview of the Achievements and Challenges*. Retrieved from http://www.imf.org/external/pubs/ft/fandd/1999/06/nsouli.htm

Organization for Economic Co-operation and Development (OECD). (n.d.). OECD Factbook 2015-2017 Economic, Environmental and Social Statistics [Data file]. Retrieved from: http://www.oecd-ilibrary.org/economics/oecd-factbook_18147364

Parsons, T. (1937). *The Structure of Social Action*. Moscow: Ecsmo.

Parsons, T. (1949). *Social Theory and Social Structure*. New York: Free Press.

Prigogine, I., & Nicolis, G. (1989). *Exploring complexity.* Saint-Petersburg: Piter.

Prigogine, I., & Stengers, I. (1994). *Time, Chaos, Quantum. Saint-Petersburg*: Piter.

Schumpeter, J. (1942). Capitalism. In *Socialism and Democracy*. Moscow: Ecsmo.

Schumpeter, J. A. (1949). *Theory of Economic Development.* Cambridge, MA: Harvard University Press.

Sorokin, P. (2006). *Social & Cultural dynamics: A Study of Change in Major Systems of Art, Truth, Ethics, law and Social Relationships*. Academic Press.

Spencer, H. (1852). *The Development Hypothesis.* Moscow: Ecsmo.

Spengler, O. (1918, 1922). The Decline of the West. Academic Press.

UN General Assembly. (n.d.). *Press Release United Nations Division for Sustainable Development.* Author.

United Nations Conference of Trade and Development. (n.d.). World Investment Report, 2016 [Data file]. Available from: http://unctadstat.unctad.org/wds/TableViewer/tableView.aspx?ReportId=96740

United Nations General Assembly. (1987). *Report of the World Commission on Environment and Development: Our Common Future*. Transmitted to the General Assembly as an Annex to document – Development and International Co-operation: Environment. Author.

Warnecke, H.-J. (1993). *Revolution der Unternehmenskultur. Das Fraktale Unternehmen.* Academic Press.

KEY TERMS AND DEFINITIONS

Corporate Systemogenesis: A sustainable development of socioeconomic systems as nonlinear systems in accordance with evolutionary laws.

Cyclic-Wave Theories: A theory development based on the reversibility of system changes and consistently repeating cycles of growth and decline.

Equilibrium: A condition of a system in which all competing forces are balanced.

Homeostasis: A stable state of a system in its interaction with the external environment.

Stationary Mode: A mode of functioning of the system in which the basic system parameters remain unchanged.

Sustainability: An ability of the system to keep its movement on the given trajectory, or the ability of the systems to support the given mode of functioning, in spite of external and internal disturbances.

Sustainable Development: A development is based on continuity of disorder and order, and on the creation of new through the chaos (in the context of social economic systems through the crises).

Theories of the Dialectical Evolution: Theories of development which are based on the irreversibility of the system changes in the process of evolution: dialectics, synergetics, social physics.

Chapter 3
Reaching Community Consensus on Reforms for More Sustainable Urban Water Management Systems:
The Case of Kapiti, New Zealand

Robyn M. Moore
Victoria University of Wellington, New Zealand

Victoria J. Mabin
Victoria University of Wellington, New Zealand

ABSTRACT

Reaching community consensus on water reforms was the motive—and the challenge—for this operational research. While water is comparatively abundant in New Zealand, a pattern of decline in the quality of water resources has persisted. Living up to its "clean, green" image is a significant goal for New Zealand, with high economic value derived from the effects of its globally recognized environmental credentials on key exports like tourism and agriculture. A 2009 government task force suggested that a "business as usual" approach is untenable, and water reform should be a priority. This systems study examines the challenges—and opportunities—facing Kāpiti, a rapidly growing coastal community, with water scarcity and quality constraints that had long prevented them from meeting their sustainable development objectives. The authors used a stakeholder typology to identify system stakeholders, Theory of Constraints (TOC) to examine their perspectives, causal loop diagrams to highlight potential negative outcomes, and TOC to design appropriate interventions.

DOI: 10.4018/978-1-5225-5996-2.ch003

INTRODUCTION

Careful management of natural resources, including water, is of prime concern globally, as described in articles such as Bjørndal, Herrero, Newman and Romero (2011) and Udias et al. (2012). For Māori, the indigenous people of Aotearoa New Zealand, their intrinsic relationship with particular rivers (*awa*) and lakes (*roto*), is central to their identity (Parliamentary Commissioner for the Environment (PCE), 2015a). Māori as *tangata whenua*, "the people of the land", have authority in a particular place, according to their ancestors' relationship to it (Royal, 2012). Their interest in the well-being (*mauri* or life force) of water and other natural systems (PCE, 2015a) is enshrined into government legislation designed to enable New Zealanders to participate in decision-making to "manage or protect" their "natural and physical resources" (Williams, 1997). As such, themes of environmental protection or guardianship (*kaitiakitanga)* are woven into the fabric of New Zealand's national identity (see Young, 2001; 2004). The motivation for this study was to consider how a community might take a more integrated, systematic, and participatory approach to meeting the challenges of water management, and achieve healthier, more sustainable freshwater systems.

The New Zealand landscape, a narrow set of islands, intersected by mountain ranges subject to continual uplift and erosion, is due to a combination of tectonic plate activity and weather (PCE, 2015; Young, 2004). More than half of New Zealand's total land area is in pasture and arable land, and more than a quarter is under forest cover (NZ Treasury, 2016). New Zealand has a "highly urbanised" population, with nearly three quarters of its people living in urban environments of more than 30,000 people (NZ Treasury, 2016:5). "Nowhere in our island nation is far from water, and most of us live within a few kilometres of the sea" (PCE, 2015). Just over a third of the country's 4.6 million people live in the Greater Auckland area, the country's fastest growing region (NZ Treasury, 2016:5).

Primary production produces the country's biggest export earnings, with agriculture accounting for more than half those earnings, and dairy being the predominant agricultural activity (NZ Treasury, 2016:17). Dairy farming is land use that has continued to expand rapidly, and is the primary cause of increased nutrient stress on waterways (PCE, 2015a). Of the two nutrients of main concern – nitrogen and phosphorus – nitrogen is the greater water quality challenge (PCE 2016).

New Zealand has made progress in recognising and reducing point source pollution (from a single source, like a storm water pipe) since the introduction of the *Resource Management Act* (1991). Other, complementary, environmental reforms are also underway (Land and Water Forum, 2010). The signing of a Clean Streams Accord in 2003 (MfE, 2003) for example, was a collaborative effort by the rural sector aimed at decreasing pastoral runoff. This was replaced by the Sustainable Dairying Water

Accord in 2013 (PCE, 2015a). On the bridging and fencing of rivers and streams, significant progress has been made, with nearly 24,000 kilometres of waterways fenced by May 2014 (PCE, 2015a). On the other hand, progress in keeping stock out of wetlands (nature's filtration system) has been minimal (PCE, 2015a). Thus, whilst success should be acknowledged in some areas, non-point discharges, from diffuse sources, like pasture or road runoff, continue to increase unsustainably, and urban waterways remain polluted to an unacceptable level (Land and Water Forum, 2010; PCE, 2015a).

Efforts in the area of water reform include New Zealand's *Fresh Start for Fresh Water* programme (2009), with its government task force concluding that a 'business as usual' approach was undesirable. The related multi-stakeholder Land and Water Forum was set up in 2009, delivering three reports containing a range of recommendations (2010; 2011; 2012). The country's first *National Water Policy Statement* (2011) was released largely in response to the Forum's recommendations. The level of participation in the Forum nationwide, is suggestive of a willingness on the part of communities of interest, to be part of water reform. However, the means to undertake widespread reform remains the subject of debate, as the overall decline in the quality of water in our rivers, lakes and streams, and related losses in indigenous biodiversity, remain disappointingly evident (Land and Water Forum, 2010; PCE, 2015a; and see NZ Treasury, 2016:7).

New Zealand's urban and rural communities face several challenges, due to increasing population density and attendant competition for scarce natural resources. In common with the global experience, the pressures and impacts of climate change are increasingly pervasive (PCE, 2015). Indeed, New Zealand and its Pacific neighbours are becoming accustomed to increasing frequency and extremity of flooding and drought, with an ensuing range of physical, social and economic impacts (PCE, 2015; 2016).

A period of unprecedented population growth in some New Zealand cities, combined with more severe weather, is testing the limits of existing urban infrastructure (PCE, 2015). Urban water systems are under particular pressure to meet the expectations of residents and businesses (see KCDC, 2009; Swinnen, 2017). Fragmented approaches to water policies and plans persist, despite ground-breaking environmental legislation like the RMA (1991) intended to promote sustainable development and protect community interests in environmental, social, economic and cultural outcomes (for history and context see Williams, 1997; Sheppard, 2010; Young, 2001). Under the RMA (1991), regional councils have the responsibility of regional sustainability, managing the effects of using freshwater and coastal waters, through regional policy statements and consent processes. Territorial (local/district) authorities (Councils) are responsible for sustainable district wellbeing. A territorial authority's responsibilities include:

- The provision of local infrastructure for water, sewerage, storm water, transport.
- Environmental safety and health, district emergency management and preparedness, public health and environmental health matters.
- Controlling the effects of land use (including hazardous substances, natural hazards and indigenous biodiversity), noise, and the effects of activities on the surface of lakes and rivers (RMA, 1991).

There are significant challenges ahead for regional and territorial authorities aiming to deliver on their sustainable development goals. With up to 40% of urban rates typically spent on water and wastewater management in New Zealand (Kāpiti Coast District Council, 2012; Kāpiti Observer, 2013), affordability is a key concern for communities. In addition, there are competing interests and perspectives vary. Consensus is not a foregone conclusion. Thus, this study is concerned with improving the quality of resource management decisions, increasing awareness of the critical constraints to achieving healthier, more sustainable and affordable water systems. A multi-method approach is used (Mingers & Brocklesby, 1997; Mingers & Rosenhead, 2004), with the methodology demonstrated through the real-world case of Kāpiti, a coastal settlement which has been debating water quality and security issues for more than a decade. The Kāpiti Coast District has a current estimated population of just over 49,000 (KCDC, 2013) and lies approximately 50km north of New Zealand's capital city, Wellington. In 2003, Kāpiti Coast District Council had adopted a *Sustainable Water Use Strategy - Water matters* on the basis that there was an urgent need to reduce per capita demand for water (demand management) and promote a better balance between residential growth and the use of water resources (KCDC, 2013b). Preventing future resource consent limit breaches was a primary consideration. A resource consent (or 'consent') is an authorisation given by an appropriate legislative body to certain activities or uses of natural and physical resources, as required under the New Zealand Resource Management Act (RMA, 1991). The following excerpt from Kāpiti's 2009 water supply review, presents some context, in respect of the water security dilemma facing the community six years on from release of the strategy, and the pressing need for reform:

Effectively, the community is likely to be in breach of consent limits within the next five years.

Breaching the consent or running parts of the supply system out of water are not acceptable options (KCDC, 2009:5/19).

The traditional approach to the management of urban water systems which splits the various parts and manages each separately is outdated, and a more integrated, lifecycle approach is required (PCE, 2001: 50). Despite legislative requirements requiring such an integrated approach (RMA, 1991), water management systems are still being dealt with on a largely *ad hoc* basis (Land and Water Forum, 2010, 2011, 2012). Clearly a different approach is needed.

Operational Research has been used increasingly in natural resource sectors like agriculture, fisheries, forestry and mining (Bjørndal et al., 2011). In the context of these industries, the focus is primarily on meeting productivity objectives and optimising profitability (Midgley & Ochoa-Arias, 2004). In a departure from the more commonly applied mathematical programming models of 'hard' OR, this *community-focused* Operational Research (COR) acknowledges the 'leaky' boundaries between environmental and community matters (Midgley & Ochoa-Arias, 2004) to concentrate on people and their perspectives. Stainton (2010:11) has urged the OR community to pay due attention to human behaviour and values, suggesting that the core set of mathematical techniques are not sufficient, and that OR requires more than the positivist approach that has tended to characterize it (Stainton 2010:11). Trutnevyte, Stauffacher and Scholz (2012:762) argue that most complex societal problems require both 'hard' OR and 'soft' OR and that both approaches make "remarkable contributions to supporting decision making" linked with "complex societal problems". The authors warn that some 'soft' OR methods used for societal decision making, where stakeholders "aspire to the ideal-type future states of a problematic system", do not sufficiently "elaborate on the feasibility and consequences of the visions" (Trutnevyte et al., 2012:762). This research explicitly seeks to include such an assessment, of both feasibility and consequences, using the problem-structuring methodology known as Theory of Constraints (Goldratt, 2010; Dettmer, 2007; Cox & Schleier, 2010) to comprehensively explore the 'boundaries of knowledge' (Midgley & Ochoa-Arias, 2004) and reveal tangible solutions to Kāpiti's dilemma.

To consider decision making as an *event,* rather than as a *process* of inquiry that unfolds over time, complete with nuances, institutional history and debate, is to overlook the social and organisational contexts that are crucial to the success of any decision (Garvin & Roberto, 2001). Inquiry is an open process that can foster the exchange of ideas to generate alternatives, and produce a well-tested solution (Garvin & Roberto, 2001). This approach presents challenges for most people, and those charged with making a decision tend to default to advocacy (Garvin & Roberto, 2001). Despite their similarities, inquiry and advocacy produce quite different results (Garvin & Roberto, 2001). What is needed, suggest Garvin and Roberto (2001), is an emphasis on inquiry, and sensible criteria for assessing the quality of the decision-making process. This is where the Theory of Constraints (TOC) can have significant impact.

The Theory of Constraints is a goal oriented approach, aimed at improving the performance of a system, however that system is defined. The methodology recognises that any complex system is made up of multiple interests, interactions and interdependencies, resulting in emergent properties of often unpredictable nature. In such contexts, reductionist approaches for managing these systems produce suboptimal results, if the interdependencies are overlooked. However, we cannot hope to capture or model the complexity of the system in detail, so we must properly select what we must focus on. Goldratt (2010:3) reasons that focusing on everything is synonymous with not focusing on anything. Thus, TOC's approach at each stage of the analysis is to focus on the key factors - the parts of the system, the relations and operations - that are constraining the system's performance. TOC provides logic protocols and systematic processes and methods that enable users to capture the parts of the system, without losing sight of the whole (Mabin & Davies, 2010). Consistent with Huberman and Miles' (1998) call for transparency in qualitative research, the qualitative data was systematically analysed using the TOC framework, presented back to participants for validation, and cross checked against secondary sources (Moore, 2009:52).

Rather than focusing on gaps in the literature, TOC methods seek to find a critical issue of concern, and to identify a key leverage point. TOC has been successfully used in production, distribution and logistics, project management, research and development, and not-for-profits (Ronen, 2005: 1). Burton-Houle (2001) describes TOC as suitable for any organisation, including not-for-profits, and that using TOC in 'for cause' enterprises can end debates regarding whether the goal is to make money, or to deliver on the cause. TOC offers a set of tools to identify the point of most resistance, and not-for-profits can leverage that resistance to meet both their cause and their financial goals (Burton-Houle, 2001). Shoemaker and Reid (2005) provide an example of TOC in a resource management setting, using the methodology to successfully uncover systemic problems and develop a process of ongoing improvement within a large water utility. Systemic problem structuring using TOC has also been applied to community involvement in water conservation, *generating valuable results for all the parties involved* (Foote at al., 2007:1). In their case study of a small township in the South Island of New Zealand, a 'multi-method' approach was found to enable the reframing of what was seen as 'appropriate policy'. The outcome was a positive intervention in a long-standing conflict between the District Council and community (Foote et al., 2007:1). Foote at al. (2007) use the theory of *boundary critique* with problem structuring to explore the inclusion, exclusion and marginalisation of people and issues. In the case of Kāpiti, the problem structuring methodology of TOC was used with a stakeholder typology to explicitly include those parts of the system that might otherwise be marginalised or ignored (Moore, 2009a; Moore, 2009:62).

This paper is structured with Section 2 introducing the methodology; Section 3 shows the research findings; Section 4 sets out the discussion, Section 5 reflects on the research contributions and looks ahead, and the final Section sets out some conclusions. Some areas for future research are suggested.

METHODOLOGY

Combining the Theory of Constraints, Stakeholder Analysis and Systems Dynamics

The research described in this paper started out as a Master's research project by the first author in collaboration with her supervisor, the second author. The paper examines community-based, collaborative decision-making for resource management, using a problem structuring methodology (Mingers & Rosenhead, 2004). Subsequent to a piloted investigation where examining the ethics of naming participants was of particular interest, a methodological framework was proposed that would integrate three complementary perspectives. The Theory of Constraints (TOC) was used with a Stakeholder Typology to identify potential system stakeholders and examine their perspectives, while Causal Loop Diagrams (CLDs) from Systems Dynamics were constructed with participants to explore and circumvent potential negative outcomes. The aim was to test the effectiveness of the combined framework to engage stakeholders to 'think through' the issues, identify the critical problems, and their potential solutions. The descriptive term 'thinking out loud' was adopted to describe the process.

The study set out to answer two research questions:

1. *Can the combined use of TOC and a stakeholder typology aid managers and the community in their decision making, to gain better outcomes from their investment in water management and*
2. *What are the necessary conditions to improve outcomes for the particular water system under pressure?*

A pilot study was undertaken as part of the university ethics committee (HEC) process. This was to test perceptions and responses to the 'thinking out loud' approach, whereby participants' views and supporting evidence for their views would be shared in a reflective and reflexive process to agree on a community goal and on the changes needed to get there. The pilot needed to demonstrate value in fostering a supportive environment where participants would openly share perspectives with others, without the usual anonymity. The desired degree of transparency must not put at risk the objective of a trustworthy and ethical research design (as per Huberman & Miles 1998). In naming the participants and sharing their perspectives in a structured way, the project departed from the more accepted approach to qualitative research, where confidentiality around the participants' viewpoints might be paramount. In this case, while most interviews were conducted individually, not only were views shared, participants had the capacity to know whose viewpoints they were. The ethical considerations of this approach were not inconsiderable, and the university's Human Ethics Committee gave due deliberation to the potential risks and benefits of granting ethical approval. Subsequent to conducting the Pilot study with two participants, approval was granted, with attendant conditions.

The Pilot study also revealed that the selection of participants might be assisted through stakeholder analysis. However, the initial stakeholder analysis process tested during the Pilot (Elias, Cavana & Jackson, 2002; Freeman, 1984, Mitchell, 1997) appeared unnecessarily complex for this community-based research. A simple model that seemed more suited to this context of Community OR, was discovered during a later literature search, in a paper by Rawlins (2006). Specific stakeholders linked with the strategic issue were identified according to the ten categories that appear in Figure 1, following Freeman (1984). Note the bidirectional arrows, illustrating the nature of the relationship between the stakeholder and the system issue.

Following the mapping of participants (Figure 1) using the generic stakeholder map (Freeman, 1984), emphasis turned to the linkages (Rawlins, 2006), or the connections between the system and the participant (Figure 2). The participant stakeholders were represented in each of the linkages, with all but two of the participants closely involved with the system under study. The 'community' and 'consumers', comprising around a third of the participants on the stakeholder grid (see Elias et al., 2002), have either a diffused or a functional (output) linkage, with normative linkages also present in some cases.

Kāpiti's District Council (KCDC) has a functional (input) linkage, while the Ministry of Health (MOH), Greater Wellington Regional Council (GWRC) and to an extent, the Department of Building and Housing (DBH), demonstrate enabling linkages. Though geographically distant from Kāpiti, the two Starborough Flaxbourne (NZ Landcare Trust, 2011) participants were selected for their situational linkage to the system problem, described in Rawlins (2006) as a 'normative' linkage. The Starborough Flaxbourne Soil Conservation Group are a collective of farm owners who have changed their farming practices to build farm systems that are healthy, resilient and profitable, in severely drought and erosion prone eastern Marlborough in New Zealand's South Island (see Avery & Thomson, 2017).

Figure 1. The stakeholder map

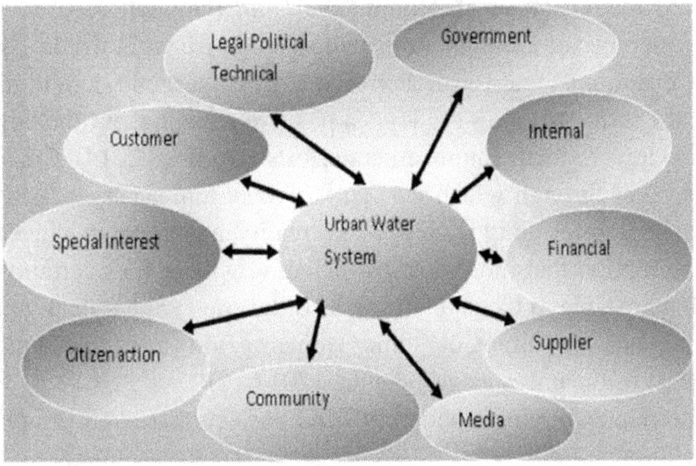

Figure 2. Stakeholder linkage model (Rawlins, 2006:4)

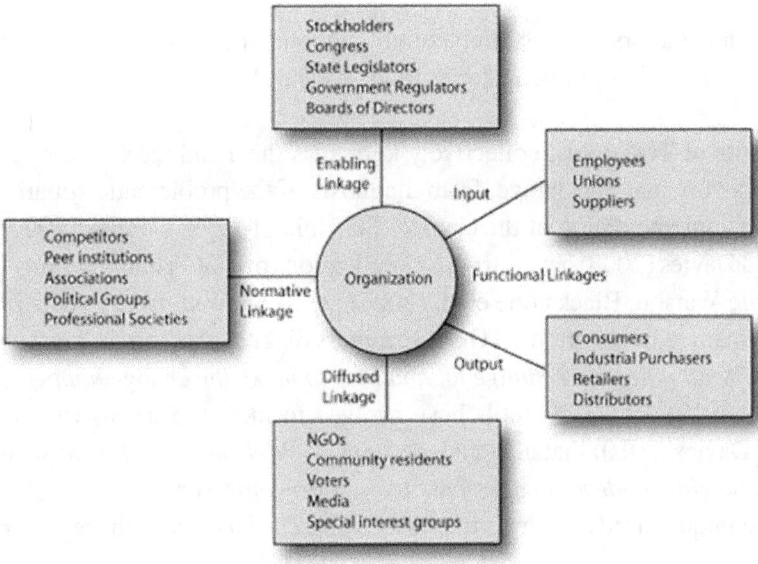

Introducing the Theory of Constraints

The TOC methodology is attributed to Dr Eliyahu Goldratt, a physicist. In 1979, Goldratt developed a successful production scheduling algorithm to resolve a bottleneck dilemma. He soon observed that challenging some basic assumptions about the way things were done could lead to breakthrough solutions, without the accompanying mathematical algorithm. *The Goal* (Goldratt & Cox, 1984) was the first of a number of Goldratt's novels to popularise what would become The Theory of Constraints (Watson, Blackstone & Gardiner, 2007). Within the TOC framework, the term 'constraint' adopts a positive meaning, as "the key indicator for where and how to focus improvement efforts" (Ashlag, 2014). Constraints are the leverage point through which we can derive system success (Ashlag, 2014). At the same time, uncertainty is viewed as a reality, and Goldratt suggests that we avoid the assumption that more detailed planning will give us certainty (Ashlag, 2014). With more time to focus on the areas of leverage, rather than putting effort into predicting what we cannot, our "improvement efforts have much bigger impact than the impact of the uncertainty" (Ashlag, 2014:18).

Three interrelated premises underpin TOC's system-based philosophy:

1. Every system has a goal and a set of necessary conditions that must be satisfied for the goal to be achieved.
2. The overall system's performance is more than just the sum of its component parts.
3. Very few factors or constraints, often only one, limit a system's performance at any given time (Shoemaker & Reid, 2005:21).

One suite of TOC tools, collectively known as the Thinking Processes, guides the process of managing change, from diagnosis of the problematic situation right through to implementation of the change (Scheinkopf 1999; Dettmer, 2007). Kim, Mabin and Davies (2008) have surveyed applications of TOC Thinking Process (TP) tools, while Watson, Blackstone et al. (2007) have reviewed twenty years of TOC. With five main logic diagrams, TOC researchers diagnose *what in the system needs to change, what it needs to change to, and how to make the change happen* (Mabin & Davies, 2010). The TOC tools have evolved to include variants (described in Mabin & Davies, 2010) that also include steps of *Why change? How to sustain the change?* and *How to develop a process of ongoing improvement?*

The techniques used to address the various phases of the research are summarised in Table 1.

By engaging with the participant stakeholders identified through the Stakeholder typology (Moore, 2009:62) and using the Thinking Processes from the TOC methodology to represent the logic underpinning their views, the problems with the most undesirable effects on the system were identified. Solutions were sought and tested using TOC's Intermediate Objectives (or Destination) Maps (also known as IO Maps and Goal Trees, and referred to in this paper as GT/IO Maps), Current Reality Tree/Branch (CRT and CRB), Evaporating Cloud conflict resolution diagrams

Table 1. Techniques used in this case study

Change Question Being Addressed	TOC Methods Used	Other Techniques Used
Why change? What is the Standard? What to change? What to change to? How to cause the change?	Goal Tree/IO Map Current Reality Branch (CRB) Evaporating Cloud (EC) Revised Goal Tree/IO Map Prerequisite Tree (PRT) Final IO Map	Literature review, Pilot study (Action research), Stakeholder Typology, Case study Individual and group interviews, Member Checking (aided by TOC GT/IO Map and EC), Fact checking (e.g. consultant engineer's MCDA analysis) Analysing Themes, Member checking (aided by GT/IO Map and EC) Causal Loop Diagrams (CLD)

(ECs), Prerequisite Tree (PRT) mapping procedures, and Causal Loop Diagrams (CLDs) from Systems Dynamics. The last in a series of GT/IO maps, a CRB and the Prerequisite Tree (PRT) are shown in this paper. A Causal Loop Diagram was used to represent a situation with potential to become a 'fix that fails' (Senge, 1994:126), and enabled the situation to be communicated clearly, helping the participants to derive their own solution - an ongoing communications strategy alongside other demand management strategies - to prevent such failure.

Goal Trees, Intermediate Objectives (IO) Maps and Their Uses

Begin with the end in mind (Covey, in Dettmer 2007). This is the premise behind using Goal Trees or IO Maps (Dettmer, 2007: Ch 3; Dettmer, 2011) at the start of a Theory of Constraints (TOC) analysis. Dettmer (2007: Ch 3) argues the Destination (or IO) map is critical to the success of the Thinking Processes and sound decision-making. The map is based on necessity logic and is readily interpreted. Thus, IO Mapping is a useful tool for Operational Researchers working on people-centred problems. Its potential for informing community and environmental research, such as in urban planning and water management, has recently begun to be explored (Moore, 2009a; Shoemaker & Reid, 2005).

The TOCICO Dictionary (Cox, Boyd, Sullivan, Reid & Cartier, 2012:68) defines the IO Map as a map of intermediate objectives needed to overcome obstacles on the path to implementation, usually generated towards the end of the analysis. It also suggests an IO Map can be used "to establish the relationships among a system's goal, critical success factors, and necessary conditions" (Cox et al., 2012:68). Dettmer (2007) was referring to this latter and less-well-known usage of an IO Map at the start of the process and his more recent term "Goal Tree" (Dettmer, 2011) helps to differentiate between these two usages. In this chapter, we will refer to the IO map as a Goal Tree/IO Map or GT/IO Map, to remind readers that this is the type of IO Map we were using.

The research began with an idea to agree a 'clear, unequivocal goal statement' (Dettmer, 2007) among the participants, using TOC GT/IO mapping. The GT/IO map is essentially a vertical hierarchy of requirements, cascading from an agreed goal. It logically describes the desired state of a system that needs improvement. It is a profoundly simple, yet effective way of identifying gaps, opportunities and critical constraints. Key to the success of this mapping procedure is that the requirements are always objective, never subjective, and they must logically connect in more or less general to specific order, down from one agreed end goal or 'goal statement' (Dettmer, 2007: Ch 3).

By simplifying the desired system view and identifying the necessary conditions and critical success factors underpinning one agreed goal, complex and perplexing system states may be readily understood by stakeholders. Critical problems and potential solutions surface by applying the knowledge of what *should be* happening (from the GT/IO map) and what is currently happening. The CRT/CRB from TOC can be used to map the Current Reality and identify critical gaps, while the PRT (Prerequisite Tree) can identify obstacles and potential solutions. A problem shared is said to be a problem halved. The value of IO mapping to Community Operational Research is in its simplicity. Specialist knowledge is not required to understand, share, and therefore, buy into, the vision that is systematically defined in an IO map, clearing the way for the stakeholders to agree the conditions required to support the goal.

Read the GT/IO map in Figure 3 from top down: *In order to...we must (ensure):*

In order to... Achieve the goal of a more sustainable urban water system for Kāpiti *We must...* Have a secure water supply *We must also...* Live within our means *In order to...* Have secure supply *and* Live within our means *We must...* Have a resilient system, *We must*...Have a flexible system Etc. In order to... Achieve reduced throughputs of potable water *We must...* Have superior (better than average) local government employees and communication *We must...* Have superior communication within the community on water issues... Etc.

The Current Reality: What Really Is Happening?

Current Reality Trees (CRTs) are sufficiency-based (*if...then*) logic trees used to compare reality with what is happening in the system now, principally, what is stopping the system from reaching its goal, as shown in the IO Map, in order to isolate what needs changing. A Current Reality Branch (CRB) focuses on one part of a system, reflecting only the part that is unfavourable (Dettmer, 2007:92). The objective of the CRT/CRB is to single out what needs changing by helping to identify the things that are clear indicators of system deficiencies, tracing them back to one or more basic causes (Dettmer, 2007:95). Figure 3 maps the system destination (the goal and Intermediate Objectives), while Figure 4 shows the Current Reality Branch

Figure 3. The final destination: Goal Tree/Intermediate Objectives map

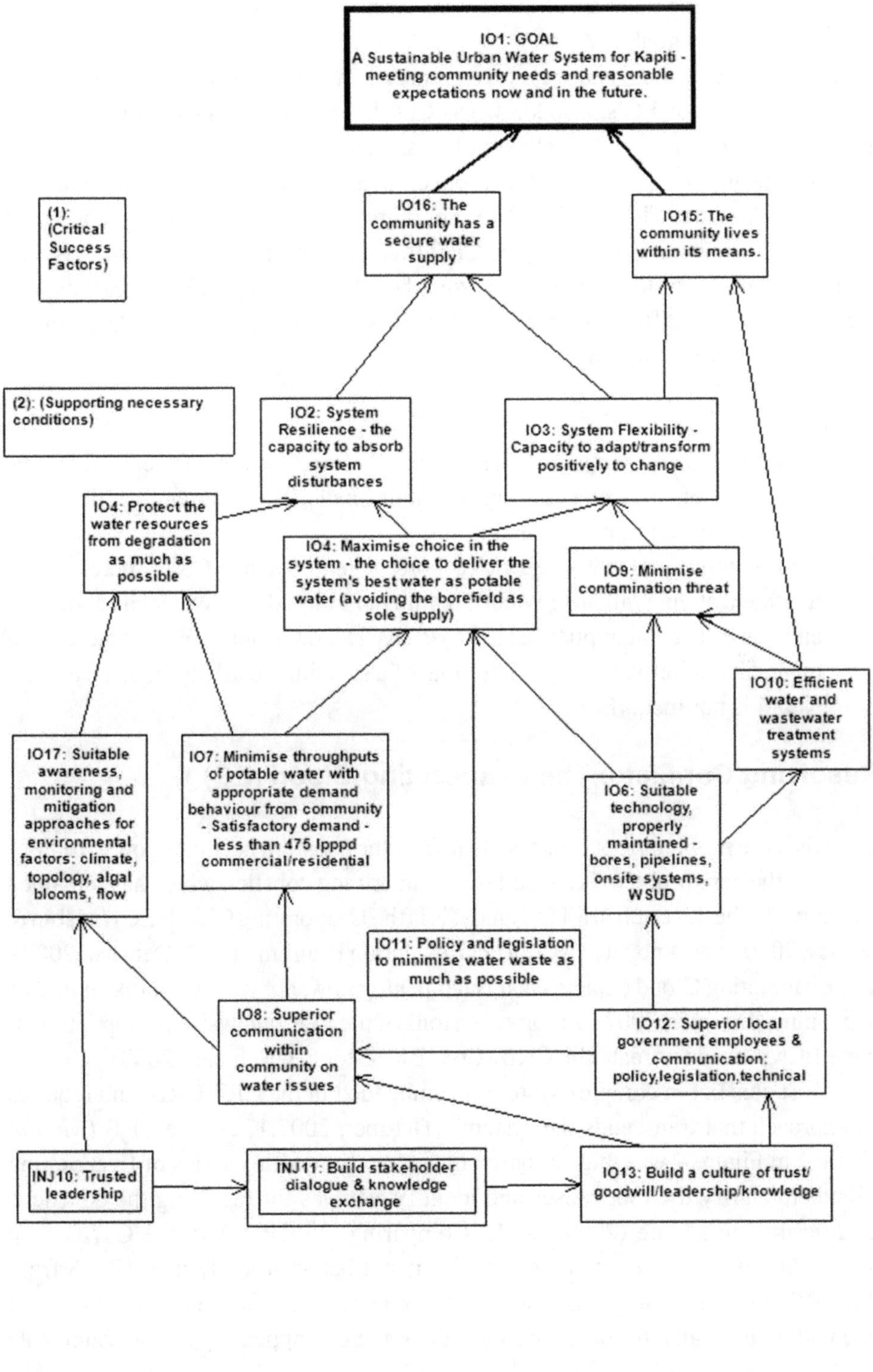

(CRB) that focuses on the dilemma facing the Waikanae/ Paraparaumu sub-region of Kāpiti, reflecting that part of the system most likely to impede the attainment of the agreed system goal.

The CRB (*what to change*) was prepared in conjunction with a series of Evaporating Clouds EC's, also known as Conflict Resolution Diagrams (Dettmer, 2007; Schragenheim, 1998). UDE signifies an undesirable effect. MAG (the bowtie connector near base of Figure 4) shows four entities combining in a magnitudinal way (Dettmer, 2007) to influence UDE2. One or more of these acts as a critical constraint to achieving the goal. Comparing the CRB (depicting participants' viewpoints on a critical part of the system) in Figure 4, with the IO map in Figure 3, the reasons why the current reality differed from the preferred system, were able to be determined.

Read the CRB bottom up:

If...then

If... Entities 1, 2, 3 and 4 exist together *Then*... Demand for water exceeds desirable limits *If*...Demand exceeds desirable limits *Then*... The community is delivered more bore water than is desirable

If... The community is delivered more bore water... *Then*... Costs increase (for treatment at the Waikanae Water Treatment Plant) AND...Water has poor taste and has effects on appliances...Etc. *If*...KCDC does not meet its objective of managing effectively *Then*...The Goal of a sustainable urban water system for Kāpiti is not met... Etc.

Resolving Conflicts: The Evaporating Cloud (EC)

Once the core problem/root cause is identified, the question becomes how to remove it? Often the core problem is a result of an underlying conflict, which has remained unresolved. The next step in a TOC analysis is the Evaporating Cloud (EC) (Mabin & Davies, 2010), also known as a Conflict Resolution Diagram, CRD (Dettmer, 2007). The Evaporating Cloud depicts what is happening now, not what we think should be happening (Dettmer, 2007). In some versions of the TOC analysis, a compilation of three ECs is used to create the CRB (Cox, Blackstone & Schleier, 2003).

"Most conflict in complex systems is embedded in the CRT/CRB, and requires an approach that transcends the system" (Dettmer, 2007:172). The CRB (*what to change*) in Figure 4 was thus prepared in conjunction with a series of Evaporating Clouds to clarify the root causes and identify avenues for resolving the conflicts, as described in Moore (2009:82-84). Comparing the CRB with the GT/IO map reveals that IO 16 [a secure water supply] is not being properly met. UDE8 from the CRB [the water supply is less and less secure] and E2 [the community must live within its means] become the catalysts for the EC, appearing as two statements

Figure 4. The Current Reality Branch (CRB) depicting what is happening in one part of the system

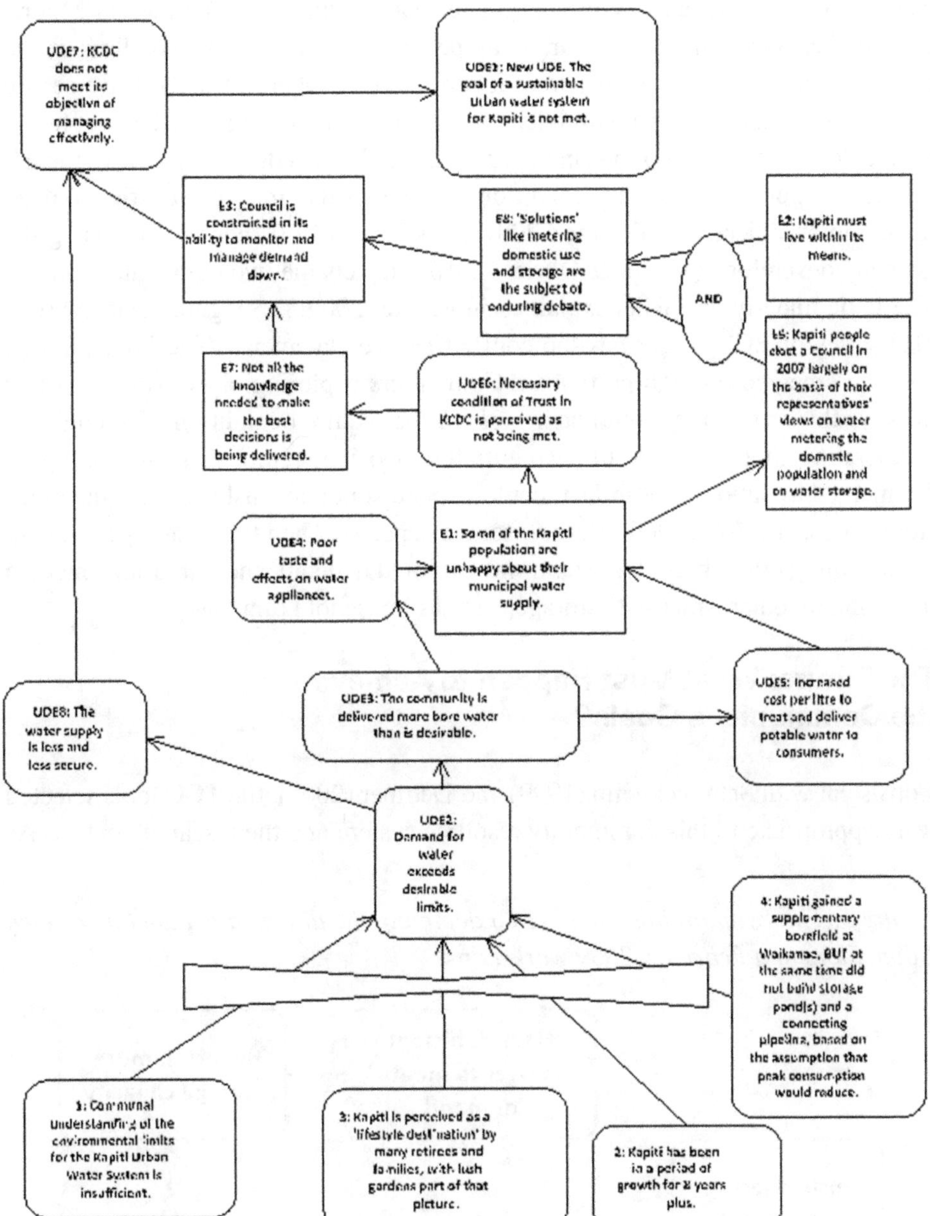

of requirements. UDE8 in the CRB showed the gap between what is required [a secure water supply] (IO16) and what is actually happening in the system [the water supply is less and less secure]. The Evaporating Cloud (shown in Moore, 2009:82) captures the core dilemma as perceived by elected Councillors at the time. The critical requirements were identified as... Kāpiti must [B. Have secure water supply] and [C. Live within its means], and were able to be used to defuse a heated debate over two options in February 2009. The needs of each side of the EC are "non-negotiable necessary conditions, outcomes that must be satisfied to meet the common objective" (Dettmer, 2007:186). The full EC, constructed using the normal conventions (Mabin & Davies, 2010), depicts the two competing options. The logic underlying this is explained in Moore, 2009:82). Figure 5, below (one of a series of EC's) describes the conflict as two statements. The breakthrough solution emerged through participant Councillors exploring objectively, a variety of ways the community could achieve *both* the required conditions. Through the Evaporating Cloud process, the two initially opposing arguments shown in Figure 5, emerged as actions that in fact, could happen simultaneously, 'Build (modular) storage capacity AND Defer (large) storage capacity. The idea of storage capacity as something other than a large dam was explored in detail, and a solution emerged involving modular, small-site storage, such as household rain tanks.

The Future: What Must Happen to Achieve the Community's Goals?

Consistent with Schragenheim (1999) and Dettmer (2007), the TOC tools selected were appropriate to this community resource system and the research context. As

Figure 5. The Evaporating Cloud (EC) depicting the dilemma debated by elected Councillors at a February 2009 workshop

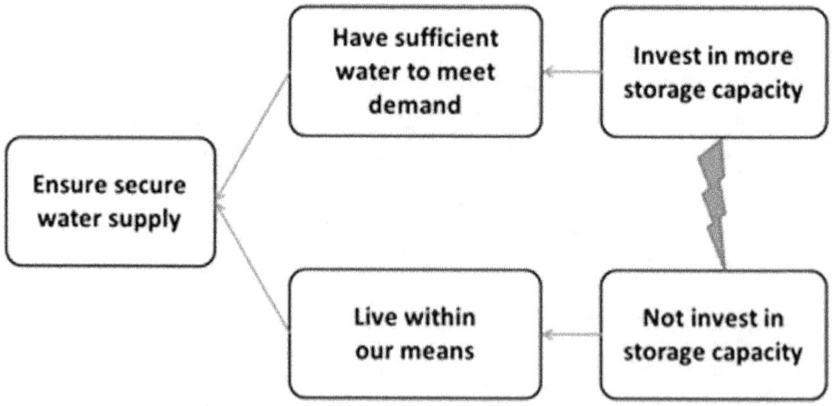

outlined above, with two elected Councillors, a key conflict regarding water storage was explored, and broken, using the Evaporating Cloud (EC). Causal Loop Diagrams (CLDs) from Systems Dynamics were then used to capture the dynamics of proposed system solutions and identify potential interactions and unintended consequences.

The CLDs were used to explore with participants how a past 'quick fix' had turned into a 'fix that fails' (Senge, 1994). Agreement was then reached on necessary interventions to prevent 'fixes from backfiring' (Senge, 1994:126). Using the EC and CLD in combination helps to devise robust solutions to avoid such negative side effects, as has been described in Mabin, Davies and Cox (2006). The original CLDs are shown in Moore (2009:92/94). An updated CLD is shown in Figure 6, next.

The Causal Loop Diagram in Figure 6 portrays the dynamic nature of water supply and demand, arising from typical approaches to managing the water supply system. We read the diagram following the arrows as follows:

Starting with the 'problem symptom' of insecurity of water supply (evident with system health and capacity being at risk), the usual response (*fix*) was to invest in more storage capacity within the reticulated water supply, increasing the water available for use. This in turn reduces the insecurity of water supply, reducing the 'problem symptom'; this chain of symptoms, actions and effects forms a balancing loop (B), bringing the system back into balance. However, increased water availability also

Figure 6. The Fix that fails (Senge, 1994) and the injection/corrective action to resolve it

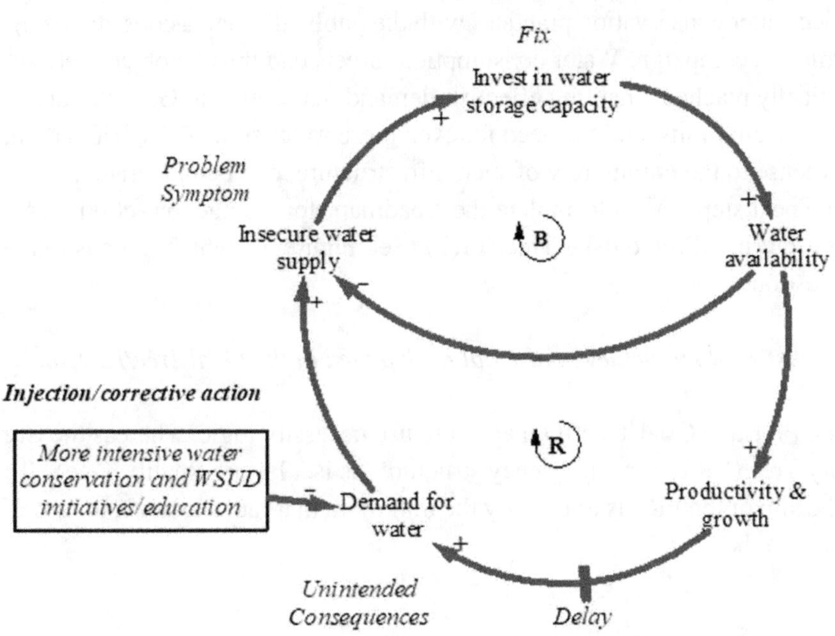

promotes productivity and growth, which after a delay, leads to an increase in the demand for water (an unintended consequence) and, the insecurity of water supply 'problem symptom' returns, in this case, magnified by climatic conditions and failing infrastructure. This chain of cause-effect forms a reinforcing loop (R), signalling that the problem symptom gets worse and worse over time.

To prevent this from happening, some corrective action (or *injection* in TOC terminology) is needed – and here it is suggested that if more intensive water conservation and water sensitive urban design (WSUD) initiatives and education are engaged in, including storage capacity outside of the reticulated system in the form of household rainwater tanks, then this would moderate the increase in demand for water, averting insecurity in the Kāpiti water supply.

The message from this is that although investing in more water storage capacity (like drilling new bores or building a dam) might appear at first glance to be the natural response to water supply insecurity, the effectiveness of this action is at risk of decreasing over time due to growth-induced levels of demand. Increasing demand, combined with the effects of a warming climate, places more and more pressure on infrastructure and elevates the risk of breaching environmental limits – this is unacceptable. It is thus imperative that investment in urban water infrastructure is also accompanied by measures such as water conservation and WSUD initiatives, and education – community-wide. Such initiatives include household rainwater tanks, investment in a successful river recharge scheme to replenish the river using borewater, rather than tapping into the poorer-tasting bore water for drinking in dry periods, rewarding the finding and fixing of household water leaks, and sharing best practice water conservation practices with the public through a consistent, ongoing information campaign. Water consumption targets can thus be objectively set and realistically reached, enabling effective demand management. Growth thus occurs within system limits, and the need for even more water storage, such as a dam, and the expense to the community of such infrastructure, can be moderated.

The next step was to formulate the 'roadmap' for change, or action plan. This was set out in a Prerequisite Tree (PRT) (See Figure 7), which intends to answer the question:

What must we do to achieve our Goal as depicted in the Goal Tree/IO Map?

The PRT and Goal Tree/IO maps both use necessity logic, whereas the Current Reality Tree (CRT) is a sufficiency structure (as is a Future Reality Tree).

The aim of the PRT is to identify the minimum that must be done, to move from one step to the next.

Figure 7. The Prerequisite Tree (PRT) implementation plan depicting the depicting the minimum that must be done to achieve the goal

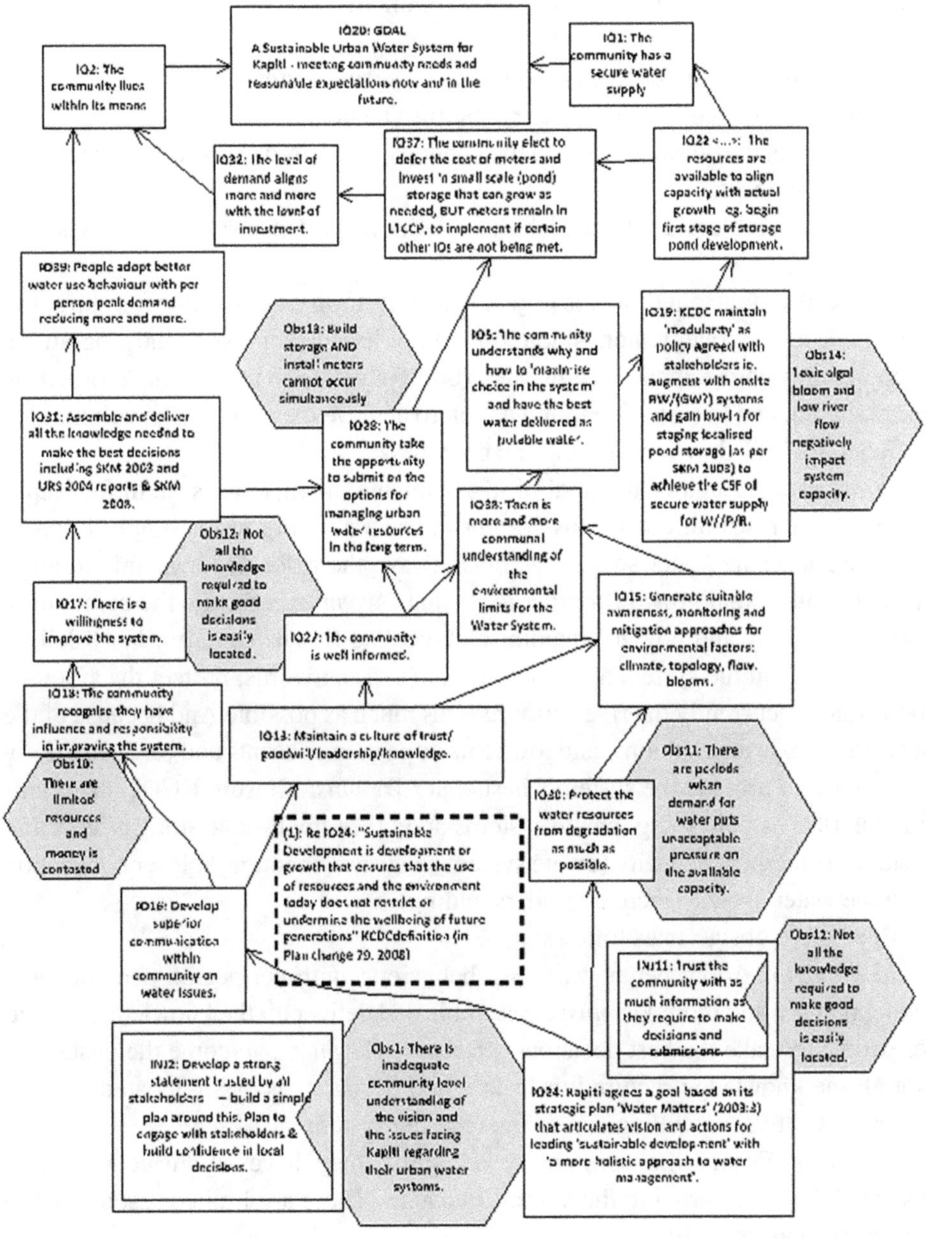

The PRT is a *necessity* structure, showing obstacles and the minimum necessary changes to move from one step to the next. In the following example, to achieve the goal of a sustainable water system for the community...

We must...Live within our means AND Have a secure water supply.
We must...Ensure people adopt better water use behaviour…
And because of the *obstacle*...Not all the knowledge required to make good decisions is easily located…
We must...Assemble and deliver all the knowledge needed to make the best decisions.

The aforementioned necessary condition took the form of a targeted communications strategy, alongside a community-led advisory panel and programme of community engagement. Its aim was to build interest and trust in the information being shared, and to foster ideas and action to achieve the goal for the community.

Read the PRT downwards from the goal, thus:

In order to... Achieve the goal of a sustainable urban water system for Kāpiti *We must*...Live within our means *We must*...Have a secure water supply In order to... Have a secure water supply *We must*...Have the resources available to align capacity with actual growth In order to do that...*We must*... Ensure the community understands why and how to 'maximise choice' so KCDC can affordably deliver the best water in the system as potable water (in other words, protect the system's 'best water' – currently the river resource – as much as possible And because of the obstacle...Toxic algal blooms and low flow impact on consents and on the capacity of the river to provide the system's best water *We must*...Ensure KCDC maintains 'modularity' as policy...eg. onsite systems and other small scale storage/recycling systems to augment capacity (onsite systems may also encourage closer connection with the water cycle to increase understanding)

Read other obstacles as follows:

In order to...Adopt better water use behaviour, with per-person peak demand reducing more and more *We must*...Assemble and deliver all the knowledge needed for people to make the best decisions... Because *We must* overcome the obstacle... Not all the knowledge required to make good decisions is easily located And, in order to...Cultivate a willingness to improve the system

We must...Ensure the community recognise they have an influence and a responsibility in improving the system Because...There are limited resources and money is contested... Etc.

RESEARCH FINDINGS

Results show the community at large accepting they have a role to play in urban water reform. The overarching message from the participants in this study is that for the urban water system to serve our communities and businesses better, the range of stakeholders must develop a deeper understanding of the system's limits and opportunities. If limits are not understood and agreed, it is difficult to live within them. If opportunities are not revealed, it is hard to grasp them. Planting drought-resistant lucerne (alfalfa) in Eastern Marlborough instead of rye-grass and clover, is one example, with its obvious parallel in the Kāpiti context, of creating landscapes and gardens that that need little or no watering, of mandatory supplementary water tanks, and other less obvious implications. A perceivable lack of resistance to a suite of measures and their associated costs, to curb water demand, is suggestive of the public accepting some responsibility for water reform in the first stages of implementation (KCDC 2013a), alongside Council staff, Councillors, and legislators. Indicators such as Annual Satisfaction Surveys (see KCDC, 2010, 2017a; CRAG, 2016) suggest that this lack of resistance has evolved into a high degree of consensus on acceptance of the reforms.

Prior to the Pilot study, the ethics of this project seemed particularly challenging, given the participants' viewpoints on possibly contentious issues were to be openly shared, albeit in a structured way. On reflection, an environment of trust between interviewer and participants was crucial to the success of this approach. TOC provides a constructive and non-threatening way to encourage a kind of deep-level reflection. In communities where many options and perspectives are raised, TOC provides a logical and collaborative framework for reviewing these, identifying the most favourable future reality and galvanising action. Causal Loop Diagrams (Senge, 1994; 2008) were part of the reflexive (Bassett, 1995) process, developed with key decision makers as a dynamic representation of the system.

In 2013, when the first author re-visited the research as preparation for a conference paper, water security had improved in Kāpiti, with steps like leak detection being credited with saving many millions of litres of treated water (Kāpiti Observer, 27 May 2013:10). Other actions included investing in the aforementioned Communications Strategy to keep the public well-informed and encourage more efficient water use behaviours. Multi-modal supply options like roofwater/greywater for outdoors, are encouraged in various meaningful ways by Council, and mandated for new buildings, while system transparency and the sharing of knowledge had been bolstered by the appointment of a Kāpiti technical advisory group (or CRAG – Charging Regime Advisory Group), led by former State Services Commissioner, Don Hunn.

With more precise information on where water is being wastefully used, and by whom, conservation education and technologies (like flow reducing tapware and pressure reduction) are able to be better targeted. The number of positive media stories, fewer negative stories and letters to the editor, and the 2010 local authority election results, suggest a turning point reached in community understanding – of the pressing need for effective water conservation and other reforms.

Recent developments include the introduction of a charging regime using water meters. The findings of this research showed two (both in a citizen action group) of the 20 participants surveyed on the question of meters, not supportive of water meters as a tool for demand management. Of those participants supporting water metering in some form, the vast majority urged no private ownership of the water utility as an attendant condition. A second condition suggested by a significant number of participants, was that businesses making use of a water resource to secure revenue, should pay by their consumption (volumetric charging). This research did not expand on this perspective, as its scope was broader than required by the thesis question.

In 2017, volumetric charging and water rights have become the subject of wider debate, and a national election issue. One side of the debate suggests that specific sectors should be required to acknowledge their actual use of water, and make a meaningful contribution, based on consumption. This contrasts with the prevailing situation, where water rights are inexpensive, and awarded on a first come first served basis. A bottling plant for example, pays a fee for a right to use the water resource and can bottle and ship offshore as much water as their allocation allows. This is seen by some as a significant benefit for an insignificant price. Others might argue that benefits accrue in the form of employment, tax revenue, and other opportunities arising from business growth, and that a unit price on business use of non-urban water would potentially slow the New Zealand economy. Articulating the conflict at the heart of this debate, using the TOC Thinking Processes, would be a useful exercise for revealing the leverage points to start with, as a means to finding where the consensus lies, in regard to reforming the parts of the system that are problematic. This could be an area for future research, with the impact of water meters and volumetric charging considered in both urban and rural contexts, taking account of interdependencies with other water conservation/demand management strategies.

DISCUSSION

A common barrier to using the kind of formula for change provided through the TOC Thinking Processes is the time-consuming nature of the diagrams. Dettmer's mapping techniques (2007) offer simplified alternatives to the traditional mapping processes of Goldratt (1997; 2004). The Goal Tree/IO Map (described as the IO Map,

'Destination' or 'Standard' in Dettmer, 2007 and revised to Goal Tree in Dettmer, 2011) is mapped at the outset. The traditionally complex and time-consuming Current Reality Tree is then able to be constructed with reference to the singular part of the system that was crucial to the Goal, identifying the one or two constraints (from the interviews) that impede system success. While TOC is regarded first and foremost as a management theory, useful as a means to articulate the nature of a situation, and therefore not always front of mind for action research, our case-study adds to a growing body of work that suggests it is effective as an action-planning approach to achieving consensus on change. Examples of TOC's use in a programme of behavioural change include Pongsart (2015), where TOC was found to be effective in helping students construct and follow a change process to ensure they completed their master's thesis on time and to meet the desired quality. The use of the Goal Tree/IO Map prior to the other TOC tools proved to be invaluable in that research, as well as in other projects such as Kimani (2015) and Mirzaei, (2016). More recently, Hohmann (2017) has shown how a simple device of colour-coding the critical success factors and necessary conditions can help to focus effort on the critical factors that are constraining performance.

The TOC Thinking Processes allow solutions to be logically tested across a field of participants. Every opinion is openly acknowledged and can be assessed against the system goal, in order to reach a consensus that can be verified. Its application in this instance, suggests it has a place as a methodology for conducting community-focused research that leads to constructive, far-reaching solutions.

CONTRIBUTIONS

A community-focused approach to dealing with resource management matters is implied in the *Treaty of Waitangi* (1840, see Young, 2001; Williams, 1997), and has been enshrined in New Zealand's contemporary environmental legislation (RMA 1991). A relevant provision is Part 2, section 7(a), which mandates having particular regard to *kaitiakitanga* (Williams, 1997; Sheppard, 2010; Young, 2001; Royal, 2012), defined in the legislation as "the exercise of guardianship by the *tangata whenua* of an area in accordance with tikanga Māori in relation to natural and physical resources [...]". By extension, this acknowledges the responsibility of the local community for the health of its natural resources, including freshwater, and is a compelling motive to investigate Community Operations Research (COR) in this country.

This study has its focus firmly on people and their perspectives and is a departure from the more commonly applied operational research models using mathematical programming. Instead, a series of conversations with diverse participants drawn (using stakeholder analysis) from an entire community, was developed into an action plan that was verified and tested using TOC's Thinking Process tools and Causal Loop Diagrams from Systems Dynamics. This multi-method approach focused on identifying the key factors at each stage, in order to find ways of helping this community to move towards its goal of a sustainable urban water management system – a systemic problem requiring a systemic solution approach. The simplicity and clarity that came from having the focus on the constraining factors, enabled both researchers and participants to visualise the system and possible solutions, in ways that matter.

Notably, the goal oriented TOC framework (Goldratt, 2010) appears to provide some useful synergies with the goal orientation of sustainable development (SD), as stated in definitions from the seminal *Bruntland Report*, the *Rio Declaration* and *Agenda 21* (Bosselmann, 2002:93) and more specifically, as articulated in *Water matters*, Kapiti's Sustainable Development Plan (KCDC, 2013b; 2003).

Water reform is arguably most effective when people are prepared to champion the need for change and the conditions that underpin the change. Since 2009, post-thesis, some of this study's most critical findings have been championed. In that year, KCDC created a new position of Water Supply Project Manager, with Phillip Stroud appointed to the role. Mr Stroud read the research findings this paper describes, and met with the first author to learn more about TOC, and particularly IO mapping (Goal Trees). In addition, he invited review of a new Water Communications Strategy and appointed an expert communicator who "brought to Council's Water Supply Project, strategic thinking that improved how the project communicated with the public. Implementation of a 'core script' (see PRT, Figure 7, INJ 2) helped the project succeed where others had failed" (Stroud, personal comment, 26 May 2010).

Revisiting the project in 2017, the first author spoke to Martyn Cole, KCDC Water and Wastewater Asset Manager. Mr Cole pointed to magnitudinal water savings brought about through the community's growing engagement with the Council on the critical water system issues they needed to deal with, including wasteful use, aging, leaky pipes, and water quality. There is a deep level of understanding and regard for the sustainability "Goal", and water use targets in the Communications Strategy that is reflective of the final Goal Tree's IO8 [Superior communication within community on water issues] and the Injection INJ11 from the CLD to prevent the "Fix that fails" [Build stakeholder dialogue and knowledge exchange].

Reflecting on the issues of concern, there is a suggestion that the focus on "what to change" and "how to cause the change" - to secure IO1 [a sustainable urban water system for Kāpiti] is becoming a more intuitive process. It is helpful that when the community is meeting or exceeding their water consumption targets, the "best water" in the system will always be available for drinking and washing, so there is tangible evidence at the tap as to whether they are achieving their water quality goals, because the "best" water (from the river) tastes, looks, and feels better than the alternative, bore water.

Mr Cole observes that the benefits of installing water meters, particularly "detecting and fixing leaks in our aging water system" were immediately visible. "Volumetric charging has been the single biggest change" with the entire district exceeding or within just a few litres of achieving its water conservation targets in 2017. A policy of financial encouragement to report and fix leaks, alongside equitable charging, might be credited to this level of progress. Leak detection at street level, with meters pointing to some substantial water losses from the network, is a source of significant savings. Changing behaviours is another important contributor. Mr Cole describes the work by volunteer advisory group CRAG (Charging Regime Advisory Group formed 2011), as crucial in determining the "equitable and fair charging regime" that the community clearly has confidence in. Notably, customer satisfaction went up in the year volumetric charging was introduced, and has increased since. The standout for Mr Cole has been the readiness of the community to embrace onsite water storage systems under Plan Change 75. This is ground breaking legislation in New Zealand and combines with other innovative policy, including reimbursement to residents of water charges related to repaired leaks, so they would be motivated to report and fix leaks. The community is more resilient says Mr Cole. "Under the Plan Change, the district now has 13 megalitres (13 million litres) of stored water - that's improving community resilience by a large factor".

OUTLOOK

New Zealand's Parliamentary Commissioner for the Environment has stressed that "the significance of an environmental issue cannot be judged without looking ahead" (PCE, 2016). The outlook for water in New Zealand varies by region - context is everything (PCE, 2016). With some regions experiencing the combined impacts of climate change, earthquakes and intensified land use, the scale of the problem of water quality and quantity is accelerating for them, while in other regions, there are water system improvements. The effects of climate change, accumulated with the longstanding impacts of erosion, pathogens and nutrient loading are known challenges.

A decade on from when this study began, and the outlook for the subject community is an increasingly resilient one, especially in light of recent earthquake events in the region. Evidence for consensus-building can be seen through a pattern of improvement in community satisfaction surveys and residents' responses to some significant policy changes (KCDC, 2015). The introduction of household water metering saw an increase in customer satisfaction the following year, despite this proposition being the cause of marked negative media attention for the authority when first proposed (Moore, 2009a; KCDC 2010). In the year 2017, 82% of ratepayers are satisfied with the overall performance of their Council, and 81% are satisfied with the water supply. By comparison, in 2010, just 35% were satisfied with water quality and less than half felt the supply was reliable, while at the same time these two issues rated most important to respondents (KCDC, 2017a, 2017b).

The Charging Regime Advisory Group (CRAG) was formed at the instigation of KCDC in 2011 to design an efficient and equitable system for the charging of water supply to the Kāpiti community (CRAG, 2016). CRAG presented 12 recommendations to the Kāpiti Coast District Council on 5 April 2012. All were in large part implemented and indications are that the charging regime has achieved a considerable measure of public acceptance (CRAG, 2016). One such CRAG recommendation was that all profits remain within water operations, rebutting a perception that water metering could be a "cash cow for KCDC" (term used by some submitters during consultation and by more than one participant in this study). This contributes to the necessary condition IO13 described by participants as [Build a culture of trust/goodwill...]. Under Plan Change 75, millions of litres of stored rainwater is available for non-potable use like toilet flushing and garden watering. With treatment options like chemical dosing with stabilised hydrogen peroxide (H2O2), householders can make their stored rainwater suitable for drinking and washing, should an emergency necessitate it. A side effect of significant quantities of detained rainwater is to ease pressure on stormwater systems, thus the policy performs a role in flood mitigation.

The local authority in this case reached a level of consensus with its community on a collection of adaptive and pre-emptive responses to the pressures and impacts of population and a warming climate. Given the current trends continue, latest reporting (KCDC, 2017a) shows significant infrastructure spending on water will be delayed by more than twenty years, with a previously tabled dam proposal off the agenda until at least 2060, a saving for the community of 36 million dollars.

CONCLUSION

This chapter was inspired by commentaries nearly a decade apart, from New Zealand's Parliamentary Commissioner for the Environment (2000) and the government's 2009 Fresh Start taskforce that suggested reaching consensus on environmental matters was one of the most pressing issues for communities. In 2015, the Commissioner released another sobering report following an investigation into the relationship between land use and two nutrient pollutants – nitrogen and phosphorus - affecting our waterways, urban and rural. On land, nitrogen and phosphorus are valuable nutrients, while in water, they can cause excessive growth of weeds, slime and algae

(PCE, 2015). Professor David Hamilton, President of the NZ Freshwater Sciences Society, wrote that the 2015 report "[...] clearly highlights the enormous challenge that we face as a nation: how to meet the government's goals of doubling the value of agricultural exports by 2025 whilst meeting the 2011 National Policy Statement for Freshwater Management which requires that the 'overall quality of freshwater' in all regions of the country be maintained or improved" (quoted in PCE, 2015).

Protecting our waterways involves questioning the focus on "ever-increasing production", and "some win-win strategic thinking occurring", according to the Parliamentary Commissioner (2015). The case of Kāpiti demonstrates a community engaging with its challenges, sparking intuitive thinking that suggests where consensus might lie, leading to creative, "win-win" ideas.

The first author used a systemic problem structuring approach to engage with a number of willing individuals, all stakeholders in a New Zealand urban water system. Participants were asked for their perspectives about the system destination, what to change, what to change to, and how to make the changes necessary, for more sustainable urban water systems.

The study developed into what was essentially a 'conversation' on the subject of how we might manage our freshwater for better outcomes, with the combined methodological framework providing a source of deep insights into the challenges, dilemmas, potential solutions, and side effects, facing resource managers and other stakeholders in an urban water system under pressure from population growth, lifestyle choices, and climatic/topographical conditions. Dettmer (2007) urges that *simplifying approaches is the key to finding out what we know, and that it is better to be approximately right than precisely wrong*. This premise of simplicity was fundamental to the design of this research and to the process of open engagement upon which this enquiry was constructed. The structured and transparent process of engagement around an agreed goal appears to provide a logical platform for consensus-building. It may also provide a reliable way to measure the level of consensus among stakeholders. This is an area for future research.

The authors found that reaching consensus on water reform involves a number of key components, not least among them, having a champion for the reforms and a sensible and consistent communications plan. A trustworthy process of engagement has, in this case, underpinned some effective reforms in the presence of perplexing and complex resource management issues, and the community continues to make progress towards achieving their system Goal of *a sustainable urban water system, that meets the community's needs and reasonable expectations, now and in the future*. A potential challenge arises when the TOC practitioner or champion disengages from the research, leaving stakeholder insights to be shared in a possibly more ad hoc manner; limiting ongoing improvement. It has been proposed that the volunteer advisory group CRAG continues to review the effects of the charging regime periodically (CRAG, 2017). That the community and KCDC have a high level of trust in CRAG is strongly evident and it is hoped that the present cycle of reflection and improvement continues. CRAG and KCDC staff like the Water Coordinator (who participated in this study originally) and Water Asset Manager (interviewed for this update), have a championing role to play in keeping the public engaged, keeping their focus firmly on the system Goal and the conditions necessary to meeting it. Following the Kāpiti case as it progresses, will reveal study limitations and points to reflect on. It is hoped that the insights gained from applying this multi-methodological framework might inform local authorities and others charged with reaching consensus on the policies and adaptive practices necessary for sustainable development, securing a clean, green future for the generations that follow.

ACKNOWLEDGMENT

With particular thanks to the participants of this study, to Arun Elias who provided feedback and critical suggestions, and Bob Cavana who provided expert advice on the Causal Loop Diagramming. This work won the inaugural IFORS EthOR (Ethics in OR) Award at EURO, Rome 2013. No funding was sought or received for this research.

REFERENCES

Ashlag, Y. (2014). *Removing constraints for business growth*. Great Barrington, MA: North River Press.

Avery, D., & Thomson, M. (2017). *The resilient farmer: Weathering the challenges of life and the land*. Penguin Random House.

Bassett, K. (1995). On reflexivity: Further comments on Barnes and the sociology of science. *Environment & Planning*, *27*, 1527–1538. doi:10.1068/a271527

Bjørndall1, T., Herrero, I., Newman, A., Romero, C., & Weintraub, A. (2011). Operations research in the natural resource industry. In *International Transactions in Operational Research*. International Federation of Operational Research Societies. Retrieved October 2012 from http://onlinelibrary.wiley.com/ doi/10.1111/j.1475-3995.2010.00800.x/

Bosselmann, K. (2002). The concept of sustainable development. In K. Bosselmann & D. Grinlinton (Eds.), *Environmental law for a sustainable society* (Vol. 1). Auckland: The New Zealand Centre for Environmental Law.

Burton-Houle, T. (2001). What if my organization's goal is not to make money? *TOC Times Quarterly Newsletter*. Retrieved 5 April 2011 from http://www.goldratt.com/for-cause/partnerperspsep2001toct.htm

Charging Regime Advisory Group (CRAG). (2016). *Tariff review March 2016.* Retrieved from http://www.kapiticoast.govt.nz/contentassets/3f9411b3fb4a4092a115c89c4207ff79/crag_report_2016_review.pdf

Cox, James, Boyd, Sullivan, Reid, & Cartier. (2012). *The Theory of Constraints International Certification Organization Dictionary* (2nd ed.). Retrieved from http://www.tocico.org/?page=dictionary 2

Cox, J. F., Blackstone, J. H., & Schleier, J. G. (2003). *Managing operations: A focus on excellence* (Vol. 1). Great Barrington, MA: North River Press.

Cox, J. F., & Schleier, J. G. (Eds.). (2010). *Theory of Constraints handbook.* New York: McGraw-Hill.

Dettmer, H. W. (2007). *The logical thinking process: a systems approach to complex problem solving*. Milwaukee, WI: Quality Press.

Dettmer, H. W. (2011). *Our goal is...What is our goal?* Retrieved 21 September 2014 from http://www. goalsys.com/books/documents/WhatisOurGoal- v5_000.pdf

Elias, A. A., Cavana, R. Y., & Jackson, L. S. (2002). Stakeholder analysis for R&D project management. *R & D Management*, *32*(4), 301–310. doi:10.1111/1467-9310.00262

Foote, J. L., Gregor, J. E., Hepi, M. C., Baker, V. E., Houston, D. J., & Midgley, G. (2007). Systemic problem structuring applied to community involvement in water conservation. *The Journal of the Operational Research Society*, *58*(5), 645–654. doi:10.1057/palgrave.jors.2602248

Freeman, R. E. (1984). *Strategic management: a stakeholder approach*. Boston, MA: Pitman.

Garvin, D. A., & Roberto, M. A. (2001). What you don't know about making decisions. *Harvard Business Review*, (Sept): 108–116. PMID:11550627

Goldratt, E. M. (1997). *Critical chain*. Great Barrington, MA: The North River Press.

Goldratt, E. M. (2010). Introduction to TOC – my perspective. In J. F. Cox & J. G. Schleier (Eds.), *Theory of Constraints handbook* (pp. 3–9). New York: McGraw-Hill.

Goldratt, E. M., & Cox, J. (1984). *The Goal: A process of ongoing improvement.* Great Barrington, MA: The North River Press.

Hohmann, C. (2017). *What is a goal tree?* Retrieved 30 August 2017 from https://hohmannchris.wordpress.com/goal-tree/

Huberman, A. M., & Miles, M. B. (1998). Data management and analysis methods. In N. K. Denzin & Y. S. Lincoln (Eds.), *Collecting and interpreting qualitative materials*. Thousand Oaks, CA: Sage.

Kāpiti Coast District Council. (2012). *Annual Plan 2011/12 – Part 1*. Retrieved 15 May 2013 from http:// www.kapiticoast.govt.nz/Documents/Downloads/ annual-plan/annual-plan-1112-part1.pdf

Kāpiti Coast District Council. (2012a). *Water – what will it cost?* Retrieved from http://www.kapiticoast.govt.nz/Docu- ments/NewslettersAndPeriodicals/water-project/ water-update/2012/water-update-9May2012.pdf

Kāpiti Coast District Council. (2012b). *Water meter survey*. Retrieved 27 May 2013 from http://www.ka- piticoast.govt.nz/Projects/Water-Meter-Installation/water-meter-survey/

Kāpiti Coast District Council. (2013). *Our district: Statistics and demographics*. Retrieved January 2013 from http://www.Kāpiticoast.govt.nz/Our-District/The-Kāpiti-Coast/Statistics-and-Demographics/

Kāpiti Coast District Council. (2013a) *Water supply project*. Retrieved 11 May 2013 from Error! Hyperlink reference not valid.

Kāpiti Coast District Council. (2013b). *Water matters – Kāpiti Coast District Sustainable Water Management Strategy, 2003*. Retrieved from http://www.kapiticoast.govt.nz/ Documents/Downloads/Strategies/Water-Matters- Sustainable-Water-Management-Strategy.pdf

Kāpiti Coast District Council (KCDC). (2009). *Kapiti water supply LTCCP review: water supply options for the Waikanae Paraparaumu Raumati area (2 April 2009)*. Retrieved 15 May 2009 from http:// www.kapiticoast.govt.nz/NR/rdonlyres/

Kāpiti Coast District Council (KCDC). (2010). *Residents opinion survey August 2010*. Retrieved from http://www.kapiticoast.govt.nz/globalassets/internal-page-images-and-documents/resident-opinion-surveys/residentopinionsurvey-august-2010.pdf

Kāpiti Coast District Council (KCDC). (2017a). *Residents opinion survey May 2017*. Retrieved from http://www.kapiticoast.govt.nz/globalassets/your-council/planning/documents/residents-opinion-survey---annual-report-13-07-17-v5.pdf

Kāpiti Coast District Council (KCDC). (2017b). *Kāpiti Coast water conservation report 2016/17*. Retrieved from http://www.kapiticoast.govt.nz/contentassets/d76c873551904e66bb1604424127b027/2016_17_waterconservationreport_v1bt.docx_.pdf

Kāpiti Coast District Council (KCDC). (2017c). *Annual Reports (on water)*. Retrieved from http://www.kapiticoast.govt.nz/services/A---Z-Council-Services-and-Facilities/Water/Water-Supply-Annual-Reports/

Kāpiti Observer. (2013). The 2013/14 programme: Financial strategy and rates impact. *Kāpiti Coast District Council – Summary – Draft Annual Plan 2013/14*. Retrieved 11 May 2013 from http://kapitiobserver.realviewdigital. com/?iid=75522&startpage=page0000027

Kāpiti Observer. (2013, May 27). *Council backing water meters*. Author.

Kim, S., Mabin, V. J., & Davies, J. (2008). The Theory of Constraints Thinking Processes: Retrospect and Prospect. *International Journal of Operations & Production Management*, *28*(2), 155–184. doi:10.1108/01443570810846883

Kimani, S. (2015). *Exploring Quality of Learning and Teaching Experiences in Higher Education using the Theory of Constraints*. Victoria University of Wellington.

Land and Water Forum. (2010). *Report of the land and water forum: a fresh start for freshwater*. Retrieved 8 October, 2010 from http://www.landandwater.org.nz

Land and Water Forum. (2011). *Land and Water forum recommendations – Implementation*. Retrieved 25 January 2013 from http://www.landandwater.org.nz/includes/ download.aspx?ID=118917

Land and Water Forum. (2012). *Third report of the land and Water forum*. Retrieved December 2012 from http://www.landandwater.org.nz/includes/download.aspx?ID=124767

Mabin, V., Davies, J., & Cox, J. (2006). Using the Theory of Constraints Thinking Processes to Complement Systems Dynamics' Causal Loop Diagrams in Developing Fundamental Solutions. *International Transactions in Operational Research*, *13*(1), 33–57. doi:10.1111/j.1475-3995.2006.00532.x

Mabin, V. J., & Davies, J. (2010). The TOC thinking processes: Their nature and use – reflections and consolidation. In J. F. Cox & J. G. Schleier (Eds.), *Theory of Constraints handbook* (pp. 631–669). New York: McGraw-Hill.

Midgley, G., & Ochoa-Arias, A. (Eds.). (2004). *Community Operational Research: OR and Systems Thinking for Community Development*. New York: Kluwer; doi:10.1007/978-1-4419-8911-6

Mingers, J., & Brocklesby, J. (1997). Multi-methodology: Towards a framework for mixing methodologies. *Omega-International Journal of Management Science*, *25*(5), 489–509. doi:10.1016/S0305-0483(97)00018-2

Mingers, J., & Rosenhead, J. (2004). Problem structuring methods in action. *European Journal of Operational Research*, *152*(3), 530–554. doi:10.1016/S0377-2217(03)00056-0

Ministry for the Environment (MfE). (2011). *National Policy Statement Freshwater Management 2011*. Retrieved October 2012 from http://www.mfe.govt.nz/ publications/rma/nps-freshwater-management-2011/ docs/nps-freshwater-mgnt-2011.pdf

Ministry for the Environment (MfE). (2003). *Dairying and Clean Streams Accord between Fonterra Co-operative Group, Regional Councils, Ministry for the Environment, and Ministry of Agriculture and Forestry*. Retrieved 28 January 2013 from http://www.mfe.govt.nz/issues/land/rural/dairying- accord-may03.pdf

Mirzaei, M. (2016). *Exploring Applicability of TOC to projects: Critical Chain Project Management and Beyond*. Wellington: Victoria University of Wellington.

Mitchell, R. K., Agle, B. R., & Wood, D. J. (1997). Towards a theory of stakeholder identification and salience: Defining the principle of who and what really counts. *Academy of Management Review*, *22*(4), 853–886.

Moore, R. M. (2009). *Shaping more sustainable communities: a case study in urban water management* (Master's thesis). Victoria University of Wellington. Retrieved from http://www.j.co.nz/robyn/Robyn_Moore_%20Shap- ing_More_Sustainable_Communities.pdf

Moore, R. M. (2009a). Shaping more sustainable communities: a case study in urban water management. In *Operations Research Society of New Zealand, Proceedings of the 44th Annual Conference ORSNZ 09*. University of Canterbury. Retrieved from http:// www.j.co.nz/robyn/Robyn_Moore_ORSNZ_2009_ conference_paper.pdf

Moore, R. M. (2009b). *Shaping more sustainable communities - Developing sustainable environmental and sound business agendas. Power-Point Presentation. ORSNZ 09 conference, December 3 – 4th 2009*. New Zealand: University of Canterbury. Retrieved from http:// www.j.co.nz/robyn/Robyn_Moore_ORSNZ_2009_thesis_presentation.pdf

New Zealand Landcare Trust. (2011). *Review of the Starborough Flaxbourne Soil Conservation Group Project.* Retrieved December 2012 from http://www. landcare. org.nz/files/file/566/Starborough%20Flaxbourne%20Review%20Dec2011.pdf

New Zealand Treasury. (2016). *New Zealand Economic and Financial Overview 2016.* Retrieved from http://www.treasury.govt.nz/economy/overview/2016/nzefo-16.pdf

Parliamentary Commissioner for the Environment. (2000). *Aging pipes and murky waters. Urban water system issues for the 21st century*. Wellington: Parliamentary Commissioner for the Environment. June.

Parliamentary Commissioner for the Environment. (2001). *Whose water is it? The sustainability of urban water systems on the Kāpiti Coast.* Wellington: Parliamentary Commissioner for the Environment. May.

Parliamentary Commissioner for the Environment. (2015). *Preparing New Zealand for rising seas: Certainty and Uncertainty*. Retrieved from http://www.pce.parliament. nz/publications/preparing-new-zealand-for-rising-seas-certainty-and-uncertainty

Parliamentary Commissioner for the Environment. (2015a). *Water quality in New Zealand: Land use and nutrient pollution.* Retrieved from http://www.pce.parliament. nz/media/1008/update-report-water-quality-in-new-zealand-web.pdf

Parliamentary Commissioner for the Environment. (2016). *The state of New Zealand's environment: Commentary by the Parliamentary Commissioner for the Environment on Environment Aotearoa 2015.* Retrieved from http://www.pce.parliament. nz/publications/the-state-of-new-zealands-environment-commentary-by-the-parliamentary-commissioner-for-the-environment-on-environment-aotearoa-2015

Pongsart, G. (2015). *Theory of Constraints (TOC) and Appreciative Inquiry (AI): a comparative study of their effectiveness in improving master's thesis students' performance (Doctoral thesis).* Wellington: Victoria University of Wellington. Retrieved from http://researcharchive.vuw.ac.nz/xmlui/bitstream/handle/10063/4743/ thesis.pdf?sequence=1

Rawlins, B. L. (2006). *Prioritizing Stakeholders for Public Relations.* Retrieved from http://www. instituteforpr.org/research_single/prioritizing_stake-holders

Resource Management Act (RMA). (1991). *Resource Management Act 1991.* Public Act 1991, no. 69. Date of assent 22 July 1991.

Royal, C. (2012). Kaitiakitanga – guardianship and conservation. *Te Ara - the Encyclopedia of New Zealand.* Retrieved 18 January 2013 from http://www.TeAra. govt.nz/en/kaitiakitanga-guardianship-and-conservation/page-4

Scheinkopf, L. (1999). *Thinking for a change, putting the TOC thinking processes to use.* St Lucie Press/ CRC Press; doi:10.1201/9781420049046

Schragenheim, E. (1998). *Management dilemmas, the Theory of Constraints approach to problem identification and solutions.* St Lucie Press.

Schragenheim, E. (1999). *Management Dilemmas: The Theory of Constraints approach to problem identification and solutions.* Boca Raton, FL: CRC Press LLC.

Senge, P. M. (1994). *The fifth discipline fieldbook: strategies and tools for building a learning organization.* London: Nicholas Brealey Publishing Ltd.

Senge, P. M., Smith, B., Kruschwitz, N., Laur, J., & Schley, S. (2008). *The necessary revolution: how individuals and organizations are working together to create a sustainable world.* London: Nicholas Brealey Publishing.

Sheppard, D. F. (2010). *Reaching sustainable management of freshwater.* NZ RMLA Conference 2010: Sustainable freshwater management: Are we there yet? Christchurch, New Zealand.

Shoemaker, T. E., & Reid, A. (2005). Applying the TOC Thinking Process: A case study in the government sector. *Human Systems Management, 24*, 21–37.

Stainton, R. J. (2010). Contextualism in epistemology and the context sensitivity of 'knows'. In Knowledge and Skepticism (pp. 113–139). MIT Press. doi:10.7551/mitpress/9780262014083.003.0007

Swinnen, L. (2017). *Water bans looming for Wellington.* Retrieved from https://www.stuff.co.nz/environment/99308577/water-bans-looming-for-wellington-as-region-on-the-cusp-of-a-shortage

(2005). Theory of Constraints: practice and research. InRonen, B. (Ed.), *The Theory of Constraints: practice and research* (pp. 1–2). Amsterdam: IOS Press, Editorial.

Trutnevyte, E., Stauffacher, M., & Scholz, R. W. (2012). Linking stakeholder visions with resource allocation scenarios and multi-criteria assessment. *European Journal of Operational Research, 219*(3), 762–772. doi:10.1016/j.ejor.2012.01.009

Udías, A., Insua, D. R., Cano, J., & Fellag, H. (2012). Cost-efficient equitable water distribution in Algeria: A bi-criteria fair division problem with network constraints. *International Transactions in Operational Research, 19*(5), 675–693. doi:10.1111/j.1475-3995.2012.00846.x

Watson, K. J., Blackstone, J. H., & Gardiner, S. C. (2007). The evolution of a management philosophy: The theory of constraints. *Journal of Operations Management*, *25*(2), 387–402. doi:10.1016/j.jom.2006.04.004

Williams, D. (Ed.). (1997). Environmental and Resource Management Law (2nd ed.). Wellington: Butterworths of New Zealand.

Young, D. (2001). *Values as Law - the History and Efficacy of the Resource Management Act*. Wellington: Institute of Policy Studies.

Young, D. (2004). *Our islands, our selves – a history of conservation in New Zealand*. Dunedin: Otago University Press.

Chapter 4
An Exploration of Thinking About Complex Global Issues and Then Taking Action

Ian Roderick
The Schumacher Institute, UK

ABSTRACT

There are many problems and challenges facing the world. They go beyond problems that can be solved. They present as continual challenges that perpetually shift and transform themselves. The word we use for these situations are issues and they are by nature complex and large scale – they are global and long term. However, action does follow thinking and talking, even if that action is internalized. This chapter explores thinking about complex global issues and taking action.

INTRODUCTION

This chapter is about complex global issues, what are they, their nature and the patterns they present. It considers what it means to act – to do something about them, whether at the global level, which is almost impossible to achieve, or at some smaller, national scale right down to local action in one's own neighbourhood, which is more likely to create an immediate change, however small, and so contribute to improving the issue.

The committed writer knows that words are action. He knows that to reveal is to change ... the writer has chosen to reveal the world and particularly to reveal man to other men so that the latter may assume full responsibility before the object

DOI: 10.4018/978-1-5225-5996-2.ch004

which has thus been laid bare .... The function of the writer is to act in such a way that nobody can be ignorant of the world and that nobody may say he is innocent of what it's all about. Sartre, 1948 - What is Literature?

According to the Greek writer Plutarch, the Egyptians used to place a skeleton at the table during a feast, to remind them that they would die one day. This spectre at the feast is something that makes people feel uncomfortable, it is a symbol of impending doom. So it seems with modern discussions of policy and politics that there is an uncomfortable spectre of, say, 'climate change' at the table. If only it were just climate change. There are a host of ghosts, threatening our existence, large scale problems, not quite here yet but looming. Although mixed in with the doom are some fascinating and exciting visions of a better world than we have today. We are going to call these mixed up apparitions, these glimpses of the future, complex global issues.

These complex global issues 'play out' at a local level (action only happens in the here and now) and different systems approaches are appropriate and useful in many different parts of these situations whether it is trying to understand the large-scale driving forces or the organisational problems of trying to cope at the grassroots with practical improvements to tackle the consequences.

We will look at real problems and we will look at the generalisations of them (the concepts and nature of complex global issues), we are concerned at how we are tackling or exploring both the specific and the general issues (our concern of course is to advance systems thinking), beyond this we are looking at the level of the processes of systems thinking (that grand word, methodologies), and finally there is the level of ourselves and the question of the future (what do we want to emerge). All the time we have to consider whom we are addressing at each level (the audiences being policy makers, practitioners, academics ... and many more). The idea is to explore how systems sciences might inform and support policy making - particularly for universal or global scale problems that manifest at national and local levels.

Hopefully we will have some double-loop learning going on (Argyris, 2005), which is when we change our own mental model of the 'what' and we then gain a shift in understanding by questioning what we are doing in the first place.

The message is to care about those people who are trying to manage complexity – deal with these issues. These people are in decision-making roles and they include politicians, policy makers, and the so-called C-suite executives and senior managers and behind them are the decision support roles and behind them the advisors in organisations like think tanks and consultants but, lest we forget, it is ultimately the individual who acts. Our task is to enable this vibrant community of potential systems practitioners to explore complex global issues using a deeper understanding of their systemic nature.

COMPLEX GLOBAL ISSUES: WHAT ARE THEY?

Government, media, business and academic worlds are full of talk and opinion on the crises we face in the coming decades. These are familiar under headings such as climate change, peak energy or net energy decline, food insecurity, water shortages, economic instability, state securitisation, militarisation and so on – a disturbingly long catalogue of woe to which we can add numerous, systemic, unforeseen consequences and complications.

This is not the place to go into these crises in any great detail but some people say we are facing a collapse of civilisation, some estimate that we only have a 50% chance of lasting to the end of the century[1]. The year 2030 is considered the point of the 'perfect storm' when many of the individual crises will come to a head. Whatever may happen, it will be different to what we can imagine now but it is most likely to result in some form of what is called *The Great Disruption* (Gilding, 2011).

The shape, scale and timing of these potentially entire-civilisation-challenging crises will remain uncertain and unpredictable. However, we can take measures to mitigate and adapt and we can seek opportunities for new ways of delivering services and products within a framework of resilience and preparedness. The skill is to know how to act and prepare in ways that are best.

If we view any community as a system then it is itself a component in larger systems that contain many similar communities but of different shapes and sizes, wealth and vulnerabilities. Our community system depends on an environment made up of other communities and, of course, the ecosystems in which we are all placed. This co-dependence (other communities equally dependent on each other) is complex (unforeseen phenomena emerge) but it clearly demonstrates that any concept of systemic sustainability requires us to look after the chains of dependency on which we rely – often unknown and unseen dependencies. This mutuality is a critical factor and must be part of any preparedness for change.

At the same time of this talk of crises and despondency, others are suggesting that opportunities are growing daily to make a Great Transition to a future of enriched lives and human solidarity, where environmental sustainability is possible, technology, innovation and human ingenuity will provide and we avoid collapse. For example the Great Transition Initiative (http://www.gtinitiative.org/) is a growing international network of scholars and activists that analyses alternative scenarios and chart a path to a hopeful future.

Megatrends

There are countless issues and problems that are foreseen in future thinking. We define a trend as a change in the issue, although we sometimes use the words trend and issue interchangeably. Some trends are classified as megatrends, which are defined as long-term changes in underlying issues that affect governments, societies and economies permanently over a long period of time – they are like super tankers, they have considerable momentum.

The McKinsey Global Institute identify just four megatrends in their study called *No Ordinary Disruption* (Dobbs, Manyika & Woeetzel, 2015). These are:

1. The rise of emerging markets
2. The accelerating impact of technology on the natural forces of market competition
3. An aging world population
4. Accelerating flows of trade, capital and people

Stefan Hajkowicz from the Commonwealth Scientific and Industrial Research Organisation (CSIRO) in Australia has produced a book called *Global Megatrends,* which has a list of seven (Hajkowicz, 2015):

1. **More From Less:** Increasing demand for limited natural resources.
2. **Going, Going ...Gone:** The window of opportunity to protect biodiversity, habitats and the global climate.
3. **The Silk Highway:** Rapid economic growth in Asia and the developing world.
4. **Forever Young:** An ageing population, changed retirement patterns, chronic illness and rising healthcare expectations.
5. **Virtually Here:** Digital technology reshaping retail, office, city functions and labour markets.
6. **Great Expectations:** Consumer and societal expectations for services, experiences and social interaction.
7. **The Innovation Imperative:** Technological advancement is accelerating, creating new markets and extinguishing existing ones.

This adds in some recognition of a 'Limits to Growth' scenario or planetary boundary agenda and the potential peaking of extraction rates of natural resources and energy and the devastating toll on biodiversity. The geopolitical shift (to the East) has importance and the emergence of cultural change and expectations makes it into his megatrend list.

In *A User's Guide to the Crisis of Civilization,* Nafeez Ahmed talks about the five greatest threats to our civilization (Ahmed, 2010):

1. **Climate Change:** These dire consequences are expected to intensify throughout the 21st century.

2. **Energy:** Our dependence on coal, natural gas, and petroleum has led to global warming, but it has also started a desperate race to gather as much natural resources as possible. This rate of consumption cannot be sustained.
3. **Food:** As global warming causes additional droughts and a steady increase in the sea level, the challenge of feeding the world's population is going to become even more difficult.
4. **Economic Instability:** As resources begin to dwindle over time, this type of unchecked greed threatens to collapse economies and leave nations bankrupt.
5. **Terrorism and Foreign Policy:** There's a long tradition of governments around the globe funding terrorist organizations in order to destabilize certain regions.

In the *Future Agenda* collaborative programme the authors identify twelve common views of the future that present challenges (Dewing & Jones, 2016):

1. Too many people
2. Running out of resources
3. Pollution out of control
4. Migration is bad
5. Not enough jobs
6. Female education is key
7. Technology will solve problems
8. Solar is the answer
9. We need to rethink retirement
10. Health care spending will only go up
11. This is the Asian Century
12. GDP growth is the best way to measure progress

Although it is a mixed bag of problems and solutions they go on to introduce what they call cross topic insights: Trust, Privacy, Inequality, Transparency and Identity are all issues with trends that will become increasingly important and will shape how we transact and structure our economies and societies.

Al Gore's book *The Future* (2013) has six section that outline the megatrends that he foresees:

1. **Global Economy:** Dramatic changes, deeply interconnected.
2. **Global Mind:** How we see the world, impact of the internet, denial, (post truth world).
3. **Power in the Balance:** Geopolitics – shift from West to East.
4. **Outgrowth:** Unsustainable consumption, limits and population.

5. **Reinvention of Life and Death:** Biotechnology, medical advances, new diseases.
6. **The Edge:** Radical new relationship between humans and the planet – the Great Transformation.

Finally, the openDemocracy network (www.opendemocracy.net) - an independent global media platform covering world affairs, ideas and culture describes three key challenges for global stability in the 21st century: economy, environment and security.

- **The First Challenge:** Is the neoliberal model of free-market capitalism that is not delivering sufficient equity and emancipation. Reflected in the populist anger that is affecting so many western countries, with refugees and migrants seen as threats.
- **The Second Challenge:** Is the fundamental issue of global environmental limits, especially climate disruption. The present-day concern over migration is as nothing as to what would happen in the face of systemic, global, climate change.
- **The Third Challenge:** Is to contain an integrated culture of militarism that prescribes military solutions in almost every situation. Even after fifteen years of the "war on terror" there is little or no recognition of its manifest failures.

There are of course many more attempts to define a list of megatrends and there are many recurring complex global issues and contradictions that leave the reader half way between dystopia and utopia.

The Third or Is It Fourth, Industrial Revolution

It is easy to paint these pictures of the future that are full of crises, catastrophes and disasters. The path dependency that holds us on a course towards breaching all planetary boundaries is the defining issue. Breaking out of a channel, switching to a new path will tax governments at all scales and will require concerted and determined action from business, civil society and individuals. However, like many seemingly intractable problems and dilemmas we should look to transcend them to find a way forward. This may come about by contextual change where the environment, activities and processes undergo transformations that change the game. This is the 'great white hope' of the technologists, that we will discover new ways to create and deliver the products and services we need in ways that lessen or even cease the headlong rush to consume the planet. Indeed the rate of change in all areas of technology is astonishing and we are now in what some call the third industrial revolution, sometime called the fourth – it is moving that fast.

The 'O' technologies (Info, Nano, Bio, Geno, Cogno) married with new materials, robotics and automation driven by a complete transformation of energy systems (end of fossil fuels) built into a circular economy (waste as resource – end of pollution) could deliver the great transformation and deliver us from the crises – but the effects will be dislocations and disruptions, industrial churn, economic strains, job losses, systemic vulnerability and dislearning, increasing inequalities.

Perhaps we should recognise the danger of *technopolies* - societies that are dominated by the ideology of technological and scientific progress, to the exclusion or harm of other cultural practices, values and world-views – as a complex global issue.

STRUCTURING COMPLEX GLOBAL ISSUES

'To get a handle on' is a fairly recent idiom that is used when we grapple with difficult concepts, it is about finding the means of understanding, controlling, or approaching a situation. Getting a handle on complex global issues means seeing if there is some structure to them. We look for commonalities and patterns, we build lists and categories and create taxonomies (more accurately we should say meronomies, as we are dealing with part–whole relationships).

Dennis Meadows distinguishes two categories of issues, these are between a global problem and a universal one. The difference is subtle but generally a universal problem is one that manifests locally in a specific way but occurs, similarly, in many places across the whole world, for example 'the illegal drugs problem' is universal, while a global problem is one that affects the whole world like climate change. We do tend to use global as a short hand to describe both types of issue.

The phrase *global problematique* was used by the Club of Rome back in the 1970s in the lead up to the publication of the Limits to Growth report (Meadows et al., 1972). Inspired by C. West Churchman, who was dedicated to inquiry into what he called 'enormous problems', the methodology was to collect relevant observations about systemic problems to generate discussion and opinion, and this inter-related mass of observations and opinion is called a problematique. The goal then becomes to clarify meanings, and to cluster observations into groups with significant similarities. But as Richard and Reidy (2009) point out, its vastness cries out for transdisciplinary or systemic approaches:

It embraces a vast number of interrelated themes across the entire range of human knowledge and social development. No single subject, framework or perspective can encompass this enormous breadth, let alone encompass the depth and sophistication of thinking implied.

This early Club of Rome work has been used by Anthony Hodgson from the International Future Forum[2] to develop the World System Model, which is a representation of our vast global problematique using just 12 nodes (figure 1).

This deceptively simple representation of the whole world allows us to explore and make sense of the interconnectedness of everything. It can act as a framework for horizon scanning work and a foundation for resilience thinking (Hodgson, 2011). This basic starting point demonstrates the combinatorial problem we face – there are 66 two-way relations in the World System Model (WSM) mandala.

What the WSM provided, though, does not embrace the megatrends we have described earlier. They are the changes that create the pressures in each of the 12 areas. As an example the megatrend of demographic underlies, or drives the changes in everything from health to trade to water demand etc., building up pressures that lead to impacts elsewhere and consequent responses. The WSM acts as a structure to explore trends.

Complex global long-term issues impinge on our lives today and we formulate policies and plans to deal with them as they are perceived to affect us. Later, we will critique traditional, rational policy making and suggest that systems approaches are required to better understand the processes in play and how from that we can let decisions emerge that have a higher probability of creating system change that will work for an improvement or be able to cope better than before in adapting to fluid complex global issues.

Figure 1. World System Model
(Courtesy of Anthony Hodgson, International Future Forum, 2016)

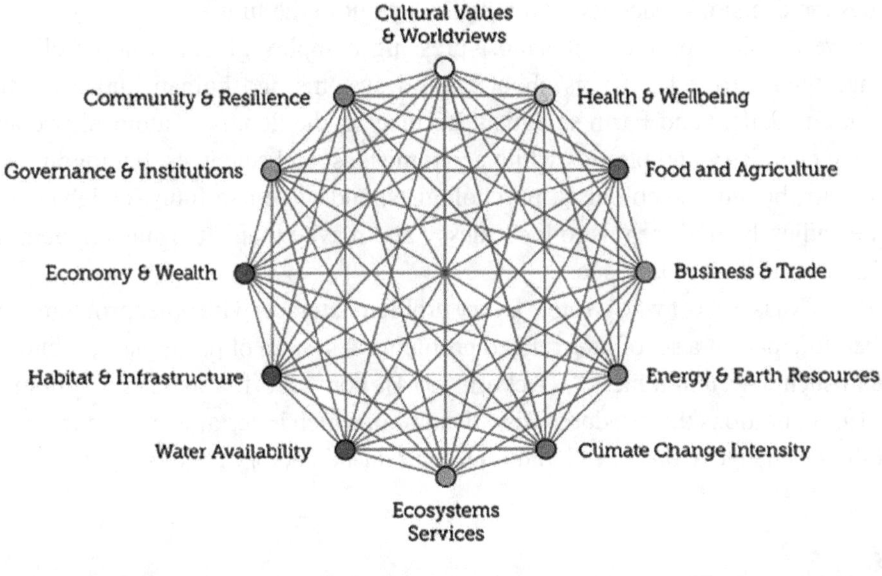

Complex global issues do not go away, but they do mutate and spawn. Often we enter games described as schismogenesis, the escalation of behaviour caused by the failure to ensure equity, these are 'arms race' process that lead to system collapse, the worst kind of system change. This exploration of complex issues is about building total resilience, maybe we could call this super-resilience – a higher level of adaptation where escalating process can be controlled and perpetually transforming challenges are held within safe boundaries.

THE NATURE OF COMPLEX GLOBAL ISSUES

Systems thinking steps into this arena of complex global issues with a little bit of humility. There are many different systems approaches soft, hard and somewhere in the middle – but all the time there is a tension between the desire for a rational method based on evidence (controlled random testing if possible) and the realisation that the complexity of the issues limits rationality. Herbert Simon (1957) describes the idea of bounded rationality where decision making is constrained by the available information (often poor quality), the tractability of the decision problem, the cognitive limitations of the decision makers (it's all too much to cope with), and the time available to make the decision is always too short. Decision-makers, in his view, act as satisficers - seeking a satisfactory solution rather than an optimal one. This resonates with what we will describe later as the garbage can model of policy making.

What we, as systems thinkers, must offer is the best way to produce the best satisfactory solution, perhaps not even a solution but just a direction to take that guides the decisions made today that will play out in the future.

If we consider probably the most pressing complex global issue of climate change then at one level it is about natural systems and human impacts – the climate modellers and Earth scientists are busy in the depths of atmospheric and oceanic chemistry and physics while economists, social scientists, demographers, biologists, botanists, ecologists, psychologists, and of course futurists debate and argue endlessly. It all begins to look messy and maybe that is the starting point to define a complex global issue.

It was Russ Ackoff who wrote: "Every problem interacts with other problems and is therefore part of a set of interrelated problems, a system of problems.... I choose to call such a system a mess", (Ackoff, 1974). For Ackoff, a mess is a system of external conditions that produces dissatisfaction, which is certainly true of the issue of climate change. Complex global issues are unsatisfactory messes.

If we are dealing with systems of problems then what types of problems are there? E.F. Schumacher describes two types, one he called convergent where over time the parties involved resolve differences, coming together with shared knowledge to arrive at a solution and eventually the problem is no more. The other type is the opposite, divergent problems are those where the parties increase their differences over time, polarising debate into fixed and self-reinforcing positions that defy any attempted solution – think of the Middle East conflict. These are problems that Bateson describes as schismogenesis. It is interesting that this phenomenon is identified in the Systems Intelligence literature where, in the everyday world, we hold back information or support to others because we feel that they are holding back - we develop an arms race of non-contribution to a problem. It is suggested that this is "the single most important key to life-decreasing, reciprocity-trivializing and vitality-downgrading mechanisms in human life" (Saarinen & Hämäläinen, 2007). Complex Global Issues are definitely messes that contain diverging problems where we all hold back.

But maybe we can add to this description by looking at Rittel and Webber's definition of a 'wicked' problem (Rittel & Webber, 1973). Wicked problems have these characteristics:

- *There is no definitive formulation.*
- *They have no stopping rule.*
- *'Solutions' are not true-or-false, but better or worse.*
- *There is no immediate and no ultimate test of a potential solution.*
- *Every attempted solution is a "one-shot operation"; you can't learn by trial and error.*
- *It is unique – no prior problem is the same as this one.*
- *Every wicked problem is a symptom of another problem.*

Indefinable, unique, never-ending, interconnected, untestable, irreversible, irreproducible - all these are characteristics of wickedness, so I think we can add the descriptor 'wicked' into the pot - complex global issues contain wicked problems.

Robert Horn picks up on this definition of wicked problems to describe what he calls social messes (Horn & Weber, 2007):

- **There is no unique "correct" view of the problem:** This goes beyond saying that there is no definitive formulation of the problem; it says that there can't be one.
- **There are many different views of the problem and contradictory solutions are offered:** Contradiction and paradox are common as people see

the problem itself as part of the problem or solutions are attempted that set up double binds, forcing other people to be spontaneously what they are not.

- **Most problems are connected to other problems:** A problem is not just a symptom of another problem but the connection may be that of causing another problem or this problem may even be part of a solution to another problem. The relationships between problems can be mysterious.
- **Data are often uncertain or missing:** We are dealing with partial information, partial in the sense of incomplete or suspect and also in the sense of filtered according to vested interest.
- **There are multiple value conflicts:** People may have deeply contrary values from one another not just different perspectives, and any attempt on one side to demonstrate the falsehood of beliefs of the other actually reinforces the other's position – a schismogenesis of beliefs.
- **Ideological, political, economic and cultural constraints:** A full understanding of the barriers, taboos and norms from a wide variety of social constructions is practically impossible.
- **Numerous possible intervention points:** If n interconnected factors are identified as pertinent in a problem then there are at least n(n-1)/2 possible intervention points involving just two factors – and so on. The combinatorial defeats you every time.
- **Consequences difficult to imagine:** Following through the causal chains in the network of interconnected factors is daunting.
- **Considerable uncertainty and ambiguity:** This is an understatement.
- **Great resistance to change:** This is often the case, a natural reaction to uncertainty, but what is really happening may be a desire for change from some and a desire to stay with the status quo from others, which again builds into an arms race of claim and counter claim just entrenching positions.
- **Problem solver(s) out of contact with the problems and potential solutions:** All too often we see that an external attempt (driven by political expediency) is imposed on the social mess. The magic bullet is sought and the seeker is unaware of the nature of the beast and unable to evaluate potential solutions.

At this point maybe we can summarise: Complex global issues are social messes of wicked, divergent problems in which we all hold back from contributing to solutions.

Let's not stop, Kelly Levin and others introduced the distinction between "wicked" and "super-wicked" problems (Levin et al., 2009). They suggest these additional characteristics:

- *Time is running out*
- *No central authority*
- *Those seeking to solve the problem are also causing it*
- *Policies discount the future irrationally*

While the characteristics of a wicked problem relate to the problem itself, these extra characteristics of super-wickedness are mostly about the agents trying to solve it. The tension between the never ending nature of a problem and the sense of urgency introduces a vicious circle of action for the sake of action with the consequences of making the situation worse rather than better and of course this heightens the urgency for a solution and so on.

The location of decision-making or authority introduces another area of tension. The dispersed nature of the problems, and attempted solutions, across space and time throws up the perceived need for co-ordination and simultaneity that is often lacking or difficult to achieve. Global, harmonised decisions for action are rare but not impossible, for example the Montreal Protocol on Substances that Deplete the Ozone Layer has been hailed as an example of exceptional international co-operation. However, many other international agreement processes get bogged down in political and economic mires of distrust and narrow interests.

It is easy to see that this 'super' prefix does bring your attention above the parapet, the problem space, to see how the decision makers are part of the problem and therefore partly the cause of it.

The more damming comment from Levin et al. is the way that decision makers, and indeed all involved, discount the future. They use the term 'hyperbolic discounting' for the irrational disregard of damage in the future. These are situations where the possibility of damage to the environment or society in general is highly likely but to the individual, in the present, the risk of damage in the distant future is uncertain and indefinite - that is when the short term, gratification decisions are made in favour over the long term, common interest. As they say, "Every scholarly tradition and set of decision tools agrees that the 'hyperbolic' discounting characteristic of super wicked problems should be avoided and is irrational."

I think it is fair to add the 'super' into our definition of Complex global issues. But let's add some more twists. Gabriele Bammer discusses complex global issues (Bammer, 2003) as having:

- Multi-party interests,
- A politically-charged history,
- Diverse cultural contexts,
- Large in geographical, temporal, hierarchical and population scales.
- Many unknowns

The unknowns are compounded by constant change; change occurring on many fronts and in manifold intended and unintended consequences of local, national and international policies and programs (Bammer, 2003).

Bammer here introduces a few more characteristics. The politically-charged history is a major factor in complex global issues, although we often talk about future generations we tend to treat them as the 'other' in our hyperbolic discounting decisions, but with a slight nod to leaving a legacy and desire to perpetuate our gene pool. But future generations are faceless while past generations are fully fleshed humans whom we admire and respect their wishes or hate and will never follow their lead. The history of issues, and the personalities contained in those histories, entrench beliefs and polarize people into many different camps. This is also the origin of the cultures in which people are set. Recognizing other worldviews is a matter of respect and understanding of a wider culture that frames the individual but this framework is not easily changed – it acts as a cage.

An important concept that Bammer uses is that of scale. She notices that, by definition, global issues are large scale problems both geographically and across populations but that these issues have a hierarchy. Global issues manifest at a spectrum of levels from the individual, through local communities, to nations and regions all the way to the totality. The nature of the issue changes as you zoom up and down, potential solutions conflict across scale boundaries, contradictions appear as interventions at one level seem counterproductive above or below. In a similar way we detect contradictions across temporal scales where we may deliberately adopt actions like increasing our squandering of fossil fuels for a few years in order to build a long lasting, renewable infrastructure for energy harvesting.

Adding all these perspectives into the pot we get:

Complex Global Issues = super-wicked, social messes with a history full of divergent problems that are large, multi-scale, political, culturally diverse and in which we all hold back from contributing to solutions, and everything is constantly changing.

There is yet another area to consider. The word 'complex' covers many things including the emergence of phenomena that are not 'in' any of the parts; this is the well-rehearsed idea that, for example, a form of transport arises from assembling the individual components of a bicycle, none of which alone can transport you. The complexity of a global issue is due to this emergence of phenomena that add more intensity to the issue and we begin to see some characteristics that produce that complexity. John Casti suggests that the characteristics of almost all complex systems are a medium-sized number of intelligent, adaptive agents interacting on the basis of local information (Casti, 1997). As a game, a complex global issue is one where the number of players and pieces is not too small and not too big but just the right number to create interesting patterns of behaviour. The players are intelligent and are able to modify and generate the rules of the game so that an ecology of rules emerges but the agents only have local information and no one has access to what all the other agents are doing. These are the conditions to generate emergent properties.

So we arrive for now with this:

Complex Global Issues = super-wicked, social messes, with a history, full of divergent problems that are large, multi-scale, political, culturally diverse and in which we all hold back from contributing to solutions, and everything is constantly changing, with a medium-sized number of intelligent agents each with local knowledge who adapt and modify the rules of the game from which new and unforeseeable phenomena emerge.

This is quite a definition. It is a work in progress as it certainly lacks beauty. Would we have it any other way? Do we want a tame and neat world? A recent project called "Enabling a Better Tomorrow" has a subtitle called: "Solving problems so they stay solved and not create new problems in the process" (Bellinger, 2012). It is a tempting idea and, as we contemplate the enormous magnitude of the Complex of Complex Global Issues and recognise the potential for tragedy that might dwarf all previous tragedies, it is unlikely that we can simplify problems enough to solve them completely and not create a host of new ones.

We are always seeking to simplify but as the saying goes: Everything should be made as simple as possible, but not simpler. We need processes that maintain complexity but in an accessible way, to over simplify is to lose the essence of the complex and trying to treat everything as just simple, local issues of the day will miss interactions across time and scale.

We can only reflect on the definition we have arrived at and treat complex global issues with humility and care, never thinking that we can find definite answers.

SCALES OF CGI'S

Systems thinking combines many disciplines to create our intellectual perspective. The areas of interest range from academic studies into the philosophies underlying the methodologies employed within systems thinking through to the minutiae of implementing participative techniques of inquiry with "real people in real situations". It is theory through to practice in a seamless weave.

This richness of perspectives requires an ability to zoom in and out; to look closely at fine granular details, the brush strokes, and then step back to see the 'big picture'. The overlapping and nested scales with which we look at the world suggest plural approaches. We need to select tools appropriate to the scale of the focus and, most importantly, appropriate to work across the boundaries of scale.

One of the values of reflection on system practice is an appreciation that systems change, indeed the reason we practice systems thinking is to affect change, directly or indirectly, in ways we hope are beneficial (the ethics of who benefits at what expense is a separate question and not addressed here). When we realise that at different times a system is different we create a time boundary and therefore a time scale. But the time scale in any systems thinking practice maybe seconds (the impact of a tweet) or it may be generations (strategies to reduce carbon dioxide emissions). We have nested scales of time that demand equal skills of zooming in and out to appreciate how a system changes from its past to its future.

Working well with time scale is vital when looking at complex global issues (CGIs). The really big picture issues like climate change, food security, resource depletion, and so on, manifest at physical scales from the individual through nations to the whole world and at temporal scales they range from daily actions to centuries in terms of the degradations of ecosystems.

One of the most important framings for CGIs, which demonstrates the need for systemic, cross scalar thinking, is the population trajectories for the rest of this century. The mid-range forecast from the UN shows global population peaking around 2050 and then declining but within that some countries exhibit much earlier peaks and declines than others thus demonstrating substantial differences across nations. The picture, at the scale of nations, brings out vital differential rates of change that will have profound effects and possible tensions among countries. Adding to that we have a time scale created by the boundary between growing and declining populations. For economies that work on the need for growth this temporal boundary indicates a phase shift to a new form of economic organisation where future income will not pay off past debt. Our scale for the period up to 2050 is likely to be decades (and years are important within that) after 2050 our scale for thinking is probably half-centuries. Of course the boundary is where interesting things happen, shifting from population growth to decline will disturb everything.

So scale is all important, what seems right at one level can be contradicted when seen from above or below – or now and then later.

Within systems thinking we attempt to sweep in as many stakeholders as possible. Identifying these stakeholders is difficult but maybe placing our focus within a scalar perspective might open the door to people whom we often consider as 'other' and not included. This way of systems thinking should bring in future generations to add their needs to what we are constructing today.

When we think about engaging citizens in thinking across timescales, there are many useful tools that can help. One such is the Three Horizons model (Sharp, 2013), as this also includes ways in which innovations can work – both to try to prop up the way things are, (by keep things going and preventing the transition to another kind of system) or to help us move to a new system that is difficult to achieve in one step.

The 1st horizon is basically business as usual which many vested interests wish to maintain, innovation may help us keep this 1st horizon as the dominant one. The 3rd horizon, currently does not look that plausible under current conditions, but is in some ways pre-figured in social, ecological and economic pilot schemes and in new thinking that has yet to be taken up, or to reach 'critical mass'.

Radical innovation can help us to reach our 3rd horizon, but there is always a danger of this being held up by those who are either wedded to the 1st horizon (whether because of perceived interests or deep conceptual and personal investment), or can only really bring themselves to contemplate 2nd horizon ideas and projects. It is this 2nd horizon that represents our best attempts at transitioning to a new future, based on our current knowledge. It paves the way and enables 3rd horizon systems to emerge.

One simple example of this model is the dominant internal combustion engine for powering vehicles, this is the 1st horizon. The 3rd horizon is fully electric vehicles and infrastructure for transport – technology that has been available for over a century. The 2nd horizon contains innovations like hybrid vehicles, public sector funded charging points, subsidised schemes for electric vehicles and so on. All of which are temporary, transitional innovations.

Figure 2. Three horizons

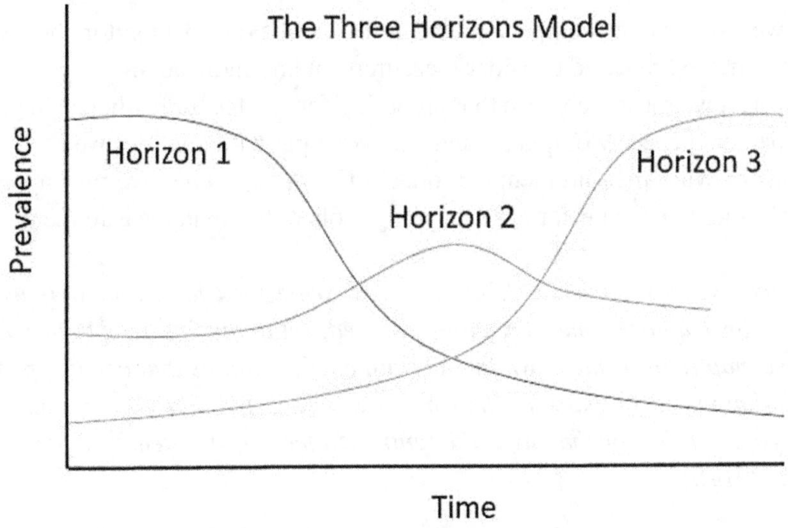

This kind of systems approach to futures can also help us to think more fruitfully about what conditions we might need to help develop, not only in order to bring people with us, but to work more effectively with those parts of society that are either more ready for change, or are actually leading pilot projects for change.

However, one aspect that has often not been fully considered is the extent to which rapid change driven by our current development model, its innovations and its crises, has hollowed out the capacity of people to personally adapt to new circumstances. In this case, to require these same people to positively engage with yet more change is a really big ask. Often it is the poorest in society that suffer from imposed structural changes, they are the ones who have to make the energy available for adaptation to new neighbours and new circumstances. It is necessary for us to understand how it feels to be told that your knowledge and life experience is now useless as the development trajectory has moved on and passed you by.

The personal psychological and conceptual skills and orientations needed to cope with change have to be a focus. It is also the case though, that complaints about change are not all to be dismissed. Some change is or has been, for the worse for some and for the better for others. Unless we can learn to listen to each other's experiences and meaning attached to change, we may be unlikely to engage in developing successful common futures. This is why it is not backward-looking to give people the chance to look back.

PATTERNS IN CGI'S

The DPSIR Framework to Consider the Future

When we think about complex global issues and how to monitor the external environment we are led to consider measurement and indicators.

However, we must recognise that preparing for change, or resilience in general, is an emergent property or quality and when dealing with qualities we easily reach the limits of what measurement can do. As Capra says when we take a systems, holistic worldview we need to use mapping tools rather than measurement:

In science, we have been told, things need to be measured and weighed. But relationships cannot be measured and weighed; relationships need to be mapped. When we map relationships, we find certain configurations that occur repeatedly. This is what we call a pattern. Networks, cycles and boundaries are examples of patterns of organisation that are at the centre of attention in systems sciences (Capra & Luisi, 2014).

One useful, and simple, framework that maps the relationships and influences and also allows the development of indicators where appropriate, is called the DPSIR framework. DPSIR stands for Drivers, Pressures, State, Impacts, and Responses.

Social, environmental and economic drivers exert pressures on the state of the system and, as a consequence, the state changes, (e.g. the provision of adequate health, resource availability, or biodiversity), this leads to impacts (on human health, ecosystems and materials) that may elicit responses that feedback on the drivers and pressures, or directly on the state or impacts, through adaptation or remedial action.

If we consider a state indicator that is concerned with limits, say carbon dioxide concentrations in the atmosphere, then the pressure indicators on this state are annual carbon emissions and the drivers are human needs for energy (heating, transport etc.), food (agriculture etc.), resources (forests, etc.). The impacts are changing climate and volatile weather, the responses are the stimulation of action at all scales to change values that are inherent in the drivers – pressure link to reduce emissions or, more likely, the responses are to just tackle the impacts (the symptoms) by adapting to extreme weather events.

In resilience thinking we talk about trends and trajectories, even megatrends and minitrends. These are phenomena that we can equate to the drivers in our framework. The need for energy, for example, has been on a steep exponential growth trajectory for many years. We also talk about stresses and shocks, and how one leads to another. The stresses are the pressures on the state of the world, which manifest in the increasing levels of carbon emissions and average temperatures, or the shifting patterns of insect inhabitation. The shocks are impacts such as volatile weather events, or a sudden collapse of an ecosystem.

In this framework, our responses are supposedly directed at the drivers, pressures, the state of the systems and the impacts. What is often the case is the we do not act until the impacts are causing significant detriment and then we only respond directly to those impact, i.e we treat the symptoms in emergency situations. Sometime we address the state of the system, thereby relieving the impact by shifting the burden on to others or into the future. Occasionally we try to reduce the pressures, for example geoengineering the planet to reduce the impact of carbon emissions. Rarely do we have the means to address the drivers, which are the megatrends like energy demand.

Preparing for change means early recognition of pressures and anticipating the impacts so that we can craft responses now to affect the drivers and pressures to reduce the harmful impacts - or encourage the beneficial ones.

DPSIR is a well-established typology but it tends to just describe the situation ‘as it is’ rather than any normative approach that attempts to say how things should be. Preparing for change opens up the discussion to ask ‘what state do we want’.

All the way around the DPSIR loop we should consider this reference to a desired state. Are we shifting the drivers to affect the pressures with our goal in mind, do the pressures change the state in the right direction so that the impacts drive responses that shift the state in the way we want.

Patterns

Trying to model a complex global issue is difficult, some attempts using Systems Dynamics have been made to tackle distinct sectors such as the food system. The whole Limits to Growth work is about modelling the world to demonstrate the impacts of 'outgrowth' and the interactions of material depletion, populations, food and pollutions - the patterns emerge.

These are valuable exercises particularly if they involve working with communities affected by the issues in model generation. In this modelling, when thinking about interactions and consequences, we look for emergent patterns, which are the similarities of behaviour that systems exhibit. These patterns suggest the future and shape our understanding of what we need to do to prepare.

A popular example of a pattern of behaviour is the Gartner hype cycle. This explains how many new ideas go through five stages. The first is the trigger or initial bursting onto the scene, this is taken up rapidly and becomes a fashion until we reach the peak of inflated expectation (the full hype) followed by a rapid collapse and out of fashion down to the trough of disillusionment (where many ideas do disappear) but that is not always the end, there is then a slow rebound up a slope of enlightenment – the real value and nature of the idea comes through – until we reach the plateau of productivity.

This pattern or shape has occurred for many different phenomena, particularly in emerging technologies like 3D printing, cloud computing, voice recognition.

Another source of patterns is the work of Peter Senge and system archetypes (Senge, 1990). He explains patterns with popular names such as the 'tragedy of the commons', 'limits to growth', 'success to the successful', 'shifting the burden', 'double bind', 'arms race' – all appearing as the result of causal loops with reinforcing or diminishing feedbacks, stocks and delays.

Looking for patterns in the past is an essential part of preparing for change and so too is imaging patterns in the future. Complex global issues are impossible to understand from only looking at the behaviour of the components, it is the patterns that emerge from the interactions of sometimes millions of components that produce the phenomena that give great concern.

Figure 3. Hype cycle

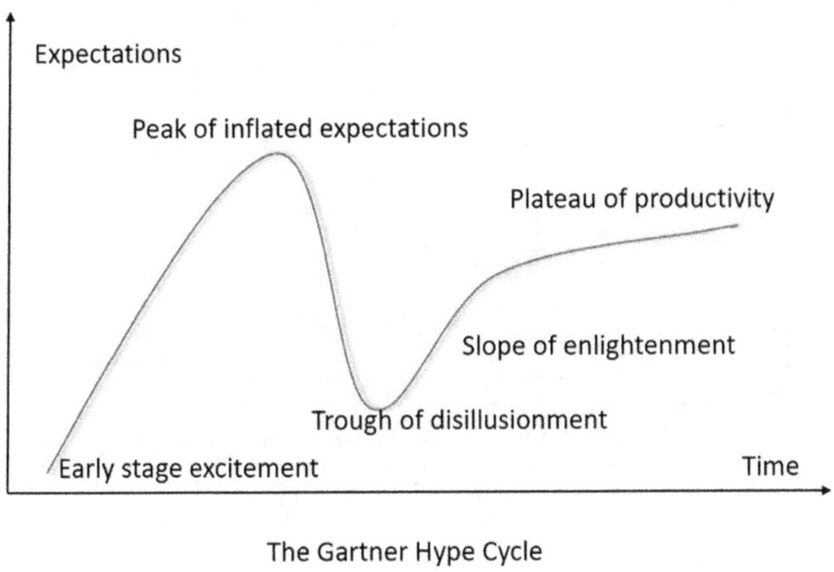

Proximity Grid

One of the major difficulties we face is that policy making and political decisions are for the *here and now*, they are for a specific locality in the present. They are not even for the global scale today as we have no decision wielding representatives of the whole world - there is no global governance. The closest we get are processes like the Global Goals (Sustainable Development Goals) that are a non-binding list of targets for the medium term (up to 2030). There are no processes of policy making that are for the global long term and, as a result, the mind-set of policymakers is not focused there.

One way to analyse this is to consider this proximity grid:

By proximity we mean closeness in both space and time. We talk about the local when we mean the vicinity in which we live and the global when we mean the whole earth. We also talk about the far distant future when we mean the long term and for the short term we often says it is 'here' and now. Drawing this out allows us to visualise the different scales of place and time - and a simple form is the 3 by 3 grid shown above.

When we think of the full consequences of climate change – probably the most complex and threatening issue we face – we are thinking in that top right-hand corner. We imagine that sea level rises will be a global concern by the end of the century, and the full effects of extreme weather will be playing out as our grandchildren enter

Figure 4. Proximity in space and time

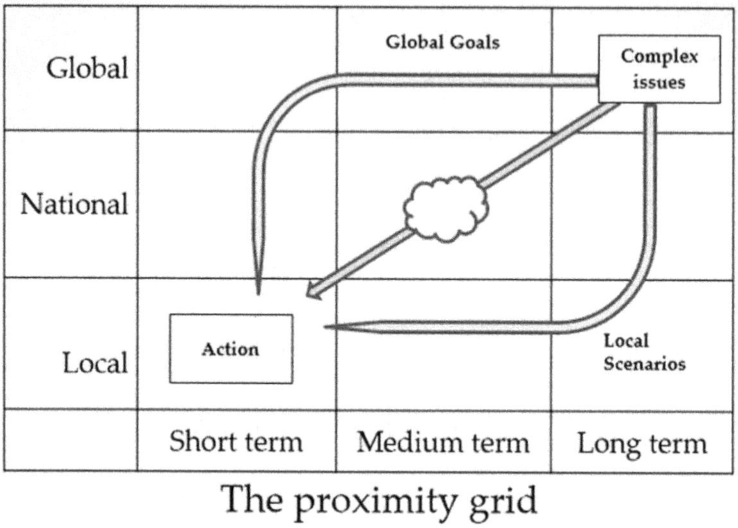

old age. The trouble is we can only act today in the short term and usually only act in our own patch, the local, which is where policy making and thinking gets translated into action. How can we improve the mind shifting that is necessary to move from the global, long term imagination to the local, short term space for action?

For the global long term, you are required to have a mind-set that sees the vision of how the world could be. It is a utopian vision of course where everyone has a fairly equivalent standard of living and opportunity and we are not trashing the planet nor running out of resources to keep the show on the road. A less utopian vision, a more realistic one, is of a world that is continually seeking to converge, it is always striving for that utopia but it is holding that idea in a dynamic of a constantly evolving complex system where the unforeseen can emerge. The proximity grid allows us to explore different ways to shift from global, long term thinking to local, short term thinking. Let's consider three routes.

The first is the direct, diagonal route from the global long term through something like a national, medium term policy framework and then down to the local, short term. This is shown as a cloud in the central box because it is often a cloudy business, opaque decision making affected by vested interest. What often happens is that global level conferences, usually organised by some prestigious body like the UN, produce global scale medium term goals, for example, the Global Goals published in September 2015 take us to 2030, they are the successor to the Millennium Development Goals.

Most countries have developed strategies and plans for this, they are often expressed as targets like an 80% cut in greenhouse gas emissions (from the 1990 baseline) by 2050, which was made law in the UK Climate Change Act of 2008. What happens next is moot. Accompanying these medium term national visions, expressed as goals, are a set of mechanisms to measure progress, some statutory instruments to enforce change, some removal of barriers, some exhortation and education to persuade companies and peoples. All of these policy making approaches and actions are top down pressures to effect local change today. The mind-set is that massive problems, like climate change, require governments to work out what to do and then command and control the implementation of plans. Too often the plans are drawn up, targets set but the delivery is lacking.

The second route is to bring our thinking back in time from the global, long term to a global perspective in the here and now – what is happening today, in the world zeitgeist, to develop answers to complex global issues – for example how do we create sustainable and fair living that *takes notice of what the long term future might be like.* This is the approach taken by organisations like the Fairtrade Foundation - their objective is to establish mechanisms for fairness at a global level in the present. From this many thousands of local, immediate initiatives spring up to deliver these practical, convergent processes today. This approach is really about the grassroots today, it allows the emergence of ideas and adaptations for action, yet it is shaped by global thinking about what to do now, which in turn has its vision derived from long term thinking.

The third route is to translate global, long term visions into local, long term ones and to then use this perspective to identify routes to achieve it, again bringing our thinking back to the present. This allows people to consider the path dependencies, the historical context, and the peculiarities of place that produce specific actions shaped by the contingencies of the local. An example is the way that cities are picking up the challenge of climate change, environmental issues and social justice and determining their own, unique path forward. This route is often cast in a framework of resilience and preparation for change, it can lead to isolation and survivalism, a sense that looking after one's own is the main focus. This approach is one of small, human scale change where complex global issue are reframed as complex local issues, *our* issues, rather than being seen as so massive and in the future that the recourse is to give up and hope that somebody else (the ever invisible 'they') will sort it out.

The value of using the proximity grid is to recognise that there are different ways to shift our thinking across scales and across time – in the 3 by 3 grid above there are nine ways to move from the top right corner to the bottom left. They have different characteristics and if we wish to make recommendations – ranging from suggesting strategies through to stating full blown rules – then a clear understanding of how we are stepping back (sometimes called back casting) to the here and now is essential so that we can better appreciate whom we need to involve and what their worldviews might be.

It is of course a generalisation that when we are concerned with the here and now and we have to make decisions for action, we tend to have a mechanical view of the world – we think in linear terms of cause and effect – there is a problem and we apply a solution. In contrast those whose thinking dwells in the far and distant future often work with a complexity perspective, recognising the non-linear interconnectedness of global issues - with a realisation that there are no simple solutions.

Long-term thinking is all very well and setting medium term goals is important but you can't do a goal. We can only do actions in the present and we need short term planning for specific activities that are the small steps along the way to a goal.

If we look at the proximity grid and consider the direct diagonal route from long term global thinking through to the national, medium term then we have intentionally placed a foggy cloud here in the area where goal setting often takes place, each country setting itself targets for 15 to 20 years hence that are deliberately well outside normal political life cycles. This is where it all gets bogged down, diluted and captured.

These insights point to the problems of moving from visionary principles to recommendations and goals for action. Yet there are problems with setting goals. A goal is an end which individuals or groups seek to achieve. For example, Global Goal #1 is: *By 2030, eradicate extreme poverty for all people everywhere, currently measured as people living on less than $1.25 a day*. It is difficult to argue with this as a target but:

- We are trying to fix the game. We establish beforehand what it is that we want versus the approach where what we want emerges and changes as we travel.
- We state the goal as a 'finite' game[3] – there is an end and it finishes with no regard for what happens next and no regard for the 'infinite' game in which it sits, nor the continuing context of other interacting, multiplayer games. We do not explore the consequences of aiming for a goal. There is no concern for the interaction of the variables in the goals with other variables and goals. It is an isolated, reductionist only way of looking at complex systems.
- We need to understand who are the individuals and groups that are pursuing the goal, who is playing, who is excluded, what are the power and control relationships involved. Global Goal #1 sounds much in the vein of 'we' are going to do something to 'them'.

The conclusion is that policy making itself is a super-wicked social mess and if systems thinkers are to have any impact they have to keep learning about the context and methods of policy making – it is a garbage can but there are many methods that give an appreciation of the mess and with luck produce higher quality decision than before.

It's the lifeboat problem. If you get too involved with too many people with too many problems they drag you under. You don't save them. They sink you.(Pirsig, 2013)

The early years of the 21st century are proving increasingly complex. We do have many challenges including a population explosion with unprecedented population migration, but also growing inequality, managing our finite resources, climate change, and the wholesale abuse of our environment. We have still to grasp the full impact of technology upon human civilisation.

If we are to develop we should reflect upon how we organise ourselves and how to create policies relevant to our survival and how we can turn them into feasible practicality. We believe that systems thinking has a significant part to play in that development thinking.

The calls for more 'joined up' government have sounded for a long time, many organisations charged with policy making or set up to do thinking (literally called think tanks) understand the nature of the complexity of global issues.

Since the birth of modern systems thinking and practice (c.1946) many theories, philosophies, methodologies, methods and tools have been developed but all are united through a concern to think of the entity as a whole and as a part of a greater whole. These systems approaches must inform and support policy making - particularly for universal or global scale problems that manifest at local levels.

The title of Cabrera book: *Systems Thinking Made Simple: New Hope for Solving Wicked Problems* (Cabrera, 2015) points to a central concern for encouraging the use of systems thinking – simplicity is key. If we are dealing with sophisticated systems thinking then we might devise elaborate and sophisticated methods to appreciate the problem, the reality is that policy makers are sophisticated thinkers but about their subject and are not versed in the systems thinking language and subtleties. We need systems thinking to be simple, accessible, and timely.

We have always lived with natural dangers ever since sabre-toothed tigers roamed Europe, but today we live in what is called the 'risk society'. Ulrich Beck defines it as *"a systematic way of dealing with hazards and insecurities induced and introduced by modernisation"* (Beck, 1986). What this means is that risk is a result of our triumphs – we have built, and will continue to build, a world where we can overcome diseases and poverty, live into our 80s or 90s, free ourselves from drudgery, have wonderful artefacts like smartphones. These are the upsides of progress but with these triumphs of technology and development have come questions:

- Can we share this abundance equitably?
- Can we avoid destroying the planet, and of course ourselves?
- Can we avoid running out of resources?
- Can we accept the consequences – the hazards and insecurities that these triumphs generate?

We have added many human manufactured dangers and hazards to our lives – these are destined to create change.

When we consider the long catalogue (the meronomy) of complex global issues there are some obvious questions. Where and when are these events going to happen, where will the effects be felt? Who will suffer, who might benefit? These are questions that we can only explore as we build resilience and prepare for what we imagine. It can sound pretty gloomy. What hope is there? Where do we look for positive things and opportunities?

There are three ways forward:

- We learn how to restructure and provide in ways that do not destroy the fabric of the earth and societies but actually enhance them. A ‘global mind’ develops – and it is benign!
- We carry on developing new technology
- Our mechanisms of exchange are succeeded by new mechanisms that prevent abuse and instead encourage equity and sharing and frugality. We change the rules of the game; this route is favoured by Hornberg.

A convergent resilience worldview recognises our dependence on others and suggests that it might it be prudent to:

- Support and care for those on whom we depend.
- Diversify, have many channels rather than one or two – small (and many) is beautiful.
- Reduce our dependence (go local etc.) taking full regard for the livelihoods of those supplying us now.

None of these pathways and directions are going to happen easily, the social justice and environmental movements have proclaimed the need for these transitions for decades without seeing major shifts.

WHY SHOULD WE CARE ABOUT CGIS?

Viable Systems Model

Every living system senses its environment through vibration, touch, sound, smell, taste, electric field, sight or whatever. The reason is simple: to find food and eat or to avoid being food and eaten. There is a particular systems model for human organisation that was developed in 1972 by Stafford Beer, (Beer, 1972). It is called the Viable Systems Model (VSM) for the organisational structure of any autonomous system capable of producing itself – this is called autopoietic, or self-creating. A viable autopoietic system is structured to meet the demands of surviving in a changing environment. One of the main features of such systems is the capability to adapt.

VSM can be boiled down to five characteristic subsystems or levels:

- **System 1:** Any viable system, from a living organism to a commercial enterprise, must be autopoietic, that is, self-producing. A body produces the cells and organs of which it is made up of; a viable business is one that insures its own continued existence. System 1 are the subsystems in a given system that produce the system in question. Such systems are viable systems themselves, and subject to recursive treatment
- **System 2:** Aids in the coordination of the activities of the different system 1's by coordinating common services:
- **System 3:** Is the management of the whole of the system's self-production. It allocates resources to the different system 1s with the welfare of the whole system in mind. I

Systems 1,2 and 3 constitute the inside-and-now of a viable system. Together, they constitute a systems' production of itself. No other subsystems would be necessary if the system had no need to grow, adapt, or make decisions in a changing environment (which is the function of system 4).

- **System 4:** Observes the anticipated future environment and its own states of adaptiveness and act to bring them into harmony" . Its role is to understand and plan for an anticipated future.

A system that has a system 4 subsystem is capable of learning; that is, of transforming itself. Such a system's identity would therefore change over time and pose a cognitive challenge to the system in question.

The role of operational closure, of producing a system's separate identity, falls to system 5:

- **System 5:** Provides logical closure to a system. It makes any decision that is not handled by any of the other subsystems. System 5 is also responsible for figuring out the system's identity (which is an ongoing task in a constantly changing and learning system); and, finally, for deciding how much resources to devote to system 3 (the day to day operation of the system) and how much to devote to system 4 (planning for the future)

In summary, at the lowest level, the components of the system perform actions, above this level of action we require a level to co-ordinate and look for synergies, a level to resolve conflicts, a level to monitor the external environment and respond to changes and finally we require a level to establish the identity and integrity of the system.

VSM also scales up and down, each system contains components, which are systems themselves similarly structured into levels, and each system is itself a component in a larger system – the same principles of organisation apply at all scales. It provides a strong framework for organisations: action, synergy, conflict resolution, monitoring and identity.

One of the meanings that we can attach to globalism is that we are creating a worldwide viable system – some people might call this a global brain. With hope this system will develop its identity to avoid collapse, to restore the damage we have done and to keep us on a path of similar flow – converging to a world of equalities and within limits. At all scales below this global identity each system needs to interpret its own viable system framework.

VSM fits well with the ideas of structural coupling of a system with its environment, the co-creation of the system and its environment through cognition or sensing what is happening. The scale of the human endeavour that is creating its own environment as it creates itself is now so great that it defines what we could call the Anthropocene – the largest force shaping the planet is the environment that is co-evolving to support human life.

The key point for us in this discussion of complex global issues is the requirement in VSM of a level in the system to monitor the external environment, to think through the consequences and to therefore prepare for change.

Futurism: Or Future Studies

Preparing for change means constantly looking out for what might be coming towards us – commonly called scanning the horizon. It means identifying weak and early signals of change and systematically thinking through the consequences as all global scale issues are interrelated. It is looking for opinions on trends and movements, for signs of the big shifts in the world as well as the small advances (or retreats) that might have important consequences, or offer opportunities for improvement, or pose threats. Preparing for change is a form of future studies.

This definition is derived from Wikipedia (Futures Studies, 2017):

Futures studies' (also called futurology and futurism) is the study of postulating possible, probable, and preferable futures and the worldviews and myths that underlie them. Futures studies seek to understand what is likely to continue and what could ***plausibly change****. Part of the discipline looks for patterns and systematic understanding of past and present to determine the likelihood of future events and trends. Unlike the physical sciences where a narrower, more specified system is studied, 'futures studies' concerns a much bigger and more complex world system.*

Stories about the past and present are important when they suggest the changes that might occur.

MOVING TO ACTION

The concept of planetary limits or boundaries was introduced in 2009 (Rockström et al., 2009) and subsequently updated in 2015 (Steffen, 2015). The spider web diagram used to illustrate this idea is now one of the iconic pictures of the environmental movement. There is a set of nine thresholds and as long as we do not move outside of them then humanity can continue to develop and thrive for generations to come. Crossing these boundaries could generate abrupt or irreversible changes on a colossal scale. They define our safe operating space - at least from an earth systems science perspective.

Respecting these boundaries reduces the risks to human society of catastrophic change.

The aggregate effects of human activity are now shaping the world through our waste streams into the atmosphere and waters and through our land use such as transforming forests into fields for crops, energy and livestock. We dam rivers, gouge out minerals, build cities and generally disrupt just about every ecosystem on the planet. We have now breached at least four of the nine planetary boundaries and earth systems are seriously out of balance. This marks the transition to what is popularly called the Anthropocene epoch.

Committed: Locked in, Canalised, Path Dependent

What does the Anthropocene mean in terms of what are the full range of changes that might arise as a result of human influence on the biogeosphere of a scale that defines a new epoch? This is the difficult task we face: to look out for the changes, some of which will have a damaging effect while others may offer a chance to grasp opportunities to improve well-being and regenerate nature – move us back within the thresholds. Will our transgressions of the boundaries lead to a collapse of economies and societies, or a graceful degradation and retreat, or to a new era of opportunity to recover and restore the dynamic balance? So much depends on the history of how we got into this mess, whether we have the wisdom to escape our past and whether we can cope with, and prepare for, the changes that that implies.

We are committed to a pathway that started at the time of the first industrial revolution. After centuries of person-powered, hard labour activities, augmented with a little wind and water, came the use of coal and steam power. These early stages of change occurred at the end of the 18th century – and were hardly noticeable. At a global scale it was a few, tiny blips in a small number of locations in Europe and the US.

The beginnings of exponential growth are gentle, so it wasn't until we reached the middle of the 19th century that we 'woke up' to the intensity of energy use from fossil fuels. This period marked the start of what is called the second industrial revolution, which saw the widespread use of machinery, through greatly increased use of coal for steam power, then adding in the use of oil and gas and the beginnings of electricity and communications systems.

For someone in 1850 to imagine and predict the changes that would occur in the next few decades is unimaginable. The pace of development with the discoveries and inventions of technology, the expansion of communications and transport and the ideological shifts resulting from turmoil, made foresight impossible and the consequences of the changes were only vaguely seen. All credit though to Svante Arrhenius who, in 1896, formulated his greenhouse law "if the quantity of carbonic acid (CO_2) increases in geometric progression, the augmentation of the temperature will increase nearly in arithmetic progression". No one listened.

For someone in 1850 to come out with a suggestion that we slow down and stop this juggernaut of change that was bringing huge wealth (to a few) and held out great promise of power, would also be unimaginable. The world was locked in. Although many might have admired Thoreau's ideas of simple living and his musings on early environmentalism, few were willing to contemplate resisting the tide. We were in a self-reinforcing channel, the more we dug the easier it became to keep digging and the rewards kept getting bigger. The systems term 'canalisation' could be used for this phenomenon, it means a development pathway that is robust and returns to where it was going after any perturbation – you can't stop it or divert it from its final form. Let's remember that the major social perturbations of war and conflict actually increased the speed of industrialisation, increasing the canalisation.

The characteristics of the early stages of exponential growth show this reinforcing feedback, each technological advance opens up new ground for further advances. The major driving force in this reinforcing feedback was what Hornborg calls 'displacing problems to others' (Hornborg, 2014). It was through a systematic displacement of labour and environmental degradation from countries like Britain to colonised parts of the world where labour and land were at zero marginal cost.

What was also a major component in this flood of change was the phenomenal increase in energy availability. These early days of industrialisation saw fossil fuel energy literally rising out of the ground by itself. Once the rigs were installed and the pipes in place, the cost of the next barrel of oil was negligible – it was zero marginal cost energy.

Industrial development was on a trajectory and the initial conditions set the pattern for what followed. The path dependency was established. Economists, however, determine three types of path dependency. The first type is benign, or efficient, where the outcome of following one path rather than another is equivalent, you reach pretty much the same destination. The second type is where some paths have better outcomes than others, but you had no knowledge to determine which one was best - chance and a slight early advantage selected one path over any other and sometimes it was definitely not the best overall outcome, it was less efficient. The third type is like the second but where knowledge of the different outcomes was foreseen from the beginning, but no action was taken to select the overall better path when events were moving towards the poorer, less efficient outcome that unfortunately had the early advantage.

The classic, and probably apocryphal, example of the second type of path dependency is the QWERTY keyboard – the slight early advantage of a keyboard layout that reduced the typewriter keys sticking together set us down a self-reinforcing pathway while an alternative layout (like the Dvorak scheme) would have been dramatically better but no one foresaw this and decision making to switch was dispersed – no central authority existed to force a change of direction.

The first and second industrial revolutions exhibited this second type of path dependency. Methods and processes of foresight to appreciate the outcomes of an unfettered rush into a fossil fuel world were not sophisticated enough to present the dangers nor was there any central controlling mechanism to reset the path even if there were awareness of the outcome.

This is where we are now, it is summed up well in this quote from Dennis Meadows:

The problem that faces our societies is that we have developed industries and policies that were appropriate at a certain moment, but now start to reduce human welfare, like for example the oil and car industry. Their political and financial power is so great that they can prevent change. It is my expectation that they will succeed. This means that we are going to evolve through crisis, not through proactive change. (Meadows, 2012)

Note that Meadows is not just expressing the situation 'as is', but he recognises the unlikelihood of vested interest changing the situation. In this sense he accuses us collectively of the third type of path dependency – an almost wilful disregard of the outcome when we do have the knowledge of what will occur and we can choose a different way forward yet we follow the path of easy wins today, we take the early advantage.

The story of responding to climate change exhibits all of the characteristics of this third type of path dependency. There is ample foresight yet we are incapable of resisting short term advantages. In 1990, from the first assessment report for the Intergovernmental Panel on Climate Change (IPCC) we have this:

The predicted population explosion will produce severe impacts on land use and on the demands for energy, fresh water, food and housing, which will vary from region to region according to national incomes and rates of development. In many cases, the impacts will be felt most severely in regions already under stress, mainly the developing countries. Human-induced climate change due to continued uncontrolled emissions will accentuate these impacts.

Twenty five years later the trajectory of increase in global CO_2 emissions has not deviated from the 'business as usual' case. Year on year the pressing political need to make sure that economic growth continues, has meant we keep taking the path of early advantage – short term gains. We have poured huge resources into technologies to extract every last drop of oil rather than choosing to expand rapidly renewable energy and circular economy processes.

What this implies is in the last line of Meadow's statement above: we are going to evolve through crisis. Accepting this shifts the debate in environmental and social justice. Without wishing in any way to ease up on mitigation efforts we are forced to accept that adapting to change is vital - to sustainability we add resilience. Again, in the same interview, Meadows says it well:

The fact that the collapse hasn't occurred so far doesn't mean it won't take place in the future. There is no doubt that the world is changing, and we will have to go along with it. There are two ways to do that: One is, you see the necessity of change ahead of time and you make the change, and the second is that you don't and are finally forced to do it anyway. Let's say that you're driving a car inside a factory building. There are two ways to stop: Either you put on the brakes or you keep going and hit the wall. But stop you will, because the building is finite. And the same holds true for Earth's resources.

POLICY MAKING: TELLING STORIES

Being able to control the stories around issues that are presented in public is highly empowering for any group (a complex global issue in its own right). The stories that are created and projected by the engagement to the world are a critical thread that bridges the internal and external contexts of the engagement. These stories, how, and how well they are communicated is what helps potential participants discern whether or not they should become engaged. However, it is not always easy to control what the story is or how it is interpreted. Stories can be co-opted by other groups, and/or perceptions and interpretations of the stories can stray from the original design.

About Prediction

From Wikipedia (Prediction, 2017):

Outside the rigorous context of science, the term "prediction" is often used to refer to an informed guess or opinion. A prediction of this kind might be inductively valid if the predictor is a knowledgeable person in the field and is employing sound reasoning and accurate data. Large corporations invest heavily in this kind of activity to help focus attention on possible events, risks and business opportunities, using futurists. Such work brings together all available past and current data, as a basis to develop reasonable expectations about the future.

If we are to bring the long term global perspective into the discourse of today and shape action in the here and now, we can only tell stories. By now we should well aware that Complex Global Issues defy prediction with any certainty, the complexity involved means that we must be circumspect about our reasonable expectations of the future. Returning to our well know Soft Systems Methodology perhaps we can translate it into the future and see if that helps to tell the story.

Ingredient List for a Story Translating CATWOE Into the Future

In this table we have the six components (CATWOE) but now in the right-hand column we ask the questions of what our system might be in the future perhaps as complex global issues unfold:

Table 1. CATWOE looking forward

	Present	Future
Client	Who is affected, who gains or loses?	Who is possible in the picture? Onion rings of affected people into the future
Actors	Who takes part, who else is involved?	Ditto, who might get drawn in?
Transformation	What is happening, what activities occurs, what changes?	What are the future transformations?
Worldview	How do the people see the world, what are their perspectives, beliefs and values?	Will worldviews be different?
Owners	Who ultimate owns the story – who could make it disappear?	Will ownership change?
Environment	What is the background and context in which the story unfolds?	How does the environment change?
Root definition of the Story	What is it? What is its purpose now? By what means is it achieved?	What will it be? What purpose will it have? How will it do it?

This time dimension leads to a new root definition, it allows us to explore future scenarios in a systematic way. Some of the additional factors that contribute to the stories (there are many scenarios) would include consideration of the vulnerabilities, volatility – what might change, how fragile is it?

In a similar way of course we can overlay Ulrich Beck's emancipatory view: who will control what, who will they be, and how will they do it? Who will not be included? How will we uncover this? - the future 'ought' tense.

These stories are all part of systems thinking and are essential to cope with complex issues that lead into thinking about policy making.

The Garbage Can Model

Back in 1972, Cohen, March and Olsen wrote a critique of the rational model (Cohen et al, 1972), challenging its linear, sequential approach. They suggested that policy making works more like a garbage can into which a mix of all kinds of problems and possible solutions are poured, with the outcomes determined by the contingencies of the inputs.

They found that people operate on the basis of inconsistent and ill-defined preferences. Technology is unclear and the participants in decision making come and go and they vary in how much time and commitment they have got. This results in organised anarchy.

The garbage can model is based on the assumption that decision making is sloppy and haphazard. Decisions result from an interaction between four independent streams of events: problems, solutions, participants, and choice opportunities. These leads to two outcomes:

1. Because of the challenges on conflict resolution, policy is often made without clarifying objectives.
2. Because of limited resources, rational decision making is usually not feasible.

The whole rather disheartening situation is summed up in this quote:

[Policy making] is a collection of choices looking for problems; issues and feelings looking for decision situations in which they might be aired; solutions looking for issues to which they might be the answer; and decision makers looking for work. Problems, solutions, participants, and choice opportunities flow in and out of a garbage can, and which problems get attached to which solutions is largely due to chance. (Cohen, March, Olsen, 1972)

There are consequences that arise from the garbage can model: solutions may be proposed even when problems do not exist; choices are made without solving problems; problems may persist without being solved; and some problems are solved.

The situation described by Cohen et al. (1972) is exactly what complexity theory covers. There are five characteristics of complexity that are relevant:

1. **Systemic:** Everything is connected.
2. **Context Specific:** Each situation is unique and details matter.
3. **Path Dependent:** History matters; what happens depends on the particular sequence of events that has led to the current situation.
4. **Episodic:** Change goes in 'fits and starts' - how resilient is the current situation to events/chance/actions/wider change.
5. **There Are Limits to Knowledge:** What emerges cannot be known in advance.

This emphasises that history, context and foresight are critical aspects of setting and working with strategy and policy. It is a difficult mind-set change because it emphasises that we can know less and predict less and attribute less than we hoped. If the world is complex, then acting congruently with that complexity can be simpler and more effective than trying to control a machine that does not exist.

The vernacular questions: 'Says who?', 'So what?', and 'Who cares?', are succinct and rather cynical questions that we could apply quickly to any recommendation to explore its robustness. However, expanding these, we can say that each recommendation should include some description of:

- What is it about?
- Who is it for?
- Where does it come from (conclusions and evidence)?
- How would one fulfil it?
- Why should it be fulfilled?
- What are the consequences of following it … or not?

Critical System Thinking methodologies extend this list to question the wider context from an emancipatory perspective (who is wielding power and control in this recommendation situation, who is excluded). So we need to add:

- Who has power and control?
- Who is marginalised (and how)?

These language structuring, systems thinking and checklist approaches provide useful tools to make sure that recommendations for preparing for change are robust and meaningful and can be accepted by the people involved.

SYSTEMS THINKING: WHAT CAN THIS BRING?

Cabrera develops the idea that there are four main concepts and therefore approaches for systems thinking. He believes that people can improve their thinking skills by learning to explicitly recognize these four things: distinctions, systems, relationships, and perspectives (DSRP) and that these underlie anything you may wish to understand.

DSRP consists of patterns of two opposing elements. These patterns and their elements are:

- **Making Distinctions:** Which consist of an identity and the 'other' (not the identity)
- **Organizing Systems:** Which consist of part and whole (the holistic concept)
- **Recognizing Relationships:** Which consist of action and reaction
- **Taking Perspectives:** Which consist of point and view (zoom in and pan out)

Each of these patterns will lead us to different systems ideas and different methodologies.

Distinctions

Perhaps what we mean here is Gregory Bateson's idea of 'a difference that makes a difference'. In a static sense there is a change and we can distinguish two sets and whenever we can separate things into sets we can discern a boundary between them. Making a distinction (D) is about boundaries, which is a fundamental concept in systems thinking. The inclusion and exclusion of stakeholders defines a boundary. We can immediately relate this to migration where there is palpable sense of boundary – those with a right to be here and those without.

A conflict of values (different ethics) can arise between people within the primary boundary of a system and those in the margin (outside the primary boundary, excluded but still within a further or secondary boundary of concern and affect).

This process of marginalisation opens up questions of whose values and power creates this boundary? How do we agree on a particular boundary we seek appreciation of a problem? Critical Systems Heuristics developed by Werner Ulrich (1988) proposes twelve questions focusing on: Motivation, Control, Expertise and Legitimacy. In many of these questions, we are asked to think of the difference between what 'is' and what 'ought to be'. For example, what is the purpose of the system and what ought it to be. If we are considering a system to vet migrants we need to ask what the process that happens really is and then ask what it should be.

Relationships

The second element of Cabrera scheme is about relationships, again a fundamental concept in systems thinking. What is appropriate to explore relationship are the many different approaches for understanding complex causality looking at the feedback mechanisms producing vicious and virtuous circles and the possible consequences of intervention.

What is of great use is methods like causal loop and influence diagrams that make explicit the relationships among the components of a system. We begin to appreciate the ideas of 'stocks and flows' of physical items and of information. Eventually this can build up into a full system dynamics model, which if the parameters are of good quality we can produce computer models that allow us to try out thought experiments for intervention.

It is tempting to 'see' a systems dynamics of migration that maps the relationships among indicators such the intensity of conflict, the number of dispossessed people, the flows of refugees across international borders, the requirements for housing, food, medical care, education and so on. The scale of his modelling and the resource requirement would of course be substantial and not quick.

The first stages of a systems dynamics approach though is valuable – producing even a rudimentary causal loop diagram (CLD) opens up discussion – it allows different perspectives to emerge from a group.

System

This section of Cabrera's scheme deals with approaches for developing viable and highly responsive organisations at multiple levels (global to local). Coping with wicked problems involves many organisations that have internal issues of structure and function and external issues of co-ordination, conflict resolution and awareness of what's going on out there.

Perhaps one of the best know systems approaches is the Viable System Model. A viable system is any system organised in such a way as to meet the demands of surviving in a changing environment. One of the prime features of systems that survive is that they are adaptable. Hoverstadt (2008) gives a detailed examination of the use and power of VSM.

Perspective

The final section of this four-part scheme is about the different systems approaches for addressing conflict, exploring multiple perspectives, developing mutual understanding, and agreeing solutions that people are willing to implement.

Whereas systems dynamic methods fall into a 'hard' systems category, this section is where we find 'soft' systems methodologies – although these categories help systems people to talk about the subject they blur together when we come to actually use methods – all four sections of Cabera's scheme work together and a soft systems approach is often where we start before developing any causal diagrams. What we are after is to examine closely the different worldviews that stakeholders bring and the map out what processes or transformations are happening in the system.

If we consider migration with a soft systems method (SSM) we can start simple by asking people to draw a 'rich picture'. This is simply getting down on paper a shared perspective of what's going on, what are the components of the focus. No great artistic ability is assumed. This would draw out the main elements easily enough: the migrant population, the hosts, the agencies, the reactionary elements, the conflict areas etc. It is where thoughts on marginalisation can appear and values and ethics written onto the picture. Even the emotional aspects of migration can be symbolised.

This group activity is often developed using the full Soft Systems Methodology under the mnemonic BATWOE

- Beneficiaries (sometimes 'Clients' is used)
- Actors
- Transformation
- Worldview
- Owners
- Victims
- Environmental Constraints

Peter Checkland (1999) give a full explanation of the power of this approach to get to heart of what the systems is. The next stage concentrates on the T and the W elements, we need to:

- Identify possible transformations that different people want
- Explore what these mean to different people to ensure that people are not talking past one another

Understanding that there are different perspectives and different desired outcomes from change is a vital contribution from systems approaches – it is the starting point for asking questions and challenging the status quo.

There are three systemic principles for managing wicked problems:

1. Explore boundaries (stakeholders and issues), values and processes of marginalisation up-front, and revisit the boundaries of your work when new aspects of a wicked problem present themselves.
2. Draw upon and mix methods from across the systems approaches (and the biophysical and social sciences) to be as responsive as possible to the multiple dimensions of wicked problems.
3. If you have no previous (or limited) experience, start from where you are. Try new methods when the need arises, and build capacity for the longer term.

Systems thinking gives us a better understanding of wicked problems and how to manage them … even when we think we already know the answer!

We have to recognise that at the present stage of development of the use of systems thinking in policy and decision making we cannot expect people to take part in a large number of 'workshop' type events spread over a long period. To apply many of the methods of systems thinking requires this fairly intensive approach and if people are already in the mode of systems thinking progress is all the better. Ideally we would like people to be in a mode of systemic thinking all the time and to be happy with a continual process of inquiry (a default state).

As Judi Marshall (2004) says:

I have learnt that living systemic thinking is a long-term, emergent, never-ending activity, with any sense-making always open to re-vision as action, reflection and feedback unfold. Inquiry is key to living systemic thinking and takes many forms, being self-reflective but also systemically engaged, and means taking strategic initiatives to learn more and track emerging data, knowing that I will never fully know.

If all policy makers were in this state then perhaps systems thinking methods would have no trouble in being tried out and accepted in rapid sessions. However, we are far from that ideal state so we need to develop and use methods that elicit knowledge and share perspectives from people who may be ignorant and fearful of what we are talking about and we need to engage with them with a degree of speed and succinctness. It could almost be called 'easy, speed system thinking'.

Appreciative Inquiry Method (AIM)

The strengths of the Appreciative Inquiry Method (AIM) are that it does have an ease of use and speed (it had its origins in a research project where busy managers had little time to discuss their projects, see Stowell & Welch, 2012, p49).

One of the objectives of AIM is to be non-intrusive so that individuals involved in the process have minimum disruption to their busy lives. If they can be persuaded to take time out to think then we should try to minimise the time and to make the experience worthwhile and perhaps fun.

Some systems of purposeful activity (human activity systems) are relevant to an exploration of a situation of interest (like migration) and to help us appreciate them we need to 'say what they are'. An oblique approach is necessary because it is never possible simply to describe real-world purposeful action once and for all. We can only describe a range of interpretations which are relevant to debating the real-world processes and structures. To do this we need to develop a concise definition of the system of interest.

AIM has three phases and during the process the participants are encouraged to keep five characteristics in mind by using the mnemonic PEArL (Participants, Engagement, Authority, relationships, Learning). As a soft system method, the key element is to create a cycle of learning about the situation of interest so that there is a meaningful exchange of ideas among the participants. In this way everyone gains a deep *appreciation* of the interconnectedness of the many factors involved. The idea is for people involved to make their sense of the issues, to share those perspectives and to revise what they think – appreciating the system, opening up questions and accommodating others.

The first phase of AIM is to create a systems map either in a group session or by facilitators talking to participants on a one-to-one basis. The idea is for the different people or subgroups to draw their impressions of various subsystems involved and once individual maps have been completed to combine them to arrive a single map – this process tends to open up discussion particularly about the different other people see things.

The second phase is to dig into the subsystem in the combined map, to separate out the reality from the desired. In other words to distinguish the 'what is' from the 'what ought to be'. This phase is concerned with a description of the elements of the map as purposeful activities and the soft systems check list of CATWOE is valuable in producing this description. (Clients, Actors, Transformation, Worldview, Owners, and Environment), see Peter Checkland's work (Checkland, 1999).

The third phase of AIM involves the facilitator returning to each participant with the activity model from stage two. This is used to open up discussion for how to move forward, what could be done to improve the situation. It is the stage of deliberate intervention where it is important to understand different people's expectations of different outcomes.

Throughout the three stages of AIM there is an underlying structure or set of questions to at least consider and explore if not completely answer. This is the PEArL mnemonic that asks:

- **P – Participants:** Who is involved? Why are they involved? What is their role in the study? Can you identify the boundary between 'Participants' and 'non-Participants'? Are there any transitory participants, if so why?
- **E – Engagement:** How will the participants engage with one another? What methods will be used? What timescales are involved? What environmental / context issues influence the engagement process?
- **A – Authority:** What formal authority is associated with the various roles – (e.g. control of resources)? What embedded authority do the tools for engagement have – and why? What are the influences from the environment / context (e.g. existing policies)?
- **R – Relationships:** Consider any changes to relationships occurring. Consider the 'commodities of power' that people have used in the situation - this provides insights into the ways that people have managed, or are managing, power in the situation and insights into what control strategies are being used, by different groups.
- **L – Learning:** About both theory and practice. What has been learnt about the situation? About the approach being used? About systems ideas?

CONCLUSION

The Cassandra Matrix

After we have challenged all the assumptions behind our perceptions, after we have included everyone and everything that we can, after we have recognised the history, culture and social influences, after we have explored every extraordinary option, after we have been as skeptical as possible about all knowledge and truths, how do we proceed?

Perhaps it is with trembling certainty, an image described by E.O Wilson (2002) in The Future of Life:

The biosphere creates our special world anew every day, every minute, and holds it in a unique shimmering physical disequilibrium. On that disequilibrium the human species is in thrall.

This shimmering disequilibrium applies to both the environment and our social surroundings; it has to be approached with temerity. At some point our actions must proceed with certainty and confidence (literally with faith) but as we do so we must tremble that our actions in this disequilibrium are themselves shimmering under constant critical review and are able to bend and flow.

The early years of the 21st century are proving increasingly complex. We do have many challenges including a population explosion with unprecedented population migration, but also growing inequality, managing our finite resources, climate change, and the wholesale abuse of our environment. We have still to grasp the full impact of technology upon human civilisation.

If we, the collective human race, are to develop we should reflect upon how we organise ourselves and how to create policies relevant to our survival and how we can turn them into feasible practicality. We believe that systems thinking has a significant part to play in that development thinking.

Poor Cassandra, daughter of King Priam and Queen Hecuba, when she did not return his love, the god Apollo placed a curse on her so that no one would ever believe her true predictions. Many people, who spend their lives exploring complex global issues, feel like a modern-day Cassandra. They predict catastrophic global warming or the end of civilisation as we run out of energy, but nobody seems to believe them - if they do then they find a way to pretend it is not true, to deny.

Cassandra told the truth but nobody would believe her so an anti-Cassandra is someone who always tells falsehoods but everyone believes them. Unfortunately, we cynically consider most politicians to be anti-Cassandras and we believe their falsehoods usually because they are telling us what we want to hear.

The whole aim of practical politics is to keep the populace alarmed (and hence clamorous to be led to safety) by menacing it with an endless series of hobgoblins, all of them imaginary. (H. L. Mencken)

Cassandra's dilemma arises when she asks herself the question: "What if my telling the story makes it come true? She can predict the future but should she just keep quiet because otherwise people will not believe her and so not do anything to change".

So just a couple more gems from the eminently quotable American essayist and critic H. L. Mencken:

- *The one permanent emotion of the inferior man is fear - fear of the unknown, the complex, and the inexplicable. What he wants above everything else is safety.*
- *For every complex problem there is an answer that is clear, simple, and wrong.*

REFERENCES

Ackoff, R. (1974). *Systems, Messes, and Interactive Planning, Redesigning the Future*. New York, London: Wiley.

Ahmed, N. (2010). *A User's Guide To The Crisis of Civilization*. London: Pluto Press.

Argyris, C. (2005). Double-loop learning in organizations: a theory of action perspective. In *Great minds in management: the process of theory development* (pp. 261–279). Oxford, UK: Oxford University Press.

Bammer, G. (2003). *Integration and Implementation Sciences: Building a New Specialisation*. Retrieved from http://www.anu.edu.au/iisn

Beck, U. (1986). *Risk Society: Towards a New Modernity*. Sage Publications.

Beer, S. (1972). *Brain of the Firm; Allen Lane*. London: The Penguin Press.

Bellinger, G. (2012). *Enabling a Better World*. Retrieved from http://insightmaker.com/insight/3395

Capra, F., & Luisi, P. (2014). *The Systems View of Life, A Unifying Vision*. Cambridge University Press. doi:10.1017/CBO9780511895555

Cabrera, D., & Cabrera, L. (2015). *Systems Thinking Made Simple: New Hope for Solving Wicked Problems*. Ithaca, NY: Odyssean.

Carse, J. (1986). *Finite and Infinite Games*. New York: Random House.

Casti, J. (1997). *Would-be Worlds*. John Wiley & Sons.

Checkland, P. (1999). *Systems Thinking, Systems Practice. Includes a 30 year Retrospective*. Wiley.

Cohen, M., March, J., & Olsen, J. (1972, March). A Garbage Can Model of Organizational Choice. *Administrative Science Quarterly, 17*(1), 1–25. doi:10.2307/2392088

Dewing, C., & Jones, T. (2016). *Future Agenda – Six Challenges for the Next Decade*. London: Third Millennium Publishing.

Dobbs, R., Manyika, J., & Woeetzel, J. (2015). *No Ordinary Disruption: The Four Global Forces Breaking All the Trends*. PublicAffairs.

Futures Studies. (2017). Retrieved from https://en.wikipedia.org/wiki/Futures_studies

Gilding, P. (2011). *The Great Disruption, How the Climate Crisis Will Transform the Global Economy*. London: Bloomsbury Publishing.

Gore, A. (2013). *The Future*. New York: Random House.

Hajkowicz, S. (2015). *Global Megatrends, Seven Patterns of Change Shaping Our Future*. Clayton South, Australia: CSIRO Publishing.

Hodgson, A. (2011). *Ready for Anything: Designing Resilience for a Transforming World*. Triarchy Press.

Hodgson, A. (2016). *International Futures Forum*. Retrieved from http://www.internationalfuturesforum.com/

Horn, R., & Weber, R. (2007). *New Tools For Resolving Wicked Problems: Mess Mapping and Resolution Mapping Processes*. Strategy Kinetics.

Hornborg, A. (2014). Does the Anthropocene Really Imply the End of Culture/Nature and Subject/Object Distinctions? Conference The Thousand Names of Gaia, From the Anthropocene to the Age of the Earth, Rio de Janerio, Brazil.

Hoverstad, P. (2008). *The Fractal Organization: Creating sustainable organizations with the Viable System Model*. Chichester, UK: Wiley.

Levin, K., Cashore, B., Bernstein, S., & Auld, G. (2009). Playing it forward: Path dependency, progressive incrementalism and the "Super Wicked" problem of global climate change. *IOP Conference Series: Earth and Environmental Science, 50*(6).

Marshall, J. (2004, Summer). Living systemic thinking: Exploring quality in first-person action research. *Action Research, 2*(3), 309–329. doi:10.1177/1476750304045945

Meadows, D. (1972). *The Limits to Growth; A Report for the Club of Rome's Project on the Predicament of Mankind*. New York: Universe Books.

Pirsig, R. (2013). *Lila: An Inquiry Into Morals*. New York: Bantam Books.

Prediction. (2017). Retrieved from https://ipfs.io/ipfs/QmXoypizjW3WknFiJnKLwHCnL72vedxjQkDDP1mXWo6uco/wiki/Prediction.html

Rittel, H., & Webber, M. (1973). *Dilemmas in a General Theory of Planning. In Policy Sciences* (Vol. 4). Amsterdam: Elsevier Scientific Publishing Company, Inc.

Rockström, J., Steffen, W. L., Noone, K., Persson, Å., Chapin, F. S. III, Lambin, E. F., ... Foley, J. (2009). Planetary Boundaries: Exploring the Safe Operating Space for Humanity. *Ecology and Society, 14*(2), 32. doi:10.5751/ES-03180-140232

Saarinen, E., & Hämäläinen, R. (2007). *Systems Intelligence: Connecting Engineering Thinking with Human Sensitivity*. Academic Press.

Senge, P. (1990). *The Fifth Discipline*. Random House.

Sharp, B. (2013). *Three Horizons: The Patterning of Hope*. Triarchy Press.

Simon, H. (1957). A Behavioral Model of Rational Choice. In *Models of Man, Social and Rational: Mathematical Essays on Rational Human Behavior in a Social Setting*. New York: Wiley.

Steffen, W. (2015). Planetary Boundaries: Guiding human development on a changing planet. *Science, 347*(6223). DOI: 10.1126/science.1259855

Stowell, F. A., & Welch, C. (2012). *A Managers Guide to Systems Practice*. Wiley. doi:10.1002/9781119208327

Ulrich, W. (1988). Systems Thinking, Systems Practice and Practical Philosophy: A programme of research. *Systems Practice, 1*(2), 137–163. doi:10.1007/BF01059855

Ulrich, W. (2002). Chp: *Boundary critique*. In H. G. Daellenbach & R. L. Flood (Eds.), *The Informed Student Guide to Management Science*. London: Thomson.

Wilson, E. O. (2002). *The Future of Life*. New York: Vintage Books.

ENDNOTES

1 British Astronomer Royal Sir Martin Rees writing in *Our Final Hour* (2003).

2 See The International Futures Forum - Attribution - Non-commercial - No-Derivatives 2.5 Scotland www.internationalfuturesforum.com

3 The idea of 'finite and infinite games' comes from James Carse in a book of that name (Carse, 1986).

Chapter 5
Revising Sub-Systems Models to Analyze Economic Policy and the Question of Systemic Persistence

James C. Schopf
Keimyung University, South Korea

ABSTRACT

While Easton's systems theory contributed to political science by demonstrating how the political system meets societal demands with policy outputs, he ignored the state's role in providing security in a hostile international environment. Hence, this chapter builds a sub-systemic governance model, arguing that large input generating groups require sufficient public goods to maintain the domestic political system and the state. Application to the South Korean case demonstrates that public good allocations increased along with the size of the input generating group. A functioning transmission belt, in the form of civic groups and elected local government, facilitated articulation of these demands to political leaders. Disruption of this subsystem in cases with unmet demands from large input generating groups can destabilize the state and its domestic political system. This new sub-systemic model seeks to advance understanding of the operation of the system and open up new areas of research into the persistence of the domestic political system.

DOI: 10.4018/978-1-5225-5996-2.ch005

INTRODUCTION

The systems approach has greatly contributed to the study of politics. David Easton's seminal Systems Theory drew attention to important aspects of political life and provided a critical framework with which to understand and analyze inputs into the political system and policy outputs to the social environment. The advancement of systems theory in political science was hobbled, however by methodological shortcomings. Easton failed to operationalize key concepts, and as a result, the theory was neither applied nor tested. In addition, Easton's all-inclusive system design was unable to provide insight into several systems-related questions areas of interest to social scientists, including the survival or collapse of states and their domestic political systems, regime change, and variation in the nature of policy outputs or societal inputs.

Combining Easton's policy process framework with methodologically rigorous approaches sharing key system's theory assumptions helps to deepen understanding of these issues. By narrowing Easton's system to a critical subsystem comprised of the leader and his/her supporters, it becomes evident that changes in the size of the input-generating group can markedly affect the quality of government policy outputs. This new sub systemic model yields the prediction that leader's seeking to maintain power will allocate an increased ratio of public goods to private goods, the larger the size of the input generating group. After operationalizing the size of the input-generating group and the share of public vs. private goods allocated through economic policy, this paper applies this sub systemic hypothesis to explain recent changes in economic policy making in South Korea. Modernization theory provides the added insight that the forces of industrialization and economic development are increasing the size of the input-generating group in societies throughout the world, which are calling for public policy goods, in the form of democratic political rights as well as improved overall living standards. Leadership failing to respond to these increased demands over a prolonged period not only provokes regime change, but, in certain circumstances, can destabilize and trigger the collapse of states and of domestic political systems. Research into underdeveloped institutions, economic power concentration, sectarian division and other factors impeding delivery of public goods to large input generating groups, can offer further insight into the question of systemic persistence, the central concern of Easton's systems theory.

This chapter first critiques the strengths and weaknesses of Easton's systems theory. A sub-systemic model is offered to ameliorate the methodological shortcomings of Easton's systems theory while making it applicable to questions concerning the persistence of domestic political systems and state maintenance. Applied to two cases of Korean industrial restructuring, the predictions of the sub-systemic model hold true: small input generating groups under authoritarian rule were associated with provision of private goods, whereas larger input generating groups under democracy produced policies that allocated public goods. The paper then addresses the importance of the transmission belt between societal actors and the state in order to adjust for imprecision in the systems approach's short-term predictions. The form of societal organization and political institution affect the capacity to convey demands to political leaders. The final section of the paper then explores the possible collapse of the domestic political system in cases where leadership is unable to provide public goods to large input generating groups.

Systems theory's popularity in political science grew following World War Two through the work of David Easton, the leading proponent in the field advocating application of systems theory. Easton sought to build an analytical, abstract, value-neutral, empirical theory which could explain the survival and persistence of political systems. Focusing on the macro-level, Easton's systems theory (1953) sought to understand society as a whole by studying the interrelated processes and relationships which contributed towards the system's continuity. He viewed government institutions and society as an interdependent whole, held together by elaborate systems of exchanges between the system, its environment and subsystems. His approach provides a useful framework with which to distinguish political relationships from other social relationships and study their mutual interaction.

Easton's systems theory first established a boundary between the political system and its environment, the larger society. His system described a five-stage policy-making process, consisting of input, conversion, output, feedback and environment. The social environment surrounding the political system initially generated demands for government policy goods. He termed these 'inputs' into the system. Demands were then reduced and transformed into issues. 'Gatekeepers' who "straddled the channels of admission to the system" regulated these processes, with political leaders enjoying the capacity to set the agenda for political discussion (Easton,1965a, p. 21). Demands led to decisions or policy 'outputs' of policy, directed towards a portion of society. Policy outputs then interacted with and transformed societal forces. The structure of the political system is dynamic and will adjust, enabling the system to persist.

Easton's systems theory is holistic, subsuming all levels of analysis, which enables politics to be seen in its entirety. The theory is therefore able to account for the demands of any group at any given time, the system's environmental effects, and the feedback from the environment back on decision-makers. By drawing attention to important aspects of political life, Easton has thus immensely contributed to the explanation of political behavior.

CRITIQUES OF EASTON

Easton's system's theory suffered from a few serious deficiencies, however. Critics charged that Easton's model failed to operationalize and test key variables and was not designed to address questions of interest to social scientists related to the collapse of domestic political systems, structures of domestic authority, such as regimes types, or changes in the nature of policy outputs or societal inputs. Scholars criticized Easton's model for remaining at the 'theoretical level' and neglecting to apply his key concepts. Easton had failed to follow through on his objective to design a theory capable of empirical verification. "Direct empirical application" and operationalizing Easton's variables remained as an unfulfilled "vital, separate enterprise" according to H.V. Wiseman (1965). Basically, the model remained untestable and unverifiable.

Easton had maintained that the system was dynamic and would adapt to pressure, and thus would nearly always persist. He clearly asserts that regime change would be insufficient to bring about the end or beginning of a system, and instead should be viewed as an example of the system successfully adapting to stressful disturbances. While Easton stated that termination of the system was neither impossible nor unusual, and suggested that it would entail the system's "complete breakdown and evaporation" he can find few examples throughout history and none in the last century (Easton, 1965b, pp. 82-83). Easton's extremely broad definition of the system was responsible for this finding of persistence. As long as individuals participated in some sort of political activity, the political community, and thereby the political system would persist. Even civil wars, revolutions, or military defeats would be unable to destroy a society's political system, as long as a political community remained. (Easton, 1965b, pp. 82-85).

One drawback with Easton's overly broad and all-encompassing system design is its inability to answer many questions of interest to contemporary social scientists, such questions regarding state power and collapse. Many contemporary political scientists would most likely regard state authority as a necessary condition for the existence of a domestic political system, since only the state is able to provide a secure environment to ensure the community's survival. If considered as an essential element, it is clear that structures of state authority have indeed collapsed in some cases, such as the former Soviet Union, and have been serious compromised in many others, such as Libya, Iraq, and Somalia. On the other hand, structures of state power and their accompanying domestic political systems in successfully democratizing nations, such as South Korea, have adjusted to changing conditions and have thrived. As designed, Easton's theory is thus unable to explain the survivability of distinct domestic political systems within states, or link survivability to the nature of policy outputs or the societal groups generating inputs into the system.

The Easton system's rejection of the political science concept of 'levels of analysis' partly accounts for this inability to address the issue of survivability of domestic political systems within states. Political science scholars have long agreed with the view that "there are different ontological layers in the world," (Patomaki & Wright, 2000). Explanations of events can be found in each level of analysis, from the international system, to the state, society, government and individual. Easton's system, however, makes no distinction between factors originating in the different levels and instead includes all factors within the category of 'environment.' In particularly, he rejects the notion of the state, which others have viewed as both a type of system[1] and as a unit within the international system of states.[2]

In contrast, David Singer's systems model incorporates different levels of analysis, modeling the world political system as "a hierarchy of nested sets of subsystems, each embraced by those at the next higher level of analysis and embracing those at all lower levels." Singer (1971, pg.12)

This distinction between different levels is important, because norms and principles by which systems operate can vary drastically depending on their level. As Kenneth Waltz (1979) has explained, the way in which states behave in the international system radically differs from how individuals and groups behave within states. This divergence results from the condition of anarchy existing in the international system, which fundamentally differs from structures of hierarchy which characterize domestic political systems. In short, under conditions of anarchy in the international system, self-interested states are primarily concerned with maximizing security in order to ensure basic survival, while individual citizens within a domestic political system rely on their states' government for basic security. Cooperation between actors is therefore more easily attained within the domestic political arena than within the international system.

Easton's rejection of this distinction between international and domestic levels of analysis overlooks the competitive and dangerous conditions existing within the international system, and thus fails to consider the possibility that such an environment could threaten the survival of a state, and thereby, its domestic political system. As Neil E. Harrison (2006, pg. 10) points out, "states which consistently act in ways that are selected by their...environment as suboptimal face eradication" in the international system.[3]

The Easton model also overlooks key domestic factors at the core of the domestic political system which must function properly to enable adjustment of output to satisfy societal demands or inputs. His approach can tell us nothing about the nature of desired policy outputs,[4] and on the input side, the model lacks a theory to explain changing societal demands.

The modernization approach, in fact, offers a theory of changing societal demands and regime change which in many ways resembles Easton's description of systemic adjustment. Seymour Lipset (1960) describes how modernization, resulting from industrialization, urbanization and economic development, can raise living standards. Better educated, more literate and wealthier citizens become more willing and able to politically participate. Growing demands for public goods including political rights on the input side brings about regime change and democratization, which responds by expanding political participation and increasing output of public policy goods. Although Easton's model allows for all variety of regime change,[5] strong empirical evidence backs the modernization prediction that economic development leads to democratization.[6]

A MODIFIED 'SUBSYSTEMIC' APPROACH TO POLITICS

To persist and survive, domestic political systems must respond to changes in societal demands or inputs. To better explain and predict system persistence, this paper modifies the Easton's policy process framework, and constructs a sub-systemic model able to predict, given a certain amount of input, the form of policy output needed to maintain the system.

The first step is to modify Easton's systems approach by redefining the system as the domestic political system with clear boundaries to enable scholars to operationalize variables, generate testable predictions, and address important questions of concern related to the nature of policy output and societal inputs and the overall survivability of domestic political systems. The model must be sufficiently parsimonious with a few, clearly defined variables, to enable measurement and the testing of predictions.[7]

Systems designers face a daunting challenge in delineating boundaries to determine what belongs inside the system, and what should be placed outside within the category of environment.[8] Werner Ulrich's (1983) 'creative holism' critical systems approach provides guidelines to help scholars select boundaries which will be considered acceptable and fair. For the purpose of this academic exercise, which seeks to deepen scholarly understanding of politics, the opinion of Political Science and International Relations experts is important to determining system boundaries.

In constructing a model, this paper focuses on a key sub-systemic component of Easton's system, situated within the domestic political system of the state, relying on noted International Relations scholar Kenneth Waltz's (1959) justification of the state as an important actor which exercises monopoly over the legitimate use of coercive force within its borders. Although Easton insists that the mere existence of a community is sufficient for persistence of the system, Waltz convincingly demonstrates that state provision of security is prerequisite for the survival of any community.

While seeking to minimize the number of causal variables for the sake of parsimony, the model also recognizes the domestic political system's placement within Singer's 'overall hierarchy of nested sets of subsystems,' and the effects of the international system on domestic political systems.

The model analyzes societal inputs and demands, and predicts the nature of policy outputs from the political leader, who occupies arguably the most important subsystem within Easton's systems theory. The types of policy goods which leaders choose to distribute as outputs (private goods vs. public goods) can determine whether they maintain support and their rule. Understanding the functioning of this critical core subsystem can help to advance understanding, not only of the operations and survivability of certain regime types, but also of the state and the domestic political system as well.

DEFINING THE BOUNDARIES OF THE SUBSYSTEM

Here we examine a subsystem comprised of three primary actors: the leader, the ruling coalition, and a broader group termed the selectorate.

This subsystem sits within the larger domestic political system, which includes all of society and is bounded by the state, which in turn is nested within the international system.

Figure 1. Subsystemic model showing size of input generating group and nature of policy output

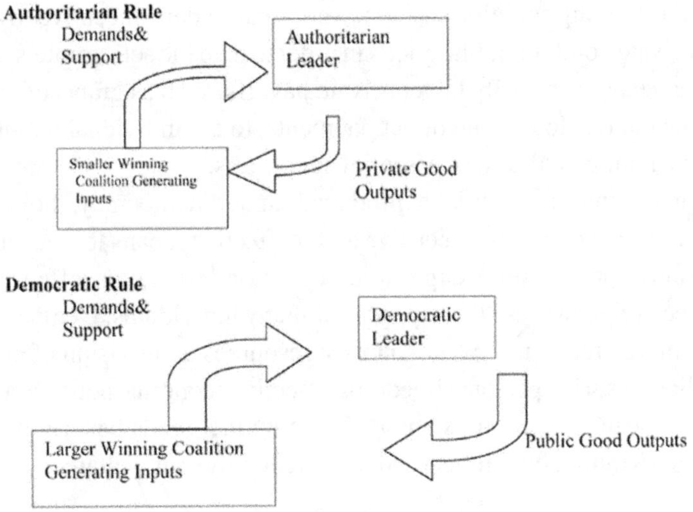

The selectorate is defined as the subset of the population that includes all those with an institutionally granted right or norm to influence the selection of the nation's leader. All other citizens of the polity are disenfranchised (Bueno de Mesquita, Smith, Siverson, & Morrow, 2003). To secure authority, a ruler must gain the support of a coalition of some minimum number of supporters within the selectorate (Roeder, 1993; Shirk, 1993). This group is termed the winning coalition.[9] As in most systems theory, the actors in this theoretical model are assumed to be rational utility maximizers. A political crisis will result if followers are not rewarded sufficiently in return for loyalty. The selectorate, therefore generates the demands, the inputs, which the leader must address in order to maintain power.

INPUTS INTO THE SUBSYSTEM

Regime type is important, however, in determining the size of the ruling coalition and selectorate to which the ruler is accountable. Authoritarian leaders normally restrict the size of the selectorate to small groups of loyalists centered on family, tribe, military ties or home region. In contrast, democratic electoral rules expand participation to all eligible voters, who use elections to hold leaders accountable. Democratic ruling coalitions, which normally include a plurality of the electorate, therefore comprise a much larger portion of the population than their authoritarian counterparts.

OUTPUTS FROM THE SUBSYSTEM

The size of the winning coalition shapes the political leader's incentives over whether to provide private goods or public goods in order to gain the selectorate's support. An authoritarian leader can easily target private payoffs, such as grants of monopolies, access to low interest loans, and direct payments, to the individual members of his/her smaller winning coalition of essential supporters.

The larger winning coalitions produced under democracy, however, make provision of private goods an expensive and ineffective means for leaders to secure support. Thinly spread over a large ruling coalition, private payoffs would be too small to effectively purchase the sup- port of many individuals. Democratic leaders are thereby more greatly motivated to invest resources in provision of public goods through policy, ensuring personal freedoms, effective economic policies, and national security as a means of gaining support. Since public goods benefit everyone in a society, larger winning coalitions produces greater overall welfare.

In fact, in studies spanning hundreds of countries and several hundred years of data, scholars have found significant correlation between the size of winning coalitions and selectorates and 31 different types of public goods, ranging from economic growth, government spending and taxation, numerous indictors of public education and public health, to civil liberties and political rights (Bueno de Mesquita, Smith, Siverson, & Morrow, 2003).[10]

APPLYING THE SUBSYSTEMIC MODEL TO THE KOREAN CASE

This paper applies this hypothesis in order to examine the relationship between expansion of the input-generating group and economic policy outputs in South Korea. After examining how democratization in 1987 expanded the Korean selectorate and ruling coalition, the paper compares the content of policy output produced under authoritarian rule with a small input generating group with policies produced by democratic leaders with larger input generating groups. The analysis gauges the allocation of private good outputs in each of the two cases.

VARIATION IN THE SIZE OF THE INPUT GENERATING GROUP

Prior to Korea's democratization, from 1980 to 1987, Chun Doo Hwan ran a brutal and ruthless dictatorship, with a very narrow ruling coalition and input generating group, reflected in the Polity Index's -8 rating of Korean political freedom. Chun relied on coercion to restrict political participation. He abolished national and regional elections and censored and shut down the press, closing 172 periodicals, and dismissing 711 journalists (Billet, 1990, pg. 300). The Chun regime barred 587 politicians from political activities, closed universities, and sent 40,000 students and protestors to be 'reeducated' in the countryside. Chun dominated the judiciary, and utilized the Korean National Security Planning Agency (NSPA) to silence dissidents through intimidation, torture and assault.[11] Chun's regime also restricted the political rights of workers by prohibiting the organizing of a union or strike without employer approval.

Chun initially restricted the size of the ruling coalition to an 81-member Legislative Council for National Security which he appointed to enact legislation.[12] Chun then revised the constitution in order to become indirectly elected President by an electoral college of 5,278 members, selected through a managed vote in which Chun's party determined the candidates (Shim, 1981). Throughout Chun's rule, then, he only needed to gain the support of a relatively small group of supporters.

KOREA'S DEMOCRATIZATION

Rapid economic development, however, had been transforming Korean society, producing broad societal demands for political participation and democratization, which was finally attained in 1987. In what became known as 'the Miracle on the Han,' South Korean GDP per capita grew from $80 in 1961 to over $5,000 in 1989, reaching $20,000 in 2007. Rapid industrialization, technological achievement, booming education, higher living standards and rapid urbanization transformed the destruction of the Korean War into a wealthy and developed OECD member country with the world's 15th largest economy.

Societal change, accompanying Korean modernization, led to growing calls for democratization. Modernization produced a large, politically active Korean middle and working class, which was joined by religious organizations, academics, students, and the media in support of democracy. Pro-democracy protests of one and 1.5 million demonstrators across 34 cities helped to force Dictator Chun Doo-hwan from power and installed a democratic government which became one of the strongest and most stable in Asia.

Korea's 1987 democratization increased the size of the winning coalition and of the selectorate, forcing leaders to respond to the demands of a much larger input generating group. Democratization expanded the size of the selectorate and ruling coalition through constitutionally guaranteed free and fair elections, universal adult suffrage, civil liberties, multi- party competition, and a free press. Following democratization, South Korea rated an 8 out of 10 according to the Polity Index, becoming the first East Asian third wave democracy to peacefully transfer power to an opposition party in 1997. Workers were guaranteed the right to form unions, bargain collectively, and strike.

The electoral system's lack of concurrent elections and single member districts, produced a string of divided governments, which strengthened the voice of the opposition party in the National Assembly, further expanding participation. Citizen participation in civil society also blossomed under democratization. The number of Korean NGOs doubled between 1996 and 1999 to 20,000, expanding to include a wide variety of citizen interests.[13] Leaders seeking power through democratic elections would therefore need to gain the support of a much larger input generating group.

VARIATION IN POLICY OUTPUT

Applying this hypothesis requires operationalizing variables, particularly the degree to which policy outputs are comprised of public vs. private goods. This paper focuses on the allocation of private goods. Examining corruption levels is one means by which to estimate the degree to which private goods comprise policy outputs, especially since the most common form of corruption is 'transactive,' entailing the provision of private goods in return for a private payoff to the political leader, usually in the form of bribes or loyalty. More overall corruption thus likely indicates increased provision of private goods.

The most common means of measuring corruption are surveys of perception of the prevalence of corruption and of direct experience with corruption. Popular perceptions of Korean corruption, as measured by the Transparency International Corruption Perception Index (CPI) gradually improved with democratization, from 3.93 CPI (1980 to 1985) under the Chun dictatorship to 3.5 to 5.7 in 2008 (Transparency International Corruption Perception Index).

Experience surveys imply an even greater decline in corruption and allocation of private goods with democratization. Those most likely to participate in corruption were the least likely to perceive graft. Whereas 51.6% of the public viewed Korean society as corrupt in October 2010 only 10.5% of government officials felt this way. (Anti-Corruption and Civil Rights Commission, 2010). TI's Global Corruption Barometer also reported a decline in the percentage of Koreans reporting giving bribes from 6% in 2004 to 2% in 2010, well below the OECD average of 8.7%.

DIRECTLY MEASURING CHANGES IN POLICY OUTPUT

Comparative statistical analysis of Korean industrial restructuring programs also shows a smaller share of private goods to public goods allocated through economic policy with expansion of the input generating group through democratization. Here we compare the provision of private goods to firms in exchange for kickbacks and loyalty through large-scale industrial restructuring programs. The Dictator Chun Doo-hwan's Industrial Rationalization program is compared with a democratic corporate restructuring program, known as the Big Deal, under democratically elected President Kim Dae-jung.

Private goods are evident in cases of 'transactive corruption' where the political leadership exchanges private favors in the form of rents, for private payoffs to the political leader, such as through bribes or political loyalty.[14] To be considered as private goods, favors cannot merely be offered as incentives to induce firms to provide public goods, such as by exporting, or improving efficiency in struggling industries through mergers and takeovers.[15]

The Korean government, in fact, insisted that industrial restructuring programs provided the public benefit of reviving unprofitable and debt-ridden industries. By selecting well run firms to acquire unviable ones, the government would encourage consolidation of excess capacity and increase productivity. Acquiring firms would be compensated with lower purchase prices (paid to creditors, not previous ownership) and financial assistance.

The criteria used by the government to select firms for receiving rents can help distinguish whether rents have been delivered as private-good favors or as compensation in return for providing public goods. Public good enhancing industrial restructuring programs should allocate rents to strong financial performers to help improve overall manufacturing productivity. On the other hand, political factors, such as bribes or elite connections, rather than financial performance, should be a factor if industrial restructuring programs have been manipulated to deliver private goods.

Rents in industrial restructuring programs can be gauged by comparing the target firm's purchase price, paid by the acquiring firm, with the market's estimation of the target firm's value. The market capitalization value (the share price multiplied by the number of shares) of both the acquiring and target firm, both before and after the public disclosure of government financial aid, can provide the market's estimate of any rents. Increases in firm value coinciding with receipt of financial aid can only be considered as rents when the firm's share price deviates from the share index of its industry share.[16]

PRIVATE GOODS WITH A SMALL RULING COALITION

The Chun dictatorship provided $8.23 billion in concessionary financial assistance through the 1985 Industrial Rationalization to 57 firms it designated to acquire others at well below market purchase prices, and often at no cost.[17] Analysis of takeover terms, purchase prices, and gains in share values reveals that acquiring firms received hundreds of millions in rents through government arranged take overs, ranging from $15 million to $600 million. Despite Chun government claims that the measure would spur recovery in the construction, shipping, steel, and textile industries by transferring ownership of unprofitable and debt-ridden operations to well managed firms, target firms outperformed the firms designated to acquire them in four of the five largest cases according to five key measures of financial performance of liquidity, profitability and stability (Schopf, 2001). Chun's self-kept record of bribes, released to the author under the Korean Freedom of Information Act, also reveals that the largest industrial rationalization rent recipients had provided the most bribes (overall and within formers to help improve overall manufacturing their respective industries).[18] Many of these large rent recipients also had marriage ties with Chun's elite. This evidence thus indicates that Dictator Chun Doo-hwan had manipulated his regime's Industrial Rationalization measure to dole out private goods in return for bribes from firms linked by family ties to his ruling elite. The small size of the input generating group made such a strategy possible.

PUBLIC GOODS WITH A LARGER RULING COALITION

Chun's manipulation of the Industrial Rationalization to reward cronies with private goods by reallocating assets and writing off loans, contrasts with the allocation of public goods through the democratically elected Kim Dae-jung government's 'Big Deal' industrial restructuring program from 1998 to 2000. Kim was responsive to a much larger input generating group, and was therefore more greatly motivated to provide, at minimal cost to the Korean tax payer, the public good of strengthening and restructuring key Korean industrial sectors suffering from excess capacity, stagnant profits, and onerous debt.

Under a deal negotiated with the largest Korean conglomerates, or 'Chaebol,' the Kim government promised financial assistance to groups that merged or swapped core affiliates with other groups in order to reduce overlap- ping capacity and increase each Chaebol's specialization of production. Unlike Chun's Industrial Rationalization, however, open and extended negotiations yielded the Chaebol far less government financial assistance than they initially sought. In fact, firms in only three of the five major swaps qualified for assistance, which was too little in two cases to raise share value during announcement of the deals or disclosure of assistance terms.

In contrast to the typical several hundred billion won debt write-offs, and forty-year combined deferment and repayment terms of the Chun's Industrial Rationalization, Big Deal firms were lucky to receive five year deferments, and instead of debt write-offs, they had to cede ownership and management control through debt-equity swaps. In fact, Hanhwa Energy, the only firm to record any appreciable gain in share value through Big Deal financial assistance, had to concede ownership rights. The absence of favors indicated that Kim had not wielded policy as a means of allocating private goods.

The government also stubbornly insisted that Chaebol seeking financial assistance first implement rigorous 'self-rescue' measures to shoulder most of the restructuring burden through disposal of redundant facilities and manpower, repayment of debt, and inducement of foreign capital. The Big Deal thus provides an example of the reduction in private goods provided through democratic industrial restructuring. Much less rent was provided than under Chun, and rent recipients were required to act in the public interest by burden sharing to qualify for assistance. This public goods policy output resulted from the democratically elected Kim Dae-jung administration's accountability to a much larger input generating group.

So, in the Korean case, economic development brought about modernization, which increased societal pressure on the input side for political participation and greater provision of public goods. The domestic political system dynamically responded through democratization, resulting in a shift of policy outputs in favor of public over private good allocation to meet the now larger winning coalitions' demands. Outputs responded to inputs, in apparent accordance with the description of Easton's model. The Korean domestic political system grew stronger and polls showed that support of Korean citizens for their government also rose following democratization (Inglehart, 2003).

THE IMPORTANCE OF THE TRANSMISSION BELT FOR SHORT TERM PREDICTIONS

Systems theories perform well in providing explanations and making predictions over the long term. Indeed, as the Korean case, and evidence across nations demonstrates, democracy has been shown to reduce corruption and the distribution of private goods over the long run (Kolstad & Wiig, 2004).

In the short term however, modified systems theory fails to explain why new democracies often experienced a temporary rise, rather than a fall in corruption. Rock (2009), in fact, discovered an inverted U pattern in the relationship between corruption and the age of democracy, also evident in Korea, in which corruption immediately rises following democratic transition, and then begins to decline after ten to twelve years.

The effectiveness of the transmission belt, the societal organizations and political institutions connecting society with the state, can influence whether the demands from the winning coalition reach the political leader.

While new democracies do possess larger selectorates, collective action problems may prevent voters in these countries from articulating their demands for anti-corruption institutional reform (Haggard & Kaufman, 1995). As Anthony Downs (1957) has argued, citizens are unlikely to take on the information costs of identifying candidates with appealing campaign platforms which offer public goods. Small, well-organized stake holder groups, such as bureaucrats, which enjoy collective action advantages (Olson, 1965) are also likely to mobilize to oppose anti-corruption reform efforts.

Fortunately, most mature, consolidated democracies possess institutionalized political parties which lower information costs through party brand names to overcome collective action barriers (Mainwaring & Scully, 1995) and mobilize civic groups and party members to break stakeholder obstructionism (Gunther et al., 1995).

Young developing democracy party systems, though, are more likely to have under-institutionalized parties with personalistic links to voters, which are less able to mobilize collective action (Mainwaring & Torcal, 2005) or pressure vested interests. Civic groups may be able to fill these roles Schmitter (2001) but they lack the membership, budgets and media presence of parties.

A further aspect of the transmission belt, the political institutional environment, and in particular, the existence of democratically elected local government, can improve the chances of civic groups to press for anti-corruption reform. Local government lowers communication costs by offering the lowest ratio and least physical distance between citizens and elected politicians (Dahl & Tufte, 1973).[19] Civic group members have the opportunity to obtain local council seats, participate in open policy deliberation (Phillips, 1996) and provide policy expertise. Weaker local government bureaucrats also offer less resistance to bureaucratic reform. Success at the local level can establish an example of bureaucratic reform which can be emulated by politicians throughout local and central government.

Korean Civic Groups Rather Than Parties, Serving as Transmission Belt

While Korea experienced a long-term improvement in corruption following democratization in 1988, Korea's Transparency International CPI score initially deteriorated, falling from 5.02 in 1996 to 3.8 in 1999, before steadily improving to remain around 5.5 from 2008 (Transparency International). Korea's World Bank's Control of Corruption score also dipped to .25 in 2000 before climbing to above 0.40 after 2009 (World Bank Governance Indicators). These gains coincided with improvement in the integrity indices of Seoul city (Seoul Integrity Index) and the Korean central government (Government of Korea Integrity Index) and a reduction in the percentage of citizens reporting bribes. (Korean government-directed Gallup surveys) (Transparency International's Global Corruption Barometer).

Korea's improved corruption and reduced bribery closely coincided with the 1999 and 2000 introduction of major anti-corruption administrative reforms by Seoul city and the Korean central government, such as the anti-corruption index, customer performance surveys, and use of e-government websites. Anti-corruption administrative reforms eliminated the structural causes of corruption, with rules which heightened transparency and increased citizen participation and competitive schemes which motivated officials to improve service and reduce graft.

Korea successfully introduced anti-corruption administrative reform in 1999 despite having personalistic (Im, 2004)[20], regional (Tan, 2000), non-programmatic, and ideologically indistinct parties (Dalton, 2008). Korean political parties did not appeal to voters on the basis of policy or ideology (Steinberg & Shin, 2006) and thus could not mobilize support for the anti-corruption reform effort.

Korea's dynamic civil society, however, which ranked as one of the largest among third wave democratizers (Salamon, 2004) and as Korea's most trusted institutions (Seong, 1999), strongly backed anti-corruption reform. The two largest groups, the Citizens' Coalition for Economic Justice (CCEJ) and People's Solidarity for

Participatory Democracy (PSPD), which were founded in 1989 and 1994, respectively (Kim, 2000), each created specialized anti-corruption divisions to push for reform. Still, membership (Ahn, 2001), budget (Choi, 2002) and media coverage[21] of the largest civic group, CCEJ, amounted to less than five percent of a major Korean political party, which made it more difficult to overcome stakeholder reform resistance.

The introduction of elected local government in 1995 was necessary for civic groups to successfully push through administrative reform. By successfully pushing local governments throughout Korea to adopt reforms like the Real Name Administrative System (RNAS), the CCEJ enabled voters to directly experience anti-corruption reform benefits, easing the way for other politicians to assemble coalitions promoting administrative restructuring. The RNAS's success led to Seoul Mayor Goh Kun's administrative reforms in 1999, including the anti-corruption index and E-government, which subsequently spread to President Kim Dae Jung's administration after 2000.

The CCEJ's anti-corruption administrative reform proposals combatted corruption by improving transparency, participation, and competition. The CCEJ's transparency-promoting proposals assigned responsibility for decisions and clarified rules and procedures to citizens. The RNAS attached official's names to government documents to assign responsibility at each stage of policy implementation (Kim, 1995), while the computerized 'Comprehensive Information Center' would provide citizens with information about their application's progress and related administrative rules, allowing citizens to identify procedural violations and proceed with grievances. By digitizing forms the 'Electronic Data Interchange System' would reduce informational asymmetry by force standardization and simplification of rules and procedures. Allowing citizens to file applications for services via computer shifted citizen-official interactions from bribery conducive intimate meetings to a transparent environment with recorded communication.

The CCEJ's participation-emphasizing measures brought outside experts into decision making and empowered citizen specialists to inquire into administrative wrongdoing. The ombudsman system established a committee of outside experts to receive appeals and investigate bureaucrats, while the Citizen's Court was comprised of outside academics, lawyers, and journalists to decide corruption-prone issues, such as the allocation of business licenses.

The CCEJ's competition-focused reforms, like the Service Improvement Reform Committee, reshaped bureaucrat incentives by offering rewards based on customer satisfaction, high productivity and citizen perception of low corruption.

The introduction of mayoral and governor elections in June 1995 provided civic groups with an exceptional opportunity to spread anti-corruption administrative reform. Civic groups pushed for administrative reform by providing policy guidance and holding policy debates for candidates in over a dozen cities, rating contenders and holding street campaigns to influence voters, monitoring elections for corruption, and unofficially running up to 1000 group members as local candidates (Kwon, 1995).

The CCEJ's efforts paid off as its administrative reforms helped reduce bureaucratic discretion and improve accountability and citizen services in newly elected local governments throughout Korea. Seoul city served as a model of RNAS implementation, while 72.6% of local governments implemented some form of information disclosure regulations (Bae, 1997). Several municipalities adopted ombudsman or citizen supervisory systems to expand independent, outside participation in decision making and oversight, introduced citizen e-mail systems to deter corruption, and implemented competitive incentive systems to award bureaucratic agencies based on service performance. A Hankuk Ilbo-PSPD survey, held one year after the 1995 local election, reported that 99.4%, 94.6%, 96% of local officials and 85.5%, 75.7%, and 72.7% of citizens felt that the newly elected governments had provided better or equal levels of citizen service, administrative credibility, and opportunities for citizen administrative participation, respectively (Lee, 1996).

Civic group local government reform success culminated in the introduction of anti-corruption administrative reform under Seoul Mayor Goh Kun, who had been elected in June 1998 on a clean governance platform. Goh, in cooperation with the CCEJ, introduced a comprehensive anti-corruption administrative reform policy package aimed at reducing bureaucratic graft through increased transparency, participation and competition.

Seoul's anti-corruption 'Integrity index,' and 'Citizen Evaluation of Administrative Service' systems gave agencies incentives to cut corruption and improve service performance. The 'Corruption Report Card Surveys' encouraged citizens to e-mail reports of improper official behavior, the 'Integrity Contract System' brought in citizen group oversight over city contract (Jang, 2001), and the weekly 'Saturday Date with the Mayor' program allowed citizens to report improprieties directly to the Mayor (Holzer & Kim, 2002).

Seoul's E-government program, "Online Procedures Enhancement for Civil Applications" or OPEN, allowed citizens to submit applications online, fill out and submit grievances, explained relevant regulations and enabled citizens to track their application in real time. OPEN's online format enabled the Audit and Inspection Bureau and Seoul City Inspector General to examine digitally recorded logs for illicit anomalies and its disclosure of licensing and permit records allowed citizens to uncover and publicize instances of favoritism. In August 1999, 86.4% of citizens responded that OPEN had improved transparency, (Kim, 1999) and following implementation of Goh's reforms, Seoul's newly developed Anti-Corruption Index improved from 64 in 1999 to 68.3 in 2000 and 70.4 in 2001. (Holzer & Kim, 2002).

The favorable response of Seoul citizens to Mayor Goh Kun's reforms lowered uncertainty for President Kim Dae Jung to adopt similar measures in the central government. The Kim administration enlisted prominent CCEJ members in top positions, and adopted several Goh Kun reforms, including OPEN and deregulation measures to increase transparency and reduce discretion, and stimulation of competition through an anti-corruption index and performance-based pay system.

The successful implementation of anti-corruption administrative reform by Korean local and central government branches from 1999 to 2004 coincided with the rapid reduction in perceived rates of bureaucratic corruption of 18%, 138%, and 30% according to Transparency International's CPI (Transparency International), the World Bank's Corruption Control indicator (World Bank Governance Indicators), and Seoul City's Integrity Index (Seoul City Integrity Index), respectively. From 2002 to 2004, the Korean central government ACRC's ACI and survey of bribery also improved by 30% and 63%, respectively. Usage of four main Korean E-government websites (Hometax, the civilian services site Minwon24, the e-procurement site KONEPS, and the e-patent site KIPOnet) were strongly correlated with Korea's improving TI CPI score from 2002 to 2014 (Schopf, 2017).

Korean civic groups, then, secured anti-corruption administrative reform by introducing bureaucratic reforms through the newly established elected local governments, which provided greater access to politicians who were more dependent on NGO policy guidance. Local government success demonstrated the benefits of reform, facilitating its spread from smaller municipal governments to Seoul and finally to the central administration. The Korean experience thus demonstrates the importance of the transmission belt, linking state and society. In Korea, as in many young democracies, the absence of institutionalized party systems posed a significant barrier to anti-corruption administrative reform, helping to explain why so many third wave democracies have struggled to reduce corruption. As the Korean case makes clear though, the combination of strong civic movements and systems of elected local government may provide an alternate means through which to transmit societal demands for public goods to democratic leaders.

A FURTHER STEP: EXPLORING THE COLLAPSE OF DOMESTIC POLITICAL SYSTEMS

Modernization theory predicts the overthrow of authoritarian leaders and regimes which fail to provide the public good of participation to politically mobilized societies. The selectorate hypothesis foretells a similar demise for leaders who fail to respond to large winning coalitions with distribution of public goods.

Numerous domestic factors, however, may preclude any type of leader in some nations from allocating public goods to meet heightened modernization-fueled societal demands. Failure to respond to these inputs can lead to civil strife, weakening the state and threatening the survival of the domestic political system in a hostile international environment. For instance, underdeveloped institutions, concentrations of economic power, and impassable societal divisions can all produce leadership unresponsive to societal demands. Future research into these issues could shed light on the determinants of resilience or weakness of domestic political systems.

Samuel Huntington (1968) described how political mobilization combined with under- developed institutions to produce strife and authoritarian reversals in many young democracies in the 1960s. Poorly performing institutions may be unable to aggregate and channel societal demands to political leaders, or may inhibit implementation of desirable government policies. Institutional legacy may provide a nation with institutions that are incompatible with local cultural norms or which are otherwise unable to meet a changing society's needs. The problem may not be easily corrected, since construction of institutions is costly and requires stakeholder agreement. Once in place, more time is required for actors to internalize and learn new rules and norms of the new institution. Future research can help to reveal if certain institutional legacies better are closely associated with resilience or the collapse of domestic political systems.

The concentration of economic resources in the hands of political leaders can also make them unresponsive to inputs from society. Michael L. Ross (2001) found that both oil and non-fuel mineral wealth impeded democratization by providing political leaders with access to economic resources independent of their societies. Oil rich regimes spent to build up internal security forces in an attempt at warding off societal pressure for democratization. As evident in the recent 'Arab Spring' however, societal pressure in many Middle Eastern nations built up to a level beyond which it could be contained. In some oil-rich Middle East nations, such as Libya and Iraq, the explosion of pent- up demands resulted in civil war, drastically weakening state authority and threatening the very existence of the domestic political system.

Demands for democratization and public goods can also be thwarted by deep ethnic, linguistic, religious sectarian division. Dankwart Rustow (1970) argued that national identity consensus was an important prior condition for democratization. Valerie Bunce (2003) also identified settling "the national question" as a central element for successful and peaceful democratic transition. Ethnically divided nations are often unable to determine who should be included in 'the people' and 'the nation.' Increased demands for participation resulting from modernization thus merely result in power struggle, deadlock, protest, demonstration, and instability. Economic and political discrimination and the denial of basic human rights, including press freedoms, civil liberties, and religious rights can lead excluded groups to adopt increasingly extreme strategies, to the point where civil war. Resulting conflicts, as in Ukraine, Georgia or Pakistan may weaken the state, even to the point of 'failure' as in Syria, Libya, Iraq, or Somalia.[22] When a state is unable to maintain a monopoly on the use of force in the type of threatening international system described by Waltz, its very existence becomes dubious. Further research can help reveal which types of sectarian division are most threatening to the persistence of the domestic political system, and help to devise strategies which to overcome this problem, such as Arend Lijhpart's (1977) consociationalist institutional designs.

CONCLUSION

Easton greatly contributed to the field of political science by providing a useful holistic framework that has increased understanding of the function of the political system in meeting societal demands with policy outputs. Easton was primarily interested in the survivability of the political system, and his model required the mere existence of community to consider the system as persisting. Communities, however, require the security provided by a state in order to survive in a hostile international environment. Hence, persistence of the domestic political system also requires maintenance of the state.

Understanding system persistence thus requires construction of a model of the domestic political system that can also explain the health of the state. Combining insights from modernization theory with a sub-systemic hypothesis focused on selectorate size and policy output, it becomes clear that in countries with growing input generating groups, the long-term survival of the state, and thus of the domestic political system as well, depends on sufficient allocation of public goods and on construction of the proper form of transmission belt to convey demands. Disruption of this critical subsystem and the limited distribution of public goods can increase long term pressure from unmet societal demands to the breaking point, giving rise to enduring instability often powerful enough to destroy both the state and its domestic political system. One of the goals of the sub-systemic model offered here is to advance understanding of the operation of the system and open up new areas of research into the survivability and persistence of the domestic political system.

REFERENCES

A megaton bomb, the circulation of the second list. (1997). *Joongang Ilbo*. (in Korean)

Abramo, C. (2005). How Far. *Transparencia Brasil*. Available online at www.tranparencia.org.br/ docs/Howfar.pdf

Ahn, J.H. (2001, January 5). CCEJ Says they won't take Corporate Funds. *Hankuk Ilbo*. (in Korean)

Alatas, S. H. (1990). *Corruption: Its Nature, Causes and Functions*. Brookfield, VT: Ashgate.

Almond, G. A., & Powell, G. B. Jr. (1966). *Comparative Politics: A Developmental Approach*. Boston: Little, Brown and Company.

Anti-Corruption and Civil Rights Commission. (2010). Retrieved from http://www.acrc.go.kr/eng/index.do

Axelrod, R. M. (1997). *The Complexity of Cooperation: Agent-Based Models of Competition and Cooperation*. Princeton, NJ: Princeton University Press.

Bae, H. S. (1997). Laws Related to Public Agency Freedom of Information Laws. [in Korean]. *Goshi Yungusa, 275*, 124–129.

Billet, B. L. (1990). South Korea at the Crossroads. *Asian Survey, 30*(3), 300–311. doi:10.2307/2644567

Bueno de Mesquita, B., Morrow, J. D., Siverson, R. M., & Smith, A. (2000). "Political Institutions, Political Survival and Policy Success," chap. In B. De Mesquita & H. Root (Eds.), *Governing For Prosperity*. New Haven, CT: Yale University Press.

Bueno de Mesquita, B., Smith, A., Siverson, R. M., & Morrow, J. D. (2003). *The Logic of Political Survival*. Cambridge, MA: MIT Press.

Bunce, V. (2003). Rethinking Recent Democratization: Lessons from the Post-communist Experience. *World Politics, 55*(2), 167–192. doi:10.1353/wp.2003.0010

Business International Corporation. (1984). *Introduction to the country assessment service*. New York: Business International Corporation.

Call, C. T. (2010). Beyond the 'failed state': Toward conceptual alternatives. In *European Journal of International Relations*. Washington, DC: US Institute of Peace. doi:10.1177/1354066109353137

Cederman, L. E. (1997). *Emergent Actors in World Politics: How States and Nations Develop and Dis- solve*. Princeton, NJ: Princeton University Press.

Chaebol and Chaebol Families. (1992). Seoul: Knowledge Industries Publications.

Chang, H. J. (2006). *The East Asian Development Experience: the Miracle, the Crisis and the Future*. St. Martin's Press.

Cho, Y. H., & Choi, B. (2004). E-government to combat corruption: The case of Seoul metropolitan government. *International Journal of Public Administration*, 27.

Chun Gathered 950 billion won. (1996a). *Joongang Ilbo*. (in Korean)

Cillers, P. (1998). *Complexity and Postmodernism: Understanding Complex Systems*. New York: Routledge.

Constitutional Court Ruling 89 Hunma 31, Third Public Report (cited as Yang vs. MOF Ruling).

Constitutional lawsuit concerning an infringement of property rights through execution of government authority: Yang Jung Moe vs. Ministry of Finance. (1993). (in Korean)

Dahl, R. A., & Tufte, E. R. (1973). *Size and Democracy*. Palo Alto, CA: Stanford University Press.

Dalton, R. J. (2008). The Quantity and Quality of Party Systems. *Comparative Political Studies, 41*(7). Retrieved from http://www.socsci.uci.edu/~rdalton/archive/CPS(2008.pdf

Downs, A. (1957). *An Economic Theory of Democracy*. New York: Harper & Brothers.

Easton, D. (1953). *The Political System*. Knopf.

Easton, D. (1965a). *A Systems Analysis of Political Life*. New York: John Wiley and Sons.

Easton, D. (1965b). *A Framework for Political Analysis*. Englewood Cliffs, NJ: Prentice-Hall, Inc.

Epstein, D. L., Bates, R., Goldstone, J., Kristensen, I., & O'Halloran, S. (2006). Democratic Transitions. *American Journal of Political Science*, *50*(3), 551–569. doi:10.1111/j.1540-5907.2006.00201.x

Goudie, A. W., & Stasavage, D. (2000). A Framework for the Analysis of Corruption. In Explaining Corruption. Elgar Reference Collection.

Government of Korea. (n.d.). *Measurement Results of the Integrity Index for Public Institutions*. Retrieved from http://kostat.go.kr/portal/korea/kor_ip/6/1/index.board?bmode=read&aSeq=309650&pageNo=9&rowNum=10&amSeq=&sTarget=&sTxt= (in Korean)

Government of the Republic of Korea, Ministry of Finance (MOF). (1988). *Reorganization of infeasible firms (a compilation of documents from the MOF, Bank of Korea (BOK), Bank Supervisory Board, and main banks).* Government of the Republic of Korea, Ministry of Finance (MOF).

Gunther, Diamandouros, & Puhle. (1995). The Politics of Democratic Consolidation. Baltimore, MD: Johns Hopkins University Press.

Haggard, S., & Kaufman, R. (1995). *The Political Economy of Democratic Transitions.* Princeton, NJ: Princeton University Press.

Harrison, N. E. (2006). *Complexity in World Politics.* State University of New York Press Albany.

Heidenheimer, A. J. (1989). Terms, Concepts, and Definitions: An Introduction. In A. J. Heidenheimer, M. Johnston, & V. T. LeVine (Eds.), *Political Corruption: a Handbook.* New Brunswick, NJ: Transaction.

Heo, U., & Stockton, H. (2005). The Impact of Democratic Transition on Elections and Parties in South Korea. *Party Politics, 11*(6), 675–688. doi:10.1177/1354068805057604

High Court of the Government of Korea. (1999). *Chun Doo Hwan trial documents.* High Court of the Government of Korea.

Holland, J. H. (1995). *Hidden Order: How Adaptation Builds Complexity.* Reading, MA: Perseus Books.

Holzer, M., & Kim, B. J. (2002). *Building Good Governance: Reforms in Seoul.* Seoul: Seoul Development Institute.

Huntington, S. (1968). *Political Order in Changing Societies.* New Haven, CT: Yale University Press.

Im, H. B. (2004). Faltering Democratic Consolidation in South Korea. *Democratization, 11*(5), 179–198. doi:10.1080/1351034041233l304642

Inglehart, R. F. (2003). How Solid is Mass Support for Democracy — And How Can We Measure It? *PS, Political Science & Politics, 36*(1), 51–57. doi:10.1017/S1049096503001689

Investigation of Chun's use of slush funds. (1996b). *Joongang Ilbo.* (in Korean)

Jang, J. C. (2001, February 16). Seoul and the Local Governments Establish the Roots of an Integrity Contract System. *Sehgeh Ilbo*. (in Korean)

Jervis, R. (1997). *System Effects: Complexity in Political and Social Life*. Princeton, NJ: Princeton University Press.

Johnston, M. (2005). *Syndromes of Corruption: Wealth, Power, and Democracy*. Cambridge, UK: Cambridge University Press. doi:10.1017/CBO9780511490965

Kang, D. (2002). *Crony Capitalism*. Cambridge, UK: Cambridge University Press. doi:10.1017/CBO9780511606175

Kim, B. J. (1995). Bringing an end to Faceless Administration through the RNAS [in Korean]. *Jibang Jachee*, *86*, 15–19.

Kim, S. H. (2000). Civic Mobilization for Democratic Reform. In L. Diamond & D. C. Shin (Eds.), *Institutional Reform and Democratic Consolidation in Korea*. Stanford, CA: Hoover Press.

Kim, S. M. (1999, August 12). OPEN is a Top 10 Korean Website. *Munhwa Ilbo*. (in Korean)

Kolstad, I., & Wiig, A. (2011). *Does democracy reduce corruption?* CMI Working Paper 4. Available from: http://www.cmi.no/publications/file/4315-does-democracy-reduce-corruption.pdf

Korea Stock Exchange Data Base. (1996). Seoul: Korean Stock Exchange. (in Korean)

Kwon, T. H. (1995, June 21). New Breeze Blowing in Elections with Citizen Group Candidates. *Hankyulleh*. (in Korean)

Lamsdorff, J. G. (2001). *Background paper to the 2001 corruptions perceptions index*. Transparency International and Gottingen University.

Lee, Y. S. (1996, July 1). One Year Survey of Citizens and Officials Regarding Local Government. *Hankuk Ilbo*. (in Korean)

Lijphart, A. (1977). *Democracy in Plural Societies: A Comparative Exploration*. New Haven, CT: Yale University Press.

Lipset, S. M. (1960). *Political Man: The Social Bases of Politics*. Garden City, NY: Doubleday & Company, Inc.

Mainwaring, S., & Scully, T. (Eds.). (1995). *Building Democratic Institutions: Party Systems in Latin America*. Stanford, CA: Stanford University Press.

Mainwaring, S., & Torcal, M. (2005). Party System Institutionalization and Party System Theory After the Third Wave of Democratization. *Working Paper #319.*

Miller, E. F. (1971). *David Easton's political theory*. The Political Science Reviewer.

Miller, W. L., Godeland, A. B., & Koshechnkina, T. (2002). Values and Norms versus Extortion and Temptation. In D. Della Porta & S. R. Ackerman (Eds.), *Corrupt Exchanges: Empirical Themes in the Politics and Political Economy of Corruption*. Baden-Baden: Nomos Verlagsgesellschaft.

Mo, J., & Moon, C. I. (2003). Business Government Relations under Kim Dae Jung. In S. Haggard, W. Lim, & E. Kim (Eds.), *Economic Crisis and Corporate Restructuring in Korea*. Cambridge, UK: Cambridge University Press; doi:10.1017/CBO9781139085083.008

National Assembly. (1988). *Investigation into Fifth Republic corruption*. National Assembly.

National Tax Office. (1989). *Confirmation of the record of the tax treatment of political donations*. National Tax Office.

Nisbett, R. E., & Ross, L. (1980). *Human Inference: Strategies and Shortcomings of Social Judgment* (p. 18). Englewood Cliffs, NJ: Prentice Hall.

Nye, J. S. (1967). Corruption and Political Development: A Cost-Benefit Analysis. *The American Political Science Review, 61*(2), 417–427. doi:10.2307/1953254

Olson, M. (1965). *The Logic of Collective Action*. Cambridge, MA: Harvard University Press.

Patomaki, H., & Wright, C. (2000). After Post- positivism? The Promises of Critical Realism. *International Studies Quarterly, 44*(2), 213–237. doi:10.1111/0020-8833.00156

Phillips, A. (1996). Why does local democracy matter? In Local Democracy and Local Government. London: Macmillan. doi:10.1007/978-1-349-25022-6_2

Przeworski, A., Cheibub, J. A., Alvarez, M. E., & Limongi, F. (2000). *Democracy and Development: Political Institutions and Material Well-being in the World, 1950-1990*. Cambridge, UK: Cambridge University Press. doi:10.1017/CBO9780511804946

Putnam, R. D. (1988). Diplomacy and Domestic Politics: The Logic of Two Level Games. *International Organization, 42*(3), 427–460. doi:10.1017/S0020818300027697

Richards, D. (2000). *Political Complexity: Nonlinear Models of Complexity*. Ann Arbor, MI: University of Michigan Press. doi:10.3998/mpub.16428

Riker, W. H. (1962). *The Theory of Political Coalitions*. New Haven, CT: Yale University Press.

Rock, M. T. (2009). Corruption and democracy. *The Journal of Development Studies, 45*(1), 55–75. doi:10.1080/00220380802468579

Roeder, P. G. (1993). *Red Sunset: The Failure of Soviet Politics*. Princeton, NJ: Princeton University Press.

Root, H. L. (1994). *The Fountain of Privilege: Political Foundations of Markets in Old Regime France and England*. Berkeley, CA: University of California Press.

Ross, M. L. (2001). Does Oil Hinder Democracy? *World Politics, 53*(April), 325–361. doi:10.1353/wp.2001.0011

Rustow, D. A. (1970). Transitions to democracy: Toward a dynamic model. *Comparative Politics*, 2(3), 337–363. doi:10.2307/421307

Salamon, L. M., Sokolowski, S. W., & ... (2004). *Global Civil Society: Dimensions of the Nonprofit Sector* (Vol. 2). Bloomfield, CT: Kumarian Press.

Samuelson, P. A. (1954). Diagrammatic Exposition of a Theory of Public Expenditure. *The Review of Economics and Statistics, 37*(4), 350–356. doi:10.2307/1925849

Schmitter, P. C. (2001). Parties Are Not What They Once Were. In L. Diamond & R. Gunther (Eds.), *Political Parties and Democracy*. Baltimore, MD: Johns Hopkins University Press.

Schopf, J. C. (2001). An Explanation for the End of Political Bank Robbery in the Republic of Korea. *Asian Survey, 41*(5), 693–715. doi:10.1525/as.2001.41.5.693

Schopf, J. C. (2017). The Governance Effects of Korea's Leading E-Government Websites. *Journal of Cases on Information Technology, 19*(2).

Seong, K. R. (1999). Civil Society and Democratic Consolidation in South Korea: Great Achievement and Remaining Problems. In *Consolidating Democracy in South Korea*. Retrieved from Seoul City Integrity Index (in Korean): http://data.seoul.go.kr/openinf/linkview.jsp?infId=OA-727

Shim, J. H. (1981). An Opposing Voice in South Korea. *Far Eastern Economic Review, 111*(12).

Shirk, S. L. (1993). *The Political Logic of Economic Reform in China*. Berkeley, CA: University of California Press.

Shleifer, A., & Vishney, R. (1993). Corruption. *The Quarterly Journal of Economics*, *108*(3), 599–617. doi:10.2307/2118402

Singer, J. D. (1971). *A General Systems Taxonomy for Political Science*. New York: General Learning.

Steinberg, D. I., & Shin, M. (2006). Tensions in South Korean Political Parties in Transition: From Entourage to Ideology? *Asian Survey*, *46*(4), 517–537. doi:10.1525/as.2006.46.4.517

Tan, A., Ho, K., Kang, K. T., & Yu, T. C. (2000). What If We Don't Party? Political Partisanship in Taiwan and Korea in the 1990s. *Journal of Asian and African Studies*, *35*(1), 67–84. doi:10.1177/002190960003500105

Tollison, R. D. (1982). Rent seeking: A Survey. *Kyklos*, *35*(4), 575–602. doi:10.1111/j.1467-6435.1982.tb00174.x

Transparency International Corruption Perceptions Index. (n.d.). Retrieved from http://www.transparency.org/research/cpi/overview

Ulrich, W. (1983). *Critical Heuristics of Social Planning*. Bern: P. Haupt.

Waltz, K. N. (1959). *Man, the State, and War*. New York: Columbia University Press.

Waltz, K. N. (1979). *Theory of International Politics*. Reading, MA: Addison-Wesley.

Wiseman, H. V. (1965). review of David Easton's "A Framework for Political Analysis," and "A Systems Analysis of Political Life" by David Easton. *The Canadian Journal of Economics and Political Science*, *31*, 4.

World Bank Worldwide Governance Indicators. (n.d.). Retrieved from http://info.worldbank.org/governance/wgi/index.aspx

ENDNOTES

1. Singer (1971, 17) viewed the state not as a solid body; it as a "coalition of all social entities at the individual, primary, and secondary levels."
2. This is similar to the concept elaborated in Putnam's (1988) two level game. In Holland's (1995) terminology, states are 'meta-agents' whose 'internal models' emerge from the interaction of domestic agents. State behavior then results from the interaction of internal model and external reality.

3 Putnam (1998) too, suggested that the state may not be able to move its internal model to accord with the reality of the international system.

4 Easton (1965b, p.89.) admits that "my approach to the analysis of political systems will not help us to understand why any specific policies are adopted by the politically relevant members in a system."

5 Easton (1965b) claims that "if a political system under stress transforms itself from a democratic to a totalitarian one …the capacity of the society to sustain some kind of political system has not been impaired."

6 Several political scientists including Richards (2000), Axelrod (1997) Jervis (1997) Cillers (1998) and Cederman (1997) have proposed complex systems models which can provide more accurate descriptions of complicated social systems, thus aiding scholars who are fishing for key causal variables. These models do not hold conditions constant, however, precluding tests of causality.

7 Przeworski et al. (2000) found that no democracy failed above a per capita GDP of $6055. Epstein et al. (2006) test the modernization hypothesis that economic development leads to democracy stands up well

8 As Almond and Powell note, a system implies "a boundary of some kind between it and its environment." This boundary enables an observer to determine what is and what is not included in the system.

9 Riker (1962) postulated that politicians will prefer to form a minimal sized coalition able to secure victory.

10 Bueno de Mesquita, Morrow, Siverson and Smith examined 196 countries, with over two hundred years of data.

11 The agency arrested and battered a Chosun Ilbo reporter in May 1983 for breaking a forced silence on the ongoing hunger strike of Kim Young Sam.

12 Legislation included the "Law on Political Purification," passed on November 3, 1980.

13 *Directory of Korean NGOs* (2000). Both 1996 and 1999 surveys were conducted by the Citizen's Newspaper, which published Directory of Korean NGOs in 1997 and 2000. Nonprofit citizen groups providing welfare services, promoting civil rights and the environment, and representing the interests of voters, taxpayers, disabled, women, and youth.

14 Syed Hussein Alatas (1990) defined 'transactive corruption' as a mutual arrangement between a donor and a recipient, actively pursued by, and to the mutual advantage of, both parties. Rents are defined by Tollison (1982) as "a return in excess of the resource owner's opportunity cost."

15 Samuelson (1954) A public good is a good that is both non-excludable and non-rivalrous in that individuals cannot be effectively excluded from use and where use by one individual does not reduce availability to others. Private goods are excludable and rivalrous.

16 Based on discussion with professionals in the Korean securities market, the study selects as a time horizon the interval from one week preceding a relevant event to two weeks afterward. Examining the interval before the event takes into account the possibility of news leaks, particularly plausible with regard to the release of financial statements, the preliminary drafts of which must receive firm approval prior to final release.

17 The aid included (at an exchange rate of 850 won per 1$) 4.19 trillion won in postponement and exemption of interest repayment, 1.64 trillion won in postponed principal repayment with normal receipt of interest, 986.3 billion won of principal written off, 460.8 billion won in new preferential rate loans, and 173.9 billion won in tax exemptions. There was no purchase price in four of the largest seven takeovers.

18 I filed a freedom of information request with the National Assembly Administrative Affairs Office on May 19, 1998, under the newly passed Korean Freedom of Information Act. My request covered several hundred thousand pages of sealed documents provided by various ministries to the National Assembly's 150th General Assembly's closed-session 38th Special Inquiry into Corruption, held in 1988. The information released to me came as one the first applications of the Korean Freedom of Information Act for classified National Assembly Documents. Materials I received under this request is noted as "FOI request materials.

19 Surveys report that citizens are more active in smaller districts and feel their participation matters more (Rose, 1996).

20 If a central party leader left one party to form another, they "gutted the organization, leaving it to wither on the vine" (Heo & Stockton, 2005)

21 The CCEJ was mentioned in 2,852 articles in the Kyunghyang and Donga newspapers vs. 70,823 for the ruling party and 56,717 for the opposition party.

22 As Charles T. Call (2010) describes, a failed state typically lacks the ability to effectively deliver basic goods and services to its population and is unable to provide security to its population or maintain a monopoly on the means of force within its borders.

Chapter 6
The "Intensification-Quality Paradox" in ICT Organizations

Solveig Beyza Narli Evenstad
Inland Norway University of Applied Sciences, Norway & Université Côte D'Azur, France

ABSTRACT

Paradoxes and system contradictions in organizations may expose employees to tensions and contradictions, which they can only partially resolve, and from which they may only partially escape. Exposure to double-bind situations leads to stress, anxiety, and other symptoms of behavioral, affective, and cognitive disturbances. When employees are caught between paradoxical injunctions and organizational defensive routines hinder metacommunication, some employees develop dysfunctional coping strategies and end up being burned out. During a phenomenological PhD research on the burnout experience of 14 ICT employees in France and Norway, a stress-inducing pathological communication pattern was discovered and named the "intensification-quality paradox." The research participants had experienced two conflicting demands as a double bind: "do more with less" and "be excellent." The employees with high standards and ideals spent excessive personal energy to do more with less while keeping the quality constant. They got increasingly exhausted, ineffective, and finally burnt out.

INTRODUCTION

There is broad consensus that the working climate has become more intense and demanding. Increased globalization, competition, and increased uncertainty of employment are identified as some of the causes for HR exploitation (Brödner &

DOI: 10.4018/978-1-5225-5996-2.ch006

Forslin, 2002). The emphasis of profitability and permanent restructuring create an organizational climate of pressure and insecurity in workers. These workers are more susceptible to stress and deterioration of their physical, psychological and mental well-being (Aubert & Pagès, 1989; Dejours, 1980). The negative side effects are stress symptoms, self-exploitation tendencies, blurring boundaries between work and private life, burnout, and eroding trust in management (Brödner & Forslin, 2002).

The antecedents of burnout are researched on widely and many organizational and personal factors are identified. Interestingly, the most influential models do not have communication as a single root factor behind burnout. In this paper, it is proposed that there are pathological communication patterns in the organizational culture that push some employees to burnout. One such pathological communication pattern is paradoxical communication and double bind in particular.

In Evenstad's phenomenological PhD research conducted in the period of 2012-2016, the role of communication behind work stress and burnout in the ICT sector in France and Norway was explored. It was found that the job-related stress factors that are used in the burnout models do not exist independently, but they arise out of *stress-inducing communication* processes (Evenstad, 2016). This research identified three major paradoxes that the research participants have experienced: '*acceleration-deceleration paradox*' (work fast in a slow organization); '*intensification-quality paradox*' ('do more with less' and 'be excellent'); '*autonomy-control paradox*' (be autonomous under the control of management). These paradoxes trapped them in a double bind and eventually led them to burnout.

The 'intensification-quality paradox' is a double bind where there are two conflicting demands on the employees: 'do more with less' and 'be excellent'. The injunction of 'be excellent' is the total quality and excellence ideal (Peters & Waterman, 1982) with a message that only the best performing organizations, those that produce the best quality goods and services, can survive, grow and be profitable. Later, with globalization and outsourcing jobs, the other injunction came along: 'resources are limited; you have to do more with less; if you give all, you will keep your job'.

This type of dominant discourses is based on hegemonic ideologies like free market capitalism, neoliberalism, technicism and managerialism. They shape the organizational and managerial discourses and even the intrapersonal discourses, i.e. the intrapersonal communication or the inner speech that takes place within an individual (Evenstad, 2016). Organizations have a tendency to take on self-sustaining, self-directing lives of their own, to evolve values and goals contrary to societal interests, to impose those values upon society and to "colonize the life-world" of the individuals they employ (Clegg & Dunkerley, 1980; Habermas, 1984). Thus, these discourses colonize individuals' life-worlds and establish a cultural control through the internalization of excellence demands.

Some employees may have already certain personality traits such as perfectionism that will exacerbate the latter. Thus, the employees are caught between the demands for achieving excellence and the demands for doing more with less, which in itself means that the quality will have to be lowered. The employees with high standards and ideals are unable to lower quality below their personal and professional standards. The situation becomes a double bind because they cannot escape the situation either. They cope by spending excessive energy in their attempts to fulfill the excellence ideal while resources diminish and demands increase. They experience mental, physical and emotional *exhaustion*, the first dimension of burnout (Maslach & Leiter, 1997). They get also increasingly ineffective and experience *reduced personal effectiveness* – the third dimension of burnout - (ibid). They burn out.

In this chapter, the paradoxes as pathological communication patterns will be explained, the particular paradox of "intensification-quality" will be elaborated, and how this double bind was experienced in the daily work of the ICT workers in Evenstad's research will be presented.

A NEW ACCELERATED WORLD OF WORK

An acceleration of the speed of life has effects on individuals' experience of time: it causes people to consider time as scarce, to feel under time pressure and stress. We are now experiencing an accelerating pace of change in everyday life that is producing less permanent relationships, a breakdown in bureaucratic structures and the emergence of new flexible organizational forms (Toffler, 1971).

Work Acceleration and Intensification

The acceleration seems to have increased over recent decades because of the "digital revolution" and the processes of globalization. Time and space barriers are disappearing (Aubert, 2003) and we are left with tyranny of instantaneity:

After having tamed space, speed attacks the time: the competitive advantage is not based on territorial conquest but the time savings. Since the advent of capitalism, there is continuous acceleration of the speed, and today a kind of limit is reached: instantaneity. This is the dictatorship of the real time.

According to Finchelstein (2011), the workers and the contemporary society will be 'sick' of the constant sense of emergency.

The contemporary individual is also multitasking, that is, performing various actions simultaneously, in order to squeeze in more and more experiences into the limited time s/he has. New technologies have accelerated communication processes. We are treating more data and documents that comes to us or we search ourselves in shorter time, which leads to cognitive overload. The individual employee has considerable responsible autonomy in terms of how, when and where the task is solved as long as it is solved by the deadline and results are delivered. *Hyperconnectivity* which refers to the use of multiple means of communication, such as e-mail, instant messaging, telephone is a new element of stress thanks to smartphones, which blur the boundaries between work and family domains (Derks, Duin, Tims, & Bakker, 2015; Makinson, Hundley, Feldhaus, & Fernandez, 2012). Additionally, smartphones allow users to extend their working hours beyond office hours. This means that the rhythms of work are becoming more individual, self-managed, in virtual teams and without boundaries with respect to time and space (Allvin, Aronsson, Hagström, Johansson, & Lundberg, 2006). Thus, flexibility provided by mobile technology is a double-edged sword. On the one hand, it provides flexibility that can be useful to accommodate work and family demands. On the other hand, increased availability increase workload.

Acceleration has brought about work intensification. According to Green (2006), work intensification has been associated with the tightening of 'porosity' of the working day as the time gaps between tasks shorten and employees have less opportunity to recover mentally and physically before the onset of the next demand. The outcome has been an increase in the psychological experience of overwork and work strain or work stress. He argues that this is a result of changed work organization, flexibility and the effects of ICT, and he concludes, '*The detrimental impact is unambiguous*' (Green, 2006: 174).

McGovern et al. (2007) have emphasized the increased bureaucratic discipline and the deleterious effects of ICT-based monitoring. This is very much in line with those critical studies of lean management, flexibility, and Just-In-Time (JIT) production, which have accelerated the flow of the work and increased the demands upon workers.

Continuous Change and Improvement

Working life has been subject to dramatic change over the past decades (Bourdin, 2008; Charpentier, 2004; Rifkin, 2011). Organizations have been forced to engage in various adaptive strategies in order to tackle new demands and remain vigorous in this unpredictable environment. Organizations have two options to become more profitable: they can either increase their profits or decrease their costs. Increasing profits through increased sales revenues may be a costly option with more research, development, design, marketing and so on so many organizations choose the cost reduction option. Companies employ continuous improvement methods to reduce costs (TQM[1], Kaizen[2]) but only major changes such as BPR, i.e., business process reengineering[3] (Davenport, 1993; Hammer & Champy, 1993), outsourcing, privatization for the publicly held companies, mergers and acquisitions lead to major cost savings.

The continuous change in organizations is inevitable, but the ability to rapidly change, responsiveness to the passing of time, the permanence of constant adaptability causes pain of all kinds: psychological, emotional, cognitive, structural and organizational (Heller, Huët, & Vidaillet, 2013; Pesqueux, 2008). Major change initiatives often result in personnel reductions through layoffs, offers of early retirement, and increased utilization of subcontracted workers (Burke & Cooper, 2000). The cost saving strategies that employ major change programs lead to 'job insecurity'. Job insecurity is defined as the perceived threat of job loss and the worries related to that threat (De Witte, 1999). Job insecurity may also refer to the possibility of some sort of degradation of status in a job without actually being made redundant, such as being transferred to a less interesting or less challenging post with the same employer.

Experiences of uncertainty concerning one's future employment are likely to have severe consequences for an employee's overall life situation in that economic and other highly valued aspects of life will be perceived as threatened (Ashford, Lee, & Bobko, 1989; Paugam, 2005; Sverke & Hellgren, 2002). Job stress is 'a perturbation to an individual, which can be connected to his position in the organization or a modification thereof, requiring adaptation of individual to new demands.' (Aubert & Pagès, 1989, p. 56).

EXPECTATIONS TO EMPLOYEES IN THE NEW WORLD OF WORK

Researchers in Information and Communication Sciences (ICS), particularly in the field of organizations, consider communication practices as 'temporalization practices' which implement social control through discourses of flexibility, adaptation, responsiveness and sustainability (Carayol, 2004). Employees are increasingly expected to be flexible, self-managing, responsibility-seeking, juggling between different roles and deal with the ever-increasing demands of doing more with less while chasing excellence ideals that are integrated to their performance goals.

Self-Managing Employees and Responsibility-Seeking Employees

Since the early 80s, it seems that we have witnessed the emergence of new forms of control. Certain control strategies have become more prominent and widespread as a means of disciplining employees (Beau, 2015; Wilkinson & Willmott, 1995). 'Cultural', 'normative' or 'clan' control is a key characteristic of the evolution of HRM in the late 80s.

During the 90s, Human Resources Management (HRM) and corporate culture were linked to the shaping of employee subjectivity and production of various subordinate or seduced selves (Du Gay, 1996; Townley, 1994). Normative cultural control is based upon establishing intense emotional attachment and the internalization of company values (Kunda, 1992). Under normative control, members act in the best interest of the company not because they are physically coerced, but they are driven by internal commitment and strong identification with company goals. This makes the constant surveillance and supervision of employees by managers, as a means of external control, redundant (Clegg, 1990; Jian, 2008).

Being Flexible and Having Many Roles

In today's organizational life, employees wear 'many hats'. These hats reflect the many roles and responsibilities they must take on. It is expected that employees embody a broader and more proactive approach to their roles in which they both own, or feel responsible for, work beyond their immediate operational tasks and recognize the importance of acquiring and using a wide range of skills and knowledge to enable them to contribute at that broader level. They need to 'own' organizational and team goals (cost, quality, lead time, customer satisfaction) and solve problems.

Organizational members not only wear more hats in today's challenging environment but they also more frequently transition between these hats over time. Role transitions may be untimely and warranted. Hence, we talk about role conflict that create stress.

Dealing With the Expectations

The question has been raised as to why workers have tended to 'go along with' these demands that have led to work intensification. Economic turbulence, crisis, recession and redundancies can contribute to workers insecurities, creating the sense that it is necessary to work harder and to accept greater performance targets in order to maintain job security. Employers develop policies that 'harness the motivation of insecurity" of *frightened workers* under repetitive downsizing (Cappelli 1999). Rosa

(2014) elaborates on the hamster wheel metaphor and proposes that employees, spinning like mad, have the feeling of not going or getting anywhere at all and they burn out at full speed (Rosa, 2014). He observes that the acceleration has detrimental effects for human beings such as the burnout.

ACCELERATION AND CHANGING MANAGEMENT PRACTICES IN THE ICT WORLD

The ever-faster computers, new operating systems, changing architectures, new programming languages, and new software packages have been flooding the ICT workplaces, in an ever-increasing speed. ICT companies have been striving with profitably and there has been strong competitive pressure from low labor cost countries with favorable tax schemes such as Ireland, India, former Eastern Europe, China, Mexico, North African countries, and Philippines. This has brought new challenges such as having team colleagues in different time zones and involuntary prolongation of office hours. In addition, social and intercultural communication skills including language skills are becoming important for collaboration with external partners overseas. As businesses demand extended hours, many ICT service companies operate with '24 hours x 7 days principle', which involves on call duty and hyperconnectivity for operational support personnel. ICT companies have accelerated their projects and processes in response to customer demands and competitive forces.

Acceleration of ICT-Projects

ICT organizations are project oriented, and project is a work form that is used for both software development and deliveries of operational or structural art such hardware upgrades, maintenance work and so on. Large ICT projects are notorious with delays and overtime work. Especially until the 90s, the system development projects were lengthy and slow with budget overruns. They followed a linear, sequential model, often called the 'waterfall model'. Projects usually worked to a fixed scope, such that when it was clear that some deliverables could not be completed, either the deadline slipped or more people were added to the project team who all worked overtime to finish the project. In a rapidly changing business world, one discovered sometimes that the 'frozen' requirements specification had already become obsolete and the system had to be amended. This caused a long list of change requests and handover to line organization was postponed sometimes several times.

In the 90s, rapidly changing business requirements demanded shorter project life cycles. The acceleration in the business world, the discourse of rapid value generation led to new methods in the ICT world, such as RAD (Rapid Application Development). RAD was followed by ASD (Agile software development) which required quality and time to be fixed, but flexibility is allowed in scope (Schwaber & Beedle, 2002). Delivering the most important features first leads to an earlier return on investment. ASD was followed by 'Lean software development' (LSD) which borrowed principles from lean manufacturing. Thus, the very discourses of acceleration in the business world have been 'imported' to the ICT world and they shaped ICT sector's socio-technical dispositifs.

When it comes to system development or other project-oriented work, project contracts are standardized and define cost, time, and quality. Project management systems track time use for each work package and report delays compared to the base line. Sanctions may be attached to delivery delays. Internal invoicing systems have been taken into use in order to capture and manage ICT employees' time use in developing, deploying, operating and maintaining systems. All these dispositifs function as a *panopticon* (Foucault, 1975) and represent 'a potential for social control' (Carayol, 2005). ICT profession has become more demanding and it requires more of individuals in terms of speed, connectivity, flexibility, commitment, and continuous learning.

Increased Responsiveness in the ICT-Services

While systems development processes were inspired by the 'time-based competition' paradigm, the need for professionalization of ICT services imposed a transformation of how ICT businesses are managed. Discourses of 'customer-orientation', 'on-time deliveries' were followed by new dispositifs such as ITIL[4] which has been implemented in most ICT organizations during the last twenty years. Service deliveries are agreed between customers and ICT companies in the form of Service Level Agreements (SLA) which define delivery standards and sanctions if the quality as defined in the SLA is not delivered.

The quality is often defined in terms of uptime, response time to critical incidents and so on. Problem management systems with strong workflow functionality have improved responsiveness of ICT departments to solve problems and help users. Automatic confirmation, logging and follow up functionality improved the communication between ICT departments and users. Time and problem management dispositifs capture time spent in solving a critical incident, escalate problems that take too long time automatically to the supervisor. Thus, visibility of poor or excellent service by teams or individuals and traceability has increased as well as surveillance possibilities. These developments have increased professionalization of operational ICT departments.

PARADOXES IN ORGANIZATIONAL CONTEXTS

The nature and function of paradox in human communication has been discussed at length (Bateson, Jackson, Haley, & Weakland, 1956; Watzlawick, Beavin, Jackson, & Alto, 1967). Paradox differs from simple contradiction because of this recursiveness. A paradoxical injunction consists in simultaneously requiring something and its opposite from an individual, which drives the individual to madness. As in the case of "Ignore this Statement", to obey is to disobey; to disobey is to obey. In work life, inconsistency in organizational conditions is likely to be reflected in incongruence in managerial communication. Managers may consciously or unconsciously send ambiguous messages, in two ways. First, they may send messages that are verbally incongruent. Examples are: '*You are running the show, however...*'; '*You make the decisions, but get the approval of...*'; '*That's an interesting idea, but pay attention...*' (Argyris, 1988). Second, managers may send messages that are verbally and nonverbally incongruent. Their nonverbal behavior seems to convey a different message than their verbal utterances. An example is a manager praising the work of an employee with a cynical tone of voice.

In some organizations, paradoxical communication becomes truly pathological (Evenstad, 2016). These organizational communication pathologies are for instance and not limited to incongruent communication such as paradoxes and 'double binds' (Bateson et al., 1956; Watzlawick et al., 1967), 'organizational defensive routines' (Argyris, 1988), and 'systematically distorted communication' (Habermas, 1984). These organizational communication pathologies create stress, anxiety, depression, burnout and sometimes suicides. According to (Deetz, 1982, p. 139):

Organizations can be filled with contradictions and formed in such a way as to alienate the individual. Decisions are manipulated, data are distorted, and power is inevitably distributed unequally. Political, economic, and community forces and individual interests are inscribed in organizational arrangements, social relations, and in every perception. Habermas, suggests, these inscribed interests result in blockages, repressions, and distorted communication. Meaning structures are filled with privileged interests, and from these perceptions of objects and events are formed.

Some common examples of contradictions, paradoxes and double binds in organizational life are as follows:

The Paradox of 'Do More With Less'

One employee said: 'Doing more with less, doing more with less. And, I'm thinking, how can I do that? Doing more will have to mean that I have to do what I'm doing

Table 1. Contradictions, paradoxes, double binds in organizational life

Be excellent	Do more with less
We expect you to be expert in your profession	But you should be polyvalent too
We expect you to be flexible and willing to change	But you should be stable and commit to stay
We value diversity	But it is important that you fit into the team
We need you to show initiative, be independent	But check with your team and your boss first
We aim for zero defects and quality problems have to be reported immediately	But you will be punished if you make a mistake
We think long term	But our profitability is measured quarterly
Have a holistic mind and don't sub optimize	But pay attention to details and focus on your task and results
Take responsibility	But your manager is the one with the decision making authority
We share knowledge and cooperate	But we have individual bonus schemes
We expect you to plan your projects rigorously	But you will be given the deadline and budget

less well'. It is frustrating, especially to employees who are highly engaged and want to deliver excellent results. These employees commonly respond by increasing their efforts, despite feeling overworked and overburdened. Finally, they end up feeling: 'The more I give the more they want. I'm giving less.' Thus, 'Do more with less' hurts, disengages and demotivates those giving most. This is the road to burnout. The individual will feel being trapped in a vicious circle or being stuck in a double bind situation.

Historically, organizations have always needed to do more with less, and they have developed many methods to improve speed, accuracy, and quality, while decreasing costs. In his 1758 essay, 'The Way to Wealth', Benjamin Franklin argued, "By diligence shall we do more with less perplexity. Sloth makes all things difficult, but industry all easy". In Franklin's essay, he was talking about making work simpler by just being focused and industrious. Jack Welch has promoted this motto, but he was thinking that it would be through innovation and intellectual capital.

Everything about this enterprise is doing more with less. OK? It needs reinvention all the time. Quality is the next step in the learning process. Getting rid of layers. Getting rid of fat. Involving everyone. All that did was to get more ideas. The whole thing here is to create the learning organization. ...You have just got to constantly focus on innovation. And more competitors. You've got to constantly produce more for less through intellectual capital.

Considering the drastic acceleration of work processes thanks to automation from factory floor to the administrative functions, there is no doubt that we are doing more with less today. The need to get the most out of every resource is everyone's responsibility for sustainability. To accomplish this, we must rely on productivity improvement, cost reduction, creativity, innovation and empowerment in all levels of the organization. Thus, in its right sense, 'more with less' implies sustainability and viability.

However, 'Do more with less' became a mantra that managers repeat to employees as an expression for continually raising the bar on goals and expectations. In the ICT sector, it means working longer hours or accomplishing more in the same amount of time. Common practices are assigning new roles, changing salary schemes to be inclusive overtime work, etc. In many work places, ICT employees who have retired or resigned are not being replaced and workload is redistributed to the remaining employees. Salespeople in ICT companies are selling operational capacity that does not exist and impose increased workloads on the operations. ICT projects are initiated and people are allocated to projects on top of their current responsibilities and given unrealistic deadlines. E-learning systems are introduced expecting employees to do the training in their free time. Before an organizational change has given the expected benefits, another change is launched giving more project work on top of daily responsibilities. At the same time, it is expected that ICT employees are delivering excellent products and services as always as promised in the Service Level Agreements.

The Demand of 'Be Excellent'

Peters and Waterman's '*In Search of Excellence*' (1982) have inspired and helped many organizations to improve the quality of their products and services. Many organizations have taken into use total quality models, such as the European Foundation for Quality Management's (EFQM) Excellence Model®. This model emphasizes 'succeeding through people', which is done by harnessing and releasing the talents of its people to produce results that lead to excellence. That is something different from what some managers seem to believe; that excellence is achieved by each employee making 'zero mistakes'. Technicist discourses such as 'lean thinking', 'lean management', 'zero waste', 'zero defects' incite us to be effective and efficient. Technicism obeys two fundamental norms, as if they are the two main commandments: technical perfection (or effectiveness) and efficiency (Schuurman, 1997). Mintzberg (1982) talks about 'cult of efficiency' and argues that efficiency has become an end in itself.

In the new world of work, the weakness is frowned upon and the performance must constantly be improved (Aubert & Gaulejac, 1991). The efficiency and effectiveness discourses colonize individuals' life-worlds and establish a cultural control through the internalization of excellence demands. For some employees, the excellence ideal is already an inner motivation factor as they may have perfectionist tendencies. Perfectionists often measure their own value entirely in terms of performing at an unrealistic level. Many researchers obviously viewed perfectionism in a domineeringly negative light (Bieling, Israeli, & Antony, 2004). The setting of high standards is an integral component of the construct of perfectionism, along with the self-assessment of success or failure in achieving one's goals. The downfall of perfectionists is that not only do these individuals set high and even unrealistic goals, but also they underrate their performance relative to those goals (see for instance *'performance-based self-esteem'*, Hallsten, 2005).

The 'Double Bind': 'Intensification-Quality Paradox'

Originally, the double bind was introduced in the 50's as a frame of reference to be used in understanding the environment surrounding a schizophrenic. Specifically, the 'environment' referred to by Bateson et al. (1956) were members of the schizophrenic's immediate family. According to Bateson et al. (1956) the 'double bind', a special case of paradoxical injunction is a necessary but not sufficient cause in the etiology of schizophrenia and other communication pathologies. The 'ingredients' of a double bind situation are proposed to be:

- Two or more persons in an important relationship;
- Repeated experience (rather than a single traumatic experience) so that the double bind pattern becomes an habitual expectation;
- A primary negative injunction ("Don't do this or I'll punish you," or "Do this and I'll punish you.");
- A secondary injunction conflicting with the first at a more abstract, commonly nonverbal level; and
- A tertiary negative injunction prohibiting the 'victim' from escaping the field. Further, it was postulated that the complete set of ingredients is no longer necessary once a person learns to perceive the universe in double bind patterns.

One can distinguish between paradox and double-bind by noting that "a paradox is what you see; a double-bind is what you're in." By definition of the construct, the parties 'bound' cannot conceptualize and resolve the paradoxical injunction in conventional fashion. Clinical, experimental and organizational research evidence

(Argyris, 1988) suggests that exposure to double bind situations negatively affects attitudes and behavior, leading to stress, anxiety, and other symptoms of behavioral, affective and cognitive disturbances.

Double Bind Organizations

Some organizations have an organizational culture where paradoxical communications, double binds are the rule rather than exception. In organizations, a double binding situation requires repeated experiences of contradictory communication within an *intense superior-subordinate relationship*. The superior's communication consists of two abstract messages: one message is an authoritative warning or order, and the other message is an equally authoritative warning or order that impinges somehow on the first. While the first message is often stated directly, the second message is often stated indirectly, nonverbally. The second message can deny the first message as well as expressing its opposite.

Such repeated experiences of contradictory messages are not sufficient to constitute a double bind situation. The subordinate must feel that he or she cannot leave the situation to escape the contradiction and that his or her efforts to clarify the communications are unsuccessful or blocked in some way. Often an employee can not simply resign and get away from the constant exposure to the paradoxical injunctions, thus it becomes a double bind situation.

Mixed messages, inconsistencies, contradictions, paradoxes like above would not necessarily cause major problems in organizations if they could be discussed and eliminated. However, a double bind organization has a culture that would not permit an authentic dialogical situation. The organization is not capable of having a dialogue, question the 'established facts', accept mistakes or that there may be ways of doing things. They are not capable of double-loop learning (Argyris, 1988). It becomes 'better' and 'better' at doing the same thing, because policies, norms, procedures are not questioned or altered.

In double-bind situations, managers at first consciously refuse to admit that there are problems. This approach is taken to avoid the consequences of having to deal with those problems. Then, they gradually start to lose sight of those problems, as if they did not actually exist. The victim of double bind will find it difficult to define the situation and will consequently find it difficult to know how to (re)act.

In this situation, the culture of the organization at the surface levels may be one of high levels of cooperation, although at its core there is a 'passive aggressive' culture with a fear of confrontation. The culture of many workplaces is permeated by pervasive, internalized and justified fear; '*the culture of fear*', '*management by fear*' are the expressions commonly used in France and recently in Norway[5] too.

One of the consequences of this is to avoid questioning the status quo, a desire not to 'rock the boat' in order to avoid being rejected by the 'team'.

Organizational Defensive Routines

Organizations with 'defensive mindset' develop *'defensive routines'* that make it difficult for their members detect and correct errors that are embarrassing and threatening. According to Argyris (1988) the fundamental rules of organizational defensive routines are to

- Bypass the errors, cover them up and even cover up the cover ups and act as if they didn't occur;
- Make the bypass undiscussable;
- Make the undiscussability of the bypass undiscussable.

Any member who tries to disclose the existence of the bypass and the organizational defensive routines in general, will reactivate the defensive routines that will be reinforced and proliferated.

Double bind organizations develop these defensive routines, which make it difficult for their members to detect and correct inconsistencies, contradictions, paradoxes, errors that are embarrassing and threatening. Individuals thus feel helpless about changing *'organizational defensive routines'* (see *'learned helplessness'*, Seligman, 1972) because the more they attempt to eliminate these routines, the stronger they will become and they will also experience being punished for trying to discuss the undiscussable. Learned helplessness is the frustration phase in the burnout process when feelings of powerlessness occur.

Organization members are likely to react by withdrawing and 'lying low'. Double binds can be extremely stressful and become destructive when one is trapped in a dilemma and punished for finding a way out. The punishment can take various forms from limiting rewards to incivility to mobbing. The culture of the double bind organization goes on as in Bateson's image: '*like a self-correcting system that has lost its governor, it spirals into never-ending, but always systematic distortions*' (Bateson, 1978, p.183). The chronic stress will eventually result in burnout for some employees.

STRESS AND BURNOUT IN THE ICT SECTOR

In this neoliberal era, everything is about competition and survival of the fastest to adapt to the turbulent markets and changing customer demands. Thus, the dominant organizational discourse is 'do more with less' and deliver excellent products and services as defined in Service Level Agreements and individual performance goals. Work overload is primarily a result of acceleration and intensification of operational work and the major cause of burnout in the ICT sector.

ICT Burnout

There have been some quantitative studies on stressors that lead to burnout in the ICT industry. Many mention high workload, role ambiguity, and role conflict as the most important causes of burnout in the ICT industry (Hetland, Sandal, & Johnsen, 2007; Huarng, 2001; Igbaria & Siegel, 1992; Li & Shani, 1991; M.K. McGee, 2006; Moore, 2000; Pawlowski, Kaganer, & Cater, 2007; Shropshire & Kadlec, 2012). According to this body of research, causes of burnout among ICT employees are very similar to the other vulnerable groups, but the burnout seems to evolve most along the '*exhaustion*' dimension. This is not surprising considering that overtime, too short deadlines, too little resources, being forced to renew competence in spare time, are rather the rule than the exception in this industry.

Despite the similarities with other industries, there are some special challenges that creates stress for ICT workers. The very nature of the profession creates unique challenges and strains for ICT workers. According to Hudson UK (2005) ICT employees have higher risk to burn out than other employees. To understand this statement, we must look more closely at the challenges that exist in the ICT industry which create stress for ICT-employees.

First of all, there is much overtime work in the ICT industry and a good deal of work must be done in adverse times. For example, it is common to make server upgrades in the evening, at night or on weekends, in order not to interfere with the day-to-day operations. Bjerke (2004, p.52) says 'a particular characteristic of the ICT industry is that there is no reasonable relation between the demands for the results and the allocated time. The employees are forced to very long work days to get the work done with an acceptable quality.'

According to Li and Shani (1991) work overload is the major source of perceived IS work stress, followed by role conflict, job-induced anxiety, and then role ambiguity. Deployment of increasingly new and complex technologies, long and odd hours, and unrealistic deadlines are among the main reasons for burnout (Igbaria & Greenhaus, 1992). According to Sethi, Barrier and King (1999), role ambiguity and role conflict are positively correlated with burnout.

According to McGee (2000) virtual-office technology, top management demands, and downsizing are some of the key factors leading to burnout. McGee argues that stress and burnout is more severe for IT professionals because they have to ensure that technology and systems work 24 hours x 7 days. Stress, job insecurity and burnout are included as antecedents of intention to leave the ICT field (Shropshire & Kadlec, 2012). Work overload, role ambiguity and conflict, lack of autonomy and lack of rewards are variables that influence exhaustion; work overload is pointed as the strongest contributor to exhaustion where insufficient staff and resources as the primary issue.

In a qualitative study, Pawlowski et al. (2007) found a strong association between project issues and perceived shortcomings of IT management/supervision. Their respondents mentioned poor management, dissension in the workplace, including conflicts with users and colleagues as both a cause of strain and an outcome of job pressures. Pawlowski et al. (2007) suggest that, 'the dominance of projects, as a way of organizing work, is perhaps one aspect that sets apart the IT profession from the generic professional context that underpins global burnout models.' Multitasking, skills issues (fear of obsolescence, keeping up with new technology), interaction with users, ineffective IT management and supervision, increased business competition, downturn in IT job market, IT offshore outsourcing, job insecurity, home-work interface were also important stressors.

The phenomenological study of the burnout experience of three burned out ICT employees in Norway (Evenstad, 2011) revealed some common characteristics of the organizational culture that generated high stress. Their organizations operated with excessive workloads, too short deadlines, understaffed departments or projects, unclear project mandates, and too many conflicting roles. Employees struggled with lack of control, competency issues, lack of social support and lack of management support.

ICT Sector in France and Their Challenges

In his report *Rapport sur les sociétés françaises d'ingenierie et de conseil en technologies*[6] Sartorius (2011) gives a sad and discomforting picture of the ICT sector in France. According to Sartorius, the general state of the industry has highlighted a hardship felt by ICT companies and which is based on a set of causes:

- General difficulties of subcontracting.
- The management of staff.
- The practices of purchase.
- The evolution of customer and supplier relations.
- The emergence of a ICT offshoring.
- A lack of recognition of their profession.
- Margin squeeze in the industry.

ICT Sector in Norway and Their Challenges

In their report *Fleksibelt for hvem? En undersøkelse av Negotias medlemmer i ikt-bransjen,* Nordrik and Bjerke (2011), once again scrutinize the ICT sector work conditions with a critical eye. Their main findings point out that

- ICT workers are working very hard to meet excessive demands for the company goals and client satisfaction.
- The management is giving (imposing) individual objectives and targets for each employee.
- Internal competition between employees is emerging and employees are busy working to reach their targets.
- Many reorganizations and downsizing processes are run by the management without involvement from the employees.

Sartorius' report gives the impression of a much harder business climate in the ICT sector than in Norway, which is a small country. Norwegian business practices are much more regulated than in France. For instance, it is customary to use standard contracts developed by the ICT Norway (IKT Norge, n.d.). Non-respect for contracts is not in the Norwegian business culture. It would be very visible in a small market and create very negative publicity.

METHODOLOGY

Research Epistemology

The epistemology of this research is interpretivism, being often presented as a variant of constructivism (Perret & Girod-Séville, 1999). Also, concepts from critical theory are mobilized, making the research approach '*critical-interpretive*' as Deetz calls it (Deetz, 1982, p. 138). The research method is qualitative. While quantitative research tradition has made significant progress in furthering our understanding of the burnout phenomena, an increasing number of scholars point out important limitations associated with this approach and they call for studies employing ideographic techniques (Pawlowski et al., 2007). The qualitative studies generate new information and give a profound account of the antecedents of burnout, as well as the interaction of personal and job factors.

Research Design

The research design is essentially based on the phenomenological analysis, which draws its theoretical origins in phenomenology and hermeneutics (Husserl, Heidegger and Merleau-Ponty). Phenomenological approaches are based in a paradigm of personal knowledge and subjectivity, and emphasize the personal perspective and interpretation. The particular phenomenological analysis that is chosen is Interpretive Phenomenological Analysis (IPA) which has an established process for analysis of qualitative data (Osborne, 1994; Smith, Flowers, & Larkin, 2009; Willig, 2001). IPA is similar to other phenomenological methods in its focus on understanding experience from the participant's perspective. However, it is unique in its focus on the individual's '*lifeworld*' and lived-experience of a phenomenon.

Participants

In IPA studies, selection of research participants tends to be purposive and broadly homogenous as a small sample size can provide a sufficient perspective given adequate contextualization. According to Smith et al (2009, p.49) 'IPA researchers usually try to find a fairly homogenous sample, for whom the research question will be meaningful' and they emphasize that, 'the primary concern of IPA is with a detailed account of individual experience'.

All participants were recruited through announcements in social media, unions organizing ICT people as well as using my personal network. Sample size is consistent with guidelines for conducting IPA studies for doctoral research and has in fact double as many participants as it is recommended. They all come from the ICT sector. Individuals were included if they worked in the ICT sector as ICT workers. Thus, administrative positions as HR manager or controller and so on were not included although some took contact. Furthermore, they were selected on the basis of having had received a diagnosis of burnout from their physician or occupational physician (France). In France, all participants were male except one; all in the age interval of 37-58 while in Norway, there were two female participants, all in the age interval of 34-52. All participants came from different regions in France, while most participants came from Oslo area in Norway.

Data Collection

'The main method to collect data in IPA studies is semi-structured interviews' (Smith, Flowers & Larkin, 2009, p.56). Fourteen semi-structured interviews were conducted. Seven burned out ICT workers in France and seven burned out ICT workers in Norway told about their experience of burnout, how they have experienced communication prior to their burnout. Interviews were done in French for French participants and Norwegian for Norwegian participants. One exception was a French participant who after some time offered to go over to English. The analysis of more than 250 pages of verbatim transcripts has resulted in many themes. In this chapter, we present the themes that relate to our topic.

Also, secondary data from official reports and statistics from EU-OSHA, OECD, governmental agencies in France and Norway, ICT sector agencies, blogs, discussion forums that ICT workers around the world use to discuss their stress and burnout as well web-sites of organizations, and articles from newspapers that were relevant to the research were analyzed.

Analysis

IPA is an '*idiographic*' methodology, examining in detail an individual case, then moving onto another, building up a rich, novel, valid and detailed understanding of a small sample. The first transcript was therefore read a number of times. At each reading, the transcript was annotated moving from initial thoughts and ideas raised by the text to more detailed, interpretative and structured coding. Interpretative comments and researcher thoughts were re-checked with the text allowing the emergence of specific themes for the first participant. This process was repeated for the remaining thirteen participants.

Once all fourteen transcripts had a list of themes, the emergent themes were clustered into super-ordinate themes that could then be linked together depending on their thematic relevance. Super-ordinate themes were then organized into main themes in a 'bottom-up' fashion. Throughout this process referral back to each of the individual texts kept interpretations close to the verbatim data. Themes were kept in the analysis according to prevalence within and across the fourteen transcripts. According to Smith et al. (2009, p.92) 'The themes reflect not only the participant's original words and thoughts but also the analyst's interpretation'.

RESULTS

The lived experiences of burnout of the research participants from France and Norway provide an exceptional portrait of the profound and rich relationship that exists between organization, culture and communication, stress and burnout among ICT workers. Each of the fourteen participants had a distinct experience and personal narrative, with a wealth of interconnecting descriptions of meaning attached to the experience. However, there were striking resemblances between French and Norwegian participants doing the same kind of work. For instance, a project manager in France and Norway had been given unrealistic deadlines and vague mission statements and they had parallel experiences in the communication processes prior to burnout. There were same type of organizational discourses, managerial discourses, paradoxes and deficiencies in the communication. Some of the ICT companies they came from were also similar with respect to culture, structure and processes although there were national culture characteristics that separated them.

The secondary data that were collected from anonymous blogs were also consistent with the research participants' experiences. An American peer doing software development in projects was telling the following sad story of his burnout (Logout before you burnout, 2018):

Burnout isn't related to selfconfidence,it is related to over work. I worked for a software consultancy, where the major client was a real whipcracker, but paid handsomely. I and my colleagues were told upfront that we had to work at least 90 hours a week, with 100 hours a week not being unusual. Being a fool I agreed to it.After three months of actually averaging 93 hours a week I was turning into a zombie. After more than two years of it I was unable to function socially. Then came a project that demanded even longer hours.

Another American peer who work with systems operations (Sysadmin burnout stories, 2018). told his story of hyper-connectivity that stressed him, shortly but sharply:

So many nights spent doing upgrades, so many weekends spent fixing problems, always on call, never on vacation without a laptop and cell phone to tether, never away from e-mail, never without a phone. Welcome to my life since 1999. Burnout was fast approaching, mismanaged and overspending on stupid shit tech company, and then I get laid off in January. Best thing ever.

The inefficient work processes, poor leadership and poor pay were mentioned by another system administration worker on the same blogpost:

I don't have a single thing that caused me burnout, just environments that nearly did. Worst one was being part of a tiny team in a high pressure environment where a culture of blame was promoted, the shocking lack of automation for even the most basic tasks meant that most of the day consisted of unnecessary work and the amount of afterhours work we had to do for no extra pay was insane.

These stories and many others are revealing the hardships in the ICT sector where expressions such as "logout before burnout", "from hero to zero" are common among

ICT workers. In the next section, the themes that concern the 'intensification-quality paradox' from the PhD research of Evenstad (2016) are presented.

Acceleration and Intensification

The research participants experienced the new world of work as accelerated and intensified. All organizational discourses valued speed and employees who were fast, flexible and responsive were favored. This acceleration manifested itself in the form of a constant time pressure, intensification of work processes, and an expectation for being connected and available all the time. This gave them a sense of work being without limits. They have experienced frequent organizational changes and short-term orientation from managers who relentlessly chased higher profits and increasing the shareholder value.

Time Pressure

Time pressure for the research participants were either in the form of unrealistic deadlines for projects and/or nagging for an operational delivery. Those who worked as project managers often struggled with '*death march projects*' (Yourdon, 2003) with unclear mandates, predetermined project deadlines, and understaffed project teams. For David, there was a pressure to establish a new service in his company during a major reorganization. He was transferred to a new post and given objectives to attain during very short time; in rather vague circumstances and without the necessary resources. For Anton, the time pressure was on an operational level for various service deliveries. He felt that he had to deliver fast, 'even before the service was ordered', as he ironically said. He complained about being in a 'firefighter modus' all the time and inability to be preventive so that he experienced going into vicious circles.

Work Intensification

The research participants experienced work intensification. As Edouard told, 'ICT companies have been employing less and less personnel, using more and more sub-contractors and offshoring.' He had filled two or more posts. Gudmund, who worked in the operations, experienced the same. He told that he was 'expected to fill three or may be four posts,' ironically adding a side comment, that 'this concerned, by the way, daytime.' Many participants also experienced that their teams were understaffed and employees who had resigned or retired were not replaced. The resources were kept constant while the workload was increased. This has resulted in an acceleration and intensification of work for those who remained. Anton told:

'You will hear 'yes, next year we will be hiring five new', for example, then, it might be that two persons have quit, and they employ two then, it's just a replacement for those two and we don't employ five new... Our department's revenues are increased each year with some millions without increasing the headcount.'

Hyperconnectivity and Work Without Limits

The research participants experienced that the work could never be finished; it was without limits. Sylvain, who had responsibility for five big accounts and spent much time on the road as a travelling technician, told about how he was constantly on the phone. He had two mobile phones and one was all the time charging while he was using the other. He said that he had to answer his clients 24h/24h. Anton referred to hyperconnectivity as 'these fancy smartphones make it possible to log into the system from anywhere, while hiking on the mountain in Norway or scuba diving in Thailand,' indicating that boundaries between free time and work time were erased.

Short-Term Orientation

Short termism or short-term orientation has been prevalent in all sectors in the accelerated new world of work with shorter business cycles, quick wins, frequent performance reporting. Pierre, who worked as a project director with an important client, experienced this short-term orientation of his management 'as a destructive chase for profits in the short term without taking into consideration the client's and his needs.' Per, who nearly burned out as a chief consultant with sales objectives to attain and obligation to invoice a certain amount of his time for consulting activities, experienced a lot of pressure. The owners had quarterly performance reviews; 'if he has sold enough, if he has invoiced enough.' Per was frustrated that they would not understand that sometimes it is necessary to work on a prospect for several more months before you can close a deal. Because of all this pressure, he ended up resigning and jumping into a job that turned out to be even more stressful so he burned out.

Continuous Organizational Change

The participants experienced also frequent organizational changes that happened because of a merger or acquisition, or because of the management's decision to change the organizational design. André who worked as a project manager in a public organization with a project partnership with EU, experienced *constant change*, sometimes because of appointment of a new director. Irene, who worked as a project manager in a publicly listed company, experienced 'continuous change with small adjustments but also a major reorganization' at the time she burned out. She was concerned about the impact of all these changes on the employees.

Rapid Pace of Technological Change

The research participants mentioned that ICT sector is very dynamic because the technologies evolve all the time and they needed to update their competency in accordance with the technological development. They felt that for the young generation this was interesting in itself but for older employees it becomes harder and harder to keep up with rapid pace of technological change. Serge, who was close to forty years old, worked with 'internet technologies that evolve every three months.' He experienced this as tiring and he longed for working with something that is stable and function. Irene as well found this 'relentless technological development' as demanding. As a project manager and female in a 'men's world', she felt that she 'had to master the technology that the team was working with, otherwise she would have serious problems.'

Paradoxical Communication

The research participants experienced paradoxical communication. Pierre was utterly confused in his discussion with his manager. It is so pathologic that Pierre has no chance of satisfying his manager and no chance of winning. This is a classic case of double bind (Bateson, 1956; Watzlawick, 1967) as Pierre could not escape the situation either.

- *"I won't say anything because I don't know what you're doing", um...the only thing I can say is that the top management says that it's not the way we should do it.*
- *Ah OK, so tell me how to do it!*
- *I don't know! Just find! Just work and find! And do what we want you to do!*
- *OK (laughs) OK, what should I do?*
- *So, just work it out, and we'll see if it will be OK, that's it.'*

Gudmund was offered a managerial position, which he took with the condition that the department would hire four technicians. When he started, the plan was reduced to two persons. Later on, it was reduced to none at all. Even later, he was ordered to fire one person. He experienced then that 'the management was saying one thing, but doing the completely opposite.' In the company strategy, this business area was a high priority, but in reality a 'no priority' at all.

Organizational Defensive Routines

Both groups of participants experienced there were organizational defensive routines that were invoked the moment they wanted to discuss their issues. These routines hinder discussion and also metacommunication (Argyris, 1986; Noonan, 2007). The research participants experienced that either they were met with silence or another form of non-response from their managers or the management team.

Silence or Non-Response on Management Level

Both the French and the Norwegian participants met silence, a non-response on management level. David felt that his questions were left unanswered by his manager when he sent him email with his questions and issues. He called him weekly and felt his manager listened: 'he listened to me absolutely, but he did nothing.' He experienced this as a lack of management support. Norwegian participants experienced that their 'emails were simply not answered.' Christian went all the way up to the CEO, personally: 'The CEO listened to me and thanked me for coming to talk with him and said nothing more.' He never heard anything back and there was no action to resolve his issues.

Delaying and Diversionary Tactics

The Norwegian participants experienced also diversionary tactics, so a decision was delayed as long as possible until the employee gave up. Marit was told that they 'understood her frustration,' and 'of course they couldn't continue like that, something had to be done, soon,' and so on. They thanked her for telling and then nothing happened. She felt like she had the 'sisyphean task.' Per who experienced bullying at work complained many times and demanded that something had to be done. Per's manager told him that 'he shouldn't bother himself too much with the bullies' and promised all the time that if it became very bad, he would do something about them, but nothing happened to the bullies.

Vicious Circle of Working Harder and Harder to Deliver Excellent Work

The research participants went into vicious circles of working harder and harder. Pierre described how he was exhausted of working harder and harder and became obsessed with avoiding mistakes so he would do things over and over again. Irene was obsessed with showing to her managers that she was good and deserved a wage raise. She told herself repeatedly that she would succeed. The demand to deliver excellent and do more with less were the common paradox for all research participants. They coped by working excessively and compulsively. They all pointed out that, at work there was an organizational culture of work overload especially on those employees who have high work capacity and high work moral.

The more you finish, the more workload they put on you. The challenge again, that I hear from many others today is that if you are willing to do much, you will be given much more work, be drawn into all sorts of projects, the problem is that you have already a 100% position, then you get roles that suggests you in reality end up with 140% position, right?

While they were pressured to do more with less, the demand for excellent delivery standards were there both from the higher-ups and from themselves too. Many research participants had high standards, perfectionist tendencies and a performance based self-esteem (Hallsten, 2005). They felt themselves valuable when they delivered excellently. Reducing the quality was not an option for them.

You will be measured on delivering in the shortest possible time, estimated time, with few resources....I think all the time that I was not performing well enough... not talented enough, this has to do with self-confidence.... Not good at saying no, not good at setting limits;... I am a very ambitious person...I am a person who sets high standards for myself...

Consequences of 'Intensification-Quality Paradox'

Work overload is primarily a result of acceleration and intensification of operational work. The research participants suffered from stress and burnout because of work overload. Their families too have suffered. They have felt that they were alienated and dehumanized.

Work-Family Conflict

Two thirds of the French participants have suffered a deterioration of their family relations due to prolonged work hours in the evening and on weekends. Only a third of the Norwegian participants complained from this issue. Sylvain felt that his marriage suffered from his absence, 'the relation ceased to exist' and finally ended up with divorce. His wife criticized Gudmund, because he was working on his computer all the time, 'even when visiting her parents.' In addition, the participants referred to a sense of lack of temporal limitations and inability to leave work behind at the office. Home office was more common among Norwegians and more appreciated as it gave flexibility to the dual career couples.

A Sense of Dehumanization and Alienation

The research participants experienced a gradual degradation of themselves as the work stress became chronic and burnout symptoms appeared. They no longer had job satisfaction, motivation, dedication or energy. For Serge, what he 'really loved once [IT work] was no longer an enriching experience.' The work had become alienating and he felt dehumanized in such a degree that he expressed it as if 'he was lobotomized.' Per had the sensation of 'being transformed into a robot' and 'being in a hamster wheel' as no matter how hard he worked he could not succeed to deliver all the work that had to be done with the few available resources in the organization. The bullying drained his personal energy at the same time.

DISCUSSION

Previous research on ICT burnout has emphasized work overload (Li & Shani, 1991). However, this is not unique to ICT sector. It is one of the major causes of burnout (Maslach & Leiter, 1997). The research participants experienced extreme work overload because of acceleration and intensification. They were expected to 'do more with less' and they were imposed too short deadlines and too few resources for their operational tasks and projects. They were expected to achieve 'excellence' through the personal performance goals that were imposed upon them. Furthermore, they had intrinsic motivation (Ryan & Deci, 2000) to work hard to achieve excellence because they had perfectionist tendencies and performance-based self-esteem (Hallsten, 2005).

In the new world work where '*cultural control*' has replaced managerial control in flat organizations, especially in Norway, employees are responsibility-seeking and self-managing. Our French participants have experienced both cultural control and direct and coercive managerial control as well as abusive supervision. All participants experienced that the paradoxical demands created a situation where they felt themselves 'trapped.' The organization or the particular manager invoked the '*organizational defensive routines*.' The lack of resources, the bullying, whatever the issues were, they were not discussable and the fact that they were undiscussable was undiscussable too (Argyris, 1986; Noonan, 2007). The paradoxes became a double bind and pathogenic.

The managerial discourse '*Do more with less*' may mean '*do more with less staff / managerial support/ wage raise*.' It means often '*Do more with less time.*'

The time pressure will accelerate business cycles, work processes, projects, and tasks resulting in higher efficiency. The time pressure takes various forms such as unrealistic deadlines for projects, too short deadline for business development, too short period of training, too short time for redoing a presentation for a boss, finishing a web-site revamp for a SMB customer, and racing against time for finishing maintenance during off the office hours.

The research participants have also suffered from frequent organizational changes and short-term orientation. We find these stress factors in the new world of work across all business sectors. What is unique for ICT sector is the acceleration of technological progress and ever increasing complexity of technology (King, 1995; McGee et al., 2000) which demands constant learning, often in free time.

'*Do more with less*' means that the quality will have to be sacrificed or lowered. The participants with high standards and ideals spent excessive energy in their attempts to fulfil this ideal. They have been '*consenting, useful* and *docile*' (Foucault, 1988) and worked very hard and very fast. They expected recognition of their efforts; however, they experienced lack of rewards and lack of fairness. The struggle for recognition (Honneth, 1996) was very much a struggle for speed. Long working hours, regular overtime, on call duty, having to be connected and available all the time, gave them a sense of 'work without limits'. They experienced work-family conflict, which we find in other sectors too. What has been striking was the way they described their sense of alienation and dehumanization.

In both groups, the speed had become the dominant discourse and formed their organization and its communication. '*Delivering all the time*' was the major goal of (work) life. Hence, the hamster wheel metaphor is very suitable as also referred by Rosa (2014). Nevertheless, we have discovered some significant differences between the two groups, which we relate to the cultural differences. All Norwegians except one complained about 'digital Taylorism'. They were mostly stressed by acceleration (fast deliveries and too tight project deadlines). They also complained about frequent organizational changes more than the French participants did.

The Norwegians were also experiencing being in a 'firefighting modus' most of the time. They perceived this as a systemic or organizational problem. By contrast, this was not mentioned by the French participants. According to Hofstede's country comparison (Hostede Insights, 2018), the avoidance of uncertainty is much lower in Norway than in France. In addition, the French are more long-term oriented. This may explain the more dynamic nature of work processes, the frequent organizational changes, and the firefighter style of Norwegians. French participants experienced time pressure as part of bullying strategies from their managers rather than as a general systemic characteristic.

The French participants appeared to have suffered more from work-family conflict than the Norwegians have. According to Hofstede, Norwegian culture is more 'feminine' with more gender equality and quality of life. In Norway, partners contribute to the tasks at home in a more egalitarian manner. Dual career families do not mind working out-of-office hours in exchange for flexible work hours that favor family life. Thus, the boundaries between work and free time are blurred in Norway. In France, often the husband play a dominant role as the financial provider and his long hours at the office is more annoying for his partner. His burnout and loss of job is also more detrimental for a French family.

CONCLUSION

Globalization and the relentless evolution of digital technologies transform the temporalities of organizations. ICT organizations are at the heart of this transformation. ICT workers are enablers of efficiency, effectiveness, and competitiveness and transform organizations, their communication, culture, and temporalities. However, ICT sector worldwide has also been transformed since 90s just like their business counterparts, through downsizing, M&A, and outsourcing. ICT companies have adopted the time-based competition principles, acceleration discourses and *dispositifs* in order to increase efficiency of their own operations. ICT companies and ICT workers became flexible, responsive, and agile. Their struggle for recognition became a struggle for speed. The demand for speed, flexibility and performativity produced pathologies such as burnout, depression and new forms of alienation (Rosa, 2012). ICT workers became the consenting, useful and docile enablers and victims of acceleration.

The study confirmed the results of previous research on burnout in the ICT sector.

Despite some differences between the French and Norwegian groups, the experience and the histories of burnout were almost similar: all the respondents were stressed by the struggle for an acceleration and intensification that became pathogenic. However, their experiences and sense-making were coloured by the national culture they belong to.

Based on this research it is proposed that ICT people have are both enablers and victims of the acceleration discourses and dispositifs (Evenstad, 2016). While they make business processes more automated, fast and efficient, they are expected to work in an ever-accelerated tempo according to Taylorist techniques and suffer from chronic stress. They implement *dispositifs* that help businesses 'do more with less' and 'be excellent' and they replicate these in their own organizations to survive in the tough global competition.

ACKNOWLEDGMENT

This chapter is based on an article (Evenstad, 2015) originally published in the International Journal of Systems and Society and later in '*Organizational Culture and Behavior: Concepts, Methodologies, Tools, and Applications*' as a chapter (Evenstad, 2017), both of them being IGI Global publications. That article used data obtained during the Master's thesis research of Evenstad, on the same subject at the ICT sector in Norway, but, with a smaller group of participants. The current article is a major extension of the original article which is enriched with newer literature and research findings, has a greater depth and most importantly, it uses data that is obtained during Dr. Evenstad's PhD study (Evenstad, 2016). As such, Dr. Evenstad will emphasize the research laboratory SIC.Lab Méditerranée, Université Côte D'Azur's role in her academic development, and particularly thank her supervisors Prof. Nicolas Pélissier and Prof. Sylvie-Alemanno. Dr. Evenstad thanks also her current employer Inland Norway University of Applied Sciences, Lillehammer, Norway for the courtesy of allowing her spend time on publishing from her previous research.

REFERENCES

Allvin, M., Aronsson, G., Hagström, T., Johansson, G., & Lundberg, U. (2006). *Gränslöst arbete – socialpsykologiska perspektiv på det nya arbetslivet*. Malmö: Liber.

Argyris, C. (1986). Reinforcing organizational defensive routines: An unintended human resources activity. *Human Resource Management*, *25*(4), 541–555. doi:10.1002/hrm.3930250405

Argyris, C. (1988). Crafting a theory of practice: The case of organizational paradoxes. Academic Press.

Ashford, S. J., Lee, C., & Bobko, P. (1989). Content, cause, and consequences of job insecurity: A theory-based measure and substantive test. *Academy of Management Journal*, *32*(4), 803–829. doi:10.2307/256569

Aubert, N. (2003). *Le Culte de l'urgence : La Société Malade du Temps*. Flammarion.

Aubert, N., & Gaulejac, V. (1991). *Le coût de l'excellence*. Paris: Éditions Du Seuil.

Aubert, N., & Pagès, M. (1989). *Occupational stress*. Paris: Klincksieck edition.

Ballard, D. I., & Ramgolam, D. I. (2011). *Connecting through Smartphones and open doors: A spatiotemporal perspective on communication technologies at work.* In The annual meeting of the NCA 95th Annual Convention, Chicago, IL.

Bateson, G., Jackson, D. D., Haley, J., & Weakland, J. (1956). Toward a theory of schizophrenia. *Behavioral Science, 1*(4), 251–254. doi:10.1002/bs.3830010402

Beau, P. (2015). Evaluer La Performance Individuelle Sur Les Resultats: De La Recherche D'Efficience Au Stress Professionnel. In Comptabilité, Contrôle et Audit des invisibles, de l'informel et de l'imprévisible. Academic Press.

Bieling, P. J., Israeli, A. L., & Antony, M. M. (2004). Is perfectionism good, bad, or both? Examining models of the perfectionism construct. *Personality and Individual Differences, 36*(6), 1373–1385. doi:10.1016/S0191-8869(03)00235-6

BjerkeP. (2004) Retrieved from http://www.de-facto.no/page/de-facto-kunnskapssenter-for-fagorganiserte/publikasjoner/bestille-publikasjon/det-nye-arbeidslivet/index.html

Bourdin, S. (2008). Approche communicationnelle du changement organisationnel: Organisations pédagogiques, TICE et situations de communication. *Science and Society*, (74): 123–140.

Brödner, P., & Forslin, J. (2002). *O tempora, O mores! Work intensity-why again an issue. In Creating Sustainable Work Systems. Emerging Perspectives and Practice* (pp. 15–26). London: Routledge.

Burke, R. J., & Cooper, C. L. (2000). *The organization in crisis: Downsizing, restructuring, and privatization.* Blackwell Publishing.

Cappelli, P. (1999). *The New Deal at Work: Managing the Market-Driven Workforce.* Boston, MA: Harvard Business School Press.

Carayol, V. (2004). *Communication organisationnelle: une perspective allagmatique.* Paris: L'harmattan.

Carayol, V. (2005). Principes de contrôle, communication et temporalités organisationnelles. *Études de Communication.*

Charpentier, P. (2004). La gestion du changement dans les organisations. *CAHIERS FRANCAIS-PARIS*, 29–36.

Clegg, S. (1990). Modern organizations: Organization studies in the postmodern world. *Sage (Atlanta, Ga.)*.

Clegg, S., & Dunkerley, D. (1980). *Organization, class and control*. Taylor & Francis.

Davenport, T. H. (1993). Reengineering the corporation. *MIT Sloan Management Review*, *35*(1), 103.

De Witte, H. (1999). Job insecurity and psychological well-being: Review of the literature and exploration of some unresolved issues. *European Journal of Work and Organizational Psychology*, *8*(2), 155–177. doi:10.1080/135943299398302

Deetz, S. A. (1982). Critical interpretive research in organizational communication. *Western Journal of Communication*, *46*(2), 131–149. doi:10.1080/10570318209374073

Dejours, C. (1980). *Work: Mental wear test psychopathology of work*. Centurion.

Derks, D., Duin, D., Tims, M., & Bakker, A. B. (2015). Smartphone use and work–home interference: The moderating role of social norms and employee work engagement. *Journal of Occupational and Organizational Psychology*, *88*(1), 155–177. doi:10.1111/joop.12083

Du Gay, P. (1996). Consumption and identity at work. *Sage (Atlanta, Ga.)*.

Dupuy, F. (2004). *Sociologie du changement: pourquoi et comment changer les organisations*. Dunod.

Evenstad, S. B. N. (2015). "Be Excellent and Do More with Less": A Paradox behind Job Burnout Threatening Organizational Sustainability. *International Journal of Systems and Society*, *2*(2), 52–67. doi:10.4018/IJSS.2015070104

Evenstad, S. B. N. (2016, December 12). *Pour une approche communicationnelle du burnout les apports de l'Analyse Phénoménologique Interprétative. Le rôle des communications stressantes dans l'épuisement professionnel; une analyse comparée France -Norvège de groupes d'informaticiens*. Retrieved January 20, 2017, from http://www.theses.fr/s170697

Finchelstein, G. (2011). *La dictature de l'urgence*. Fayard.

Foucault, M. (1975). *Discipline and Punish: The Birth of the Prison* (A. Sheridan, Trans.). Harmondsworth, UK: Penguin Books.

Foucault, M. (1988). Technologies of the Self. In *Hutton* (pp. 16–49). Amherst, MA: The University of Massachusetts Press.

Frykt Kultur Google Search. (2018). Retrieved from https://www.google.no/search?q=fryktkultur&ie=utf-8&oe=utf-8&client=firefox-b-ab&gfe_rd=cr&ei=4qknWKTSGMmq8wfi9bXwAQ

Green, F. (2006). *Demanding work: The paradox of job quality in the affluent economy*. Princeton University Press.

Greenwood, J. (1997). *The third industrial revolution: technology, productivity, and income inequality*. American Enterprise Institute.

Habermas, J. (1984). *The theory of communicative action; Volume one: Reason and the rationalization of society*. Boston, MA: Beacon Press.

Hallsten. (2005). Performance-based self-esteem: A driving force in burnout processes and its assessment. *Arbete och hälsa vetenskaplig skriftserie*.

Hammer, M., & Champy, J. (1993). *Reengineering the corporation. 1993*. New York: HaperCollins.

Heller, Huët, & Vidaillet. (2013). Introduction. Du sens et du rôle de la critique. In *Communication et organisation: Perspectives critiques*. Presses du Septentrion.

Hetland, H., Sandal, G. M., & Johnsen, T. B. (2007). Burnout in the information technology sector: Does leadership matter? *European Journal of Work and Organizational Psychology*, *16*(1), 58–75. doi:10.1080/13594320601084558

Honneth, A. (1996). *The Struggle for Recognition: The Moral Grammar of Social Conflicts*. Cambridge, MA: MIT Press.

Howard, A. (1995). A framework for work change. *The Changing Nature of Work*, *3*, 44.

Huarng, A. (2001). Burnout Syndrome among Information System Professionals. *Information Systems Management*, *18*(2), 1–6. doi:10.1201/1078/43195.18.2.200 10301/31272.3

Igbaria, M., & Greenhaus, J. H. (1992). Determinants of MIS employees' turnover intentions: A structural equation model. *Communications of the ACM*, *35*(2), 34–49. doi:10.1145/129630.129631

Igbaria, M., & Siegel, S. R. (1992). The reasons for turnover of information systems personnel. *Information & Management*, *23*(6), 321–330. doi:10.1016/0378-7206(92)90014-7

IKT Norge. (n.d.). *About ICT Norway*. Retrieved from https://www.ikt-norge.no/english/

Insights, H. (2018). *Compare Countries*. Retrieved from https://geert-hofstede.com/countries.html

Jian, G. (2008). Identity and technology: Organizational control of knowledge-intensive work. *Qualitative Research Reports in Communication*, *9*(1), 62–71. doi:10.1080/17459430802400365

Kamp, A., Lambrecht Lund, H., & Søndergaard Hvid, H. (2011). Negotiating time, meaning and identity in boundaryless work. *Journal of Workplace Learning*, *23*(4), 229–242. doi:10.1108/13665621111128655

King, J. (1995). Stress rattles "help!" desks. *Computerworld*, *29*, 1.

Kunda, G. (1992). *Engineering culture: Control and commitment in a high-tech organization*. Philadelphia: Temple University Press.

Li, E. Y., & Shani, A. B. R. (1991). *Stress Dynamics of Information Systems Managers: A Contingency Model*. Polytechnic State University.

Logout before you burnout. (2018). Retrieved from https://www.reddit.com/r/programming/comments/ua27q/logout_before_you_burnout/

Makinson, P., Hundley, S., Feldhaus, C., & Fernandez, E. (2012). Mobile communications anytime, anywhere: The impact on work-life balance and stress. In *2012 Frontiers in Education Conference Proceedings* (pp. 1–6). IEEE.

Maslach, C., & Leiter, M. P. (1997). *The Truth About Burnout: How Organizations Cause Personal Stress and What to Do About It*. San Francisco, CA: Jossey-Bass.

McGee, M. K. (2006). Time for a career change. *Information*.

McGee, M. K., Khirallah, D. R., & Lodge, M. (2000). Backlash. *Information Week*, *805*, 50–56.

Mintzberg, H. (1982). A note on that dirty word "Efficiency." *Interfaces*, *12*(5), 101–105. doi:10.1287/inte.12.5.101

Moore, J. E. (2000). One Road to Turnover: An Examination of Work Exhaustion in Technology Professionals. *Management Information Systems Quarterly*, *24*(1), 141–168. doi:10.2307/3250982

Noonan, W. R. (2007). *Discussing the undiscussable: a guide to overcoming defensive routines in the workplace*. San Francisco, CA: Jossey-Bass.

Nordrik, B., & Bjerke, P. (2011). *De-Facto, Oslo*. Retrieved from http://www.de-facto.no/wp-content/uploads/2015/06/193927fleksibelt-for-hvem.pdf

Osborne, J. W. (1994). Some similarities and differences among phenomenological and other methods of psychological qualitative research. *Canadian Psychology*, *35*(2), 167–189. doi:10.1037/0708-5591.35.2.167

Paugam, S. (2005). *Les formes élémentaires de la pauvreté*. Presses universitaires de France Paris.

Pawlowski, S. D., Kaganer, E. A., & Cater, J. J. III. (2007). Focusing the research agenda on burnout in IT: Social representations of burnout in the profession. *European Journal of Information Systems*, *16*(5), 612–627. doi:10.1057/palgrave.ejis.3000699

Perret & Girod-Séville. (1999). *Fondements épistémologique de la recherche*. Thiétart.

Pesqueux, Y. (2008). Un modèle organisationnel du changement ? *Communication et organisation*, *33*.

Peters, T. J., & Waterman, R. H. (1982). *In Search of Excellence: Lessons from America's Best-Run Companies*. New York: Harper & Row.

Philippe, B. (2004). *Sociologie du changement dans les entreprises et les organisations*. Le Seuil.

Rifkin, J. (2011). *The third industrial revolution: how lateral power is transforming energy, the economy, and the world*. Macmillan.

Rosa, H. (2012). *Aliénation et accélération. Vers une théorie critique de la modernité tardive*. La: Découverte.

Ryan, R. M., & Deci, E. L. (2000). Self-determination theory and the facilitation of intrinsic motivation, social development, and well-being. *The American Psychologist*, *55*(1), 68–78. doi:10.1037/0003-066X.55.1.68 PMID:11392867

Sayah, S. (2013). Managing work–life boundaries with information and communication technologies: The case of independent contractors. *New Technology, Work and Employment*, *28*(3), 179–196. doi:10.1111/ntwe.12016

Schuurman, E. (1997). Philosophical and ethical problems of technicism and genetic engineering. *Techné: Research in Philosophy and Technology*, *3*(1), 27–44. doi:10.5840/techne19973111

Schwaber, K., & Beedle, M. (2002). *Agile software development with Scrum* (Vol. 1). Prentice Hall Upper Saddle River.

Seligman, M. E. P. (1972). Learned helplessness. *Annual Review of Medicine*, *23*(1), 407–412. doi:10.1146/annurev.me.23.020172.002203 PMID:4566487

Shropshire, J., & Kadlec, C. (2012). I'm Leaving the IT Field: The Impact of Stress, Job Insecurity, and Burnout on IT Professionals. *International Journal of Information and Communication Technology Research, 1*.

Smith, J. A., Flowers, P., & Larkin, M. (2009). *Interpretative Phenomenological Analysis: Theory, Method and Research*. London: Sage Publications Ltd.

Sverke, M., & Hellgren, J. (2002). The nature of job insecurity: Understanding employment uncertainty on the brink of a new millennium. *Applied Psychology*, *51*(1), 23–42. doi:10.1111/1464-0597.0077z

Sysadmin Burnout Stories. (2018). Retrieved from https://www.reddit.com/r/sysadmin/comments/tmc28/sysadmin_burnout_stories/

Toffler, A. (1971). *Future Shock. Bantam.*

Townley, B. (1994). *Reframing Human Resource Management: Power, Ethics and the Subject at Work*. London: Sage.

Watzlawick, P., Beavin, J. H., Jackson, D. D., & Alto, M. (1967). Pragmatics of Human Communication. A Study of Interactional Patterns, Pathologies, and Paradoxes. WW Norton.

Wilkinson, A. J., & Willmott, H. (1995). *Making quality critical: New perspectives on organizational change*. Routledge.

Willig, C. (2001). Interpretative phenomenology. *Introducing Qualitative Research in Psychology: Adventures in Theory and Method*, 50–69.

Yourdon, E. (2003). *Death march*. Pearson Education.

ENDNOTES

1 Total quality management (TQM) consists of organization-wide efforts to install and make permanent a climate in which an organization continuously improves its ability to deliver high-quality products and services to customers. TQM enjoyed widespread attention during the late 1980s and early 1990s before being overshadowed by ISO 9000, Lean manufacturing, and Six Sigma.

2 Japanese for 'improvement.' When used in the business sense and applied to the workplace, *kaizen* refers to activities that continuously improve all functions and involve all employees from the CEO to the assembly line workers.

3 BPR is the fundamental rethink and radical redesign of business processes to achieve dramatic improvements in cost, quality, service, and speed.

4 ITIL, formerly an acronym for 'Information Technology Infrastructure Library', is a set of practices for 'IT service management' (ITSM) that focuses on aligning ICT services with the needs of business.

5 The term 'fryktkultur' meaning 'culture of fear' returns 19000 hits with articles on pilots, hospitals, police and so on. (Frykt Kultur Google Search, 2018).

6 See http://www.ladocumentationfrancaise.fr/var/storage/rapports-publics/114000092.pdf

Chapter 7
Expanding the Discourse on Organizational Inquiry, Power and the Metaphor Commodity

Frank Stowell
University of Portsmouth, UK

ABSTRACT

Models of organization are unsatisfactory as they assume a certain level of predictability, but research shows that each organization is unique and cannot be encapsulated by a single ontological model. Organizations exist in a turbulent environment and must change to survive, but change creates tensions that threaten its stability. Those who make up the organization cooperate only when it is in their interest to do so; in other words, it is far from being predictable. Individuals and groups use their power in an attempt to shape outcomes into one that they can accept. The way that power is used can have a significant effect upon the decisions made about the way the organization maintains its relationship with its environment. To enhance our understanding of the way organizational power is used, the notion of "commodity" as a metaphor for power is introduced as a means of surfacing the way in which individual power is used.

BACKGROUND

This chapter is an updated version of the paper published in IJSS Vol 1. No1 in 2014 and provides an opportunity to place the notion of organizational power, the subject of the paper, into context. First, we describe the Systems perspective upon which this argument is based; second we describe what we mean by organization

DOI: 10.4018/978-1-5225-5996-2.ch007

and then commodity as a metaphor for power. Finally a short section outlining the current research activities associated with the ideas expressed in this chapter

Organizational intervention is problematic as each inquirer has a different notion of what an organization is and in order to make sense of it the inquirer adopts a model or concept of 'organization'. In this paper the author revisits the way in which we set about organizational inquiry and how we can gain a greater appreciation of power and how it might be used. It is argued that a once and for all model of organization is deficient as organizations exist in a turbulent environment and must constantly change to survive. Change may bring about resistance from those most affected. The notion of Commodity as a metaphor for 'soft power' is introduced as a way of surfacing how it is used. By thinking in terms of commodity the inquirer gains insight into the effects of power within the group and opens the way for a wider range of discussions to take place that may bring out hidden and imagined manifestations of power. Understanding the way that power manifests might play a significant part in developing management strategy.

SYSTEMS AS A META DISCIPLINE

What Is an Organisation?

Organizational enquiry means intervening into it in some way, but the term 'organization' is vague with no universal definition. Organisation often relates to an entity of some kind that is given an identity by the observer. In order to address this potential difficulty we will adopt, in this paper, the Systems approach. The Systems epistemology is a meta-discipline; a way of thinking about the world unfettered by the need to adopt a particular 'discipline' driven model.

A 'systems approach' is an attempt to think in a more holistic way than does Natural Science. Its justification lies in the real existence of emergent properties. The two ways of using systems ideas (properly defined, not as in everyday language) namely hard (seeing the world as systemic) and soft (seeing the process of inquiry into the world as systemic) broadly imply two philosophical traditions: positivist allied to a sociology of functionalism, and phenomenological allied to a sociology of interpretivism, (Stowell-Checkland; U Tube 2012).

Defining a particular 'System' is problematic because everyone has a general, uninspected, idea of what organisation means; consequently it is exceptionally difficult to pin down a sharp definition. It is because each individual has a 'choice' that we need to remind ourselves of the importance of his or her world-view, or Weltanschauungen.

The Weltanschauung that is adopted by an observer determines where the observed 'system' starts and ends (for them), as what we might call a system may not be the same for someone else as it is for us. For us a 'System' is composed of a mixture of recollection and perception; it is eidetic. Attempts to describe what a particular System is immediately opens up the complexities of thinking systemically. Beer (1979, pp7-11) suggests that a System has to 'have a purpose', but adds, 'who says so?' His answer is that 'it comes from you, the observer of the system'. Churchman, (1971, p42) was of a similar view '…whether or not something is a system is regarded as a specific choice of the designer.' Capra refers to the 'fluidity' of the informal networks of communication that are an essential part of human organizations. These networks '…are embodied in the people who engage in the common practice. When new people join, the entire network may reconfigure itself…' (Capra, 2014, pp318). In other words the organization is not a fixed model but socially determined, '…The shared practice creates flexible boundaries of meaning that are often unspoken.' (Capra, 2014, p317). Patočka (2016, p53) provides something more acceptable when he speaks of a system as something, which merely enhances the '…clarity of the written account'. In a similar way referring to something as a system gives 'shape' to the phenomenon with which we are concerned: an organization could be said to be one such system.

Although this chapter is not about the complexities of boundary and Weltanschauung we should acknowledge that those who are a part might view any 'system' from different perspectives. For example, there is its formal manifestation identified by the name by which it is known and an informal one which might be meaningful to an individual. Imagine a university called "*Camford*", for this university there is both a formal definition and an individual's personal experience of it. A generic definition of a university might be, "An institution of higher education, offering courses and research facilities in mainly non-vocational subjects and having acknowledged powers and privileges, especially that of conferring degrees" (Shorter Oxford Dictionary, 2007), but it is also a place where people from different backgrounds and experiences meet and integrate (to varying degrees) into a social community. It is being a part of a social community that is often the thing that will define *their* university in a more personal way. Never-the-less if we are to intervene into an organisation the inquirer needs a model of some kind that allows helps them to understand what it is they are trying to investigate. But how does the inquirer begin to make sense of an organisation that has many dimensions? Checkland and Holwell (1998) "suggest that …in order to bring more clarity ….it does seem important to declare a defensible concept or model of organisation…" (Checkland & Holwell, 1998, p79 -84).

If we think of an organisation as an entity we can describe it as a community bound together in a contractual sense where the social relationships are individualistic and impersonal. Tönnies referred to this kind or relationship as Gesellschaft (Harris,

2009). This kind of instrumental relationship is in contrast to the informal relationship we feel as part of our family or the close-knit community to which we belong. These relationships he described as Gemeinschaft. Here we have relationships built on a sense of belonging, being a part of a tribe or community. Truzzi (1971) describes the equilibrium in Gemeinschaft as being "…achieved through morals, conformism, and exclusion - social control - while Gesellschaft keeps its equilibrium through police, laws, tribunals and prisons. … Rules in Gemeinschaft are implicit, while Gesellschaft has explicit rules (written laws)". But experience shows us that human beings do not always act in a rational way. Tönnies argued that "…Nothing happens in Gesellschaft that is more important for the individual's wider group than it is for himself. On the contrary, everyone is out for himself alone and living in a state of tension against everyone else" (Harris, 2009, p52). Checkland and Holwell, added to this view by arguing that an organisation is not "...simply a rational machine whose members willingly combine together to pursue organisational goals" (Harris, 2009, p80). This raises questions about the process of goal assimilation, adding support to the argument that we should assume that whilst its members appear willingly to conform it may be only when it serves their particular ends.

It is against this background of uncertainty about the nature of an organisation emerged a generalised concept from Lancaster University's 30 year programme of Action Research (see Checkland, 1999). The research was undertaken in *real* organisations resulting in more than 300 projects that contributed to the notion of organisation and how it might be researched. The kernel of this overall concept found, beneath the multiple concepts that can be used to describe organisation, that there are three elements:

1. A social collectivity; showing-
2. Multiple worldviews and worldviews changing over time.
3. Would-be purposeful activity

This observation provides a source of immense and changing complexity in any real world example of an organisation. The full complexity of the way in which it exists could never be described in detail, which means that only epistemological statements about organisations are justified.

FROM ONTOLOGY TO EPISTEMOLOGY

The lessons learnt from the Action Research programme at Lancaster marked a move from ontology to epistemology and the so-called hard/soft paradigm shift in systems thinking. Researchers and practitioners involved in the programme recognised that the complexity of views and aspirations of those that make up an organisation meant that intervention to achieve intended change can only be undertaken via a process approach; i.e. by defining an epistemology that can be used to investigate, explore and inquire into real world organisations.

Experience of organisational intervention reveals that change can take place at three levels. The first is Structural, that is to say changes to reporting and functional structures; the second is Process, which relates to the dynamic elements of reporting and informing, and the third, possibly the most difficult, is Attitude (Checkland, 1999, pp180-181).

An outcome from the Lancaster programme showed that in developing the required process approach it was necessary to invent ways of finding out about the real world situation that was addressed (see Stowell & Checkland, 2012, Interview, pp53-59). The question of how to gain understanding from the organisational experience by being part of that experience rather than that of an observer, exercised thinking about the method of inquiry itself. These thoughts initially comprised [i] giving attention to the intervention (which will affect the situation in question) and [ii] to the nature of the culture that is experiencing the intervention. The outcome of this was the emergence and evolution of what has become known as Soft Systems Methodology (Checkland, 1999).

Models of organisational intervention can predefine the kind of outcome from the study. Employing, a Systems Engineering approach will lead to an outcome couched in the terms of how to solve the problem taking for granted *What* that problem is. The approach is carried over to the method employed that might act as barrier between the inquirer and the subject of the inquiry. This brings us to a second and the significant contribution to organisational research made by the Lancaster programme, that is the impact of the method of inquiry itself. Studies had shown that the presence of the investigator in an intervention changes the behaviour of those that work within, "…The very act of setting up the requirements of experimental control on the lines of natural science within a business organisation may curtail 'normal' patterns of behaviour and consequently deny the possibility of anything more than limited organisational understanding" (Stowell 1989, pp 85-86), and "…cause the participants to change their behaviour and conversations so that, as a result, the researcher is not observing the natural setting, but one which is disturbed by the researcher" (McKinnon, 1988). The difficulties of observational research is well-documented elsewhere (e.g. Giddens 2001, C 20; Checkland, 1981, C5) and underlines the effect that such activities can have upon staff consequently adding to the difficulties of gaining insight into an organisation. For example Stowell, A (2010) found during some extensive field research, involving several organisations, that "The method used in the pilot study seemed to have created a barrier between the researcher and the participants…" (Stowell, 2011, p120). To overcome this she changed to a softer approach, which she found did not get in the way of learning about the enterprise. Whilst the Lancaster research showed that ways of finding out about the impact of the method to be fruitful it did not encompass one part of organisational culture, namely the nexus of power and politics found in any organisation. The intentions and expectations of the evolving methodology as being capable of radical change could be criticised as being a vehicle to preserve the status quo. Checkland himself acknowledged that the desire for radical change."…will be limited by the desire to achieve action in the real world….has led to a rather conservative use of the methodology" (Checkland, 1981/1999, p281).

It was clear that radical solutions to a problem situation generated solely by the practitioner without the cooperation of those most affected may well be resisted in favour of the status quo. A second difficulty is that if a practitioner hoped to continue to work for the client then the chances are that any recommendations that are made, if they are to be accepted, do not rock the boat. It was realised that any assessment of SSM made from the results of past projects could not, with any degree of confidence, establish if they were due to practitioner conservatism or a deficiency in SSM.

A second criticism of SSM referred to the question of its neutrality and through that its potential for radical change. Critical theorists claimed that SSM was a methodology that would always defer to authority. For example, Jackson, claimed that "…the methodology is culpable in that it is prepared to accept for implementation changes emerging from a false consensus produced by distorted communication." (Jackson, 1982, p25). Leaving to one side the improbability of proving any human communication to be 'undistorted', these were serious charges against a methodology that promoted the notion of neutrality and as being capable of engendering radical change. The response to this criticism was to point out that SSM provided a platform for debate whereas Critical theory, from which Jackson's comments emerge, seeks to engender change through advocating the process of social interaction to enable the social actors to see themselves and their social situation in a new light. It was thought possible that by creating a platform for debate it would create circumstances in which alternative worldviews were explored. Never-the-less the success of the debate was recognised as being a function of the skills and persuasion of the practitioner. It was realised that the debate could be enhanced if the agenda could include items of self-awareness and emancipation in the widest sense. Important influences such as organisational power and the effects that power might have upon the individual, the group and at an organisational level had to be made explicit and part of the discussion if real change was to be enabled.

ORGANISATIONAL POWER

The literature of organisational power reveals a number of differing views but in general there is an endorsement of its importance to achieving organisational understanding. Resort to the literatures of power and politics showed them to be model heavy (often systems models); but these were too unwieldy to be helpful. In many instances power is treated as if it exists in the real world and as such, in an equally tangible world, can be investigated through the use of a model or measured by observation. The study of power is not the neat incision into society that a model implies; it is much more involved than that. The enlightening discovery came from the very founder of the concept of politics: Aristotle. He recognised that human beings are unavoidably different and in any social grouping different interests will be being pursued (e.g. Aristotle, *The Politics*, 1981, 121b32–p108, Checkland and Scholes, 1990, p50). Latour suggests that power is manifested not as a cause to people's behaviour but as a consequence of an intense activity of enrolling, convincing and enlisting others. Latour's view is that " The obedience to an order given by someone would require the alignment of all people concerned by it who would assent faithfully without adding or subtracting anything. Such a situation is highly improbable. The chances are that the order has been modified and composed by many different people who slowly turned it into something completely different as they sought to achieve their own goals." (Latour, 1986). He is suggesting that a constant debate will rage over who obeys and who will be obeyed. If this is not to disrupt or break the collectivity, there needs to be arrangements, generally agreed to and used, which enable power play to be contained: It will always happen.

Most of us will have experienced the use of formal power, e.g. managerial decisions, Political policy and also the use of informal or Soft power e.g. Personality, appearance. To address the complex social interactions that take place in any social grouping it was clear that discussions should embrace a variety of worldviews and arguments with the aim of facilitating a meeting of minds that is reached through the consideration, by all, of the various viewpoints and arguments made. (See Stowell, 1989; Checkland, 1999)

To address the, perceived, shortcoming in SSM a means of making visible the use of informal power, but remaining true to its underpinning philosophy, was sought. We considered power not merely in terms of its formal manifestation, such as managerial power (e.g. Marabelli & Galliers, 2016; Marabelli and Newell, 2014; Hardy, 1996; Clegg, 1989), but as 'soft' commodities of power. Researchers observe that, "...the use of soft power in social groups is often used to great effect in discussions and it at times 'trumps' that of formal power" (e.g. See Cooray, 2010; Hart, 2013). The challenge was could this feature of social interaction be highlighted in such a way that it could be used to enhance discussions. The focus here is upon soft informal, commodities, of power.

THINKING OF POWER IN TERMS OF A METAPHOR

Organisational power is more difficult to define conceptually than to recognise. Subsequently a means of finding a practical method of assessing its impact "... would make an important contribution to organisational problem solving (Stowell & Allen, 1988). In order to feel secure an individual seeks to control their environment and power is a manifestation of an individual's attempt to exercise that control (Stowell, 1989).

Working within the Lancaster A/R programme Stowell (1989), was involved in an intervention in a real-world situation, which was heavily political, and suggested that the metaphor of 'a commodity' embodying power could be helpful. The expression of power through the idea of a metaphor is concerned primarily with an attempt to find a practical means of addressing its management. The notion of commodity relates to something that is considered of value by organisational members within their culture, or the value systems of a given organisation; it will vary between and within each social grouping. The language used within the group will be more readily understood in its context than by an outsider and it is through language that soft power might be exercised. It is worth noting that Patočka, discussing the phenomenon of language, remarks, '...individual subjects seek to assert themselves, and since the action of each individual always involves broader or narrower communities, linguistic communication is a primary instrument of the perpetually renewed social combat." (Patočka, 2016, pp91-92).

The metaphor *Commodity* was chosen in an attempt to provide a neutral account of perceived inputs of individuals and groups in which attempts are made to translate a situation into something else. The use of the noun commodity should, through discussion, lead to a more precise definition or explanation of what is meant. The idea embodies the Aristotelian notion of how a political discussion should be conducted, "...a variety of viewpoints and arguments are presented, and the final decision is arrived at through a consideration of the strengths and weaknesses of these viewpoints and arguments" (Clayton, 2005, section1a). Similarly the notion of commodity is a means of opening up free discussion between participants in the spirit of Gadamer's 'Fusion of Horizons' where participants work to exchange knowledge rather than seek to impose a particular opinion, (Gadamer, 2004, p304; Stowell & Welch, 2012, pp 147-148).

We see 'commodity' used to assist individuals, or groups, describe what *they* perceive to be the way *soft* power is being exercised by other members of the group.

Life's experience shows us that the employment of different commodities of power can alter the outcome of any discussion and may give "unfair" advantage to those with the skills to use them. Stowell felt that by surfacing these "commodities of power" would enable a stream of analysis to be created. In this organisation we can ask: what are these commodities? How are they obtained, exercised, defended, suffused or limited, relinquished? This idea yielded a third analysis to be incorporated within SSM at the Analysis One, Two and Three stage. In Analysis Three Checkland questions the nature of power, and asks

What is Power? It works with the fact that everyone who participates in the life of social grouping quickly acquires a sense of what you have to do to influence people. (Checkland, 1999, pA20).

In a wider sense, researchers could also use the metaphor as a way of thinking and surfacing the kinds of 'commodities' that individuals use. This has proved useful because of the difficulty in defining power, consequently when intervening in an organisation we can be left with impressions of power rather than something that can be explained to others. In a recent study Cooray observed that "...in this exercise the more technically proficient participants seemed to use their expertise as a way of influencing others. This was exacerbated because the less technically able were hesitant creating the impression that they were happy to leave this aspect to others. (N.B. It could also be an example of the use of soft power; the implication being a form of ready-made excuse if things went wrong)..." (Stowell & Cooray, 2017).

The idea of thinking in general terms about the 'process' of power and hence the notion of using a metaphor has proved useful. As Champion and Stowell point out that by "Applying the 'power as a commodity' the metaphor facilitates asking how power has been expressed within the situation and how people intend to use and maintain these 'commodities" (Champion & Stowell, 2001, p9). By thinking in terms of a commodity the researcher could gain two advantages. First, a better, more holistic, understanding of the situation and second, the metaphor can be used to discuss the effects of power within the group. Managed sympathetically it opens the way for a wider range of discussions to take place that may bring out hidden, but imagined manifestations of power.

Commodities of Power

Commodities are both tangible and intangible. The notion of controlling power seems to relate to many different attributes which are as diverse as charisma, physical strength, education, beauty, status, property ownership and as examples of a negative effect, withdrawal of affection, lack of communication, exclusion, physical threat. For example, Stowell (1989) describes a situation that manifested itself during a project to change the working practices in the commercial department of a manufacturing company. A new export contracts manager had been appointed and became involved in the project that had been on-going for several months. The project had been developed into one of participation and of shared ownership by all involved (the commercial department had 50 members, see Stowell, 1989, pp161). At his first meeting the manager, who had apparently enjoyed a reputation in his previous company for a particular style of management, joined some of the group meetings. Stowell reports that the export contracts controller seemed to want to exploit his commodities of power, described by Stowell, as [i] his expertise and [ii] his forthright opinions. "He attempted to rearrange the situation in order to make himself a valued asset but he failed to recognise the culture of his new enterprise; he appeared to want the organisation to change rather than accept there to be a new situation. He seemed to be unaware of the culture of his new company and adopted inappropriate commodities which cost him his job" (Stowell, 1989, p268). His attempts to exercise control over the project could be thought of in terms of commodities of power. But his 'power' was based on a reputation that was not recognised nor valued by the staff in the commercial department of his new company. His attempts to denigrate the project had exactly the reverse effect upon the staff that he intended. By criticising the project, he had unwittingly challenged the integrity of the staff who, by this time, were driving the project. (He incorrectly assumed that it was an outside consultant); The staff saw the new manager as attempting to undermine their efforts.

The study proved of value in helping to make sense of the way individuals reacted and used their commodities of power in the face of significant change. The projects reported by Stowell (1989) involved the participation of all members of the relevant departments in the development of new ways of working that was controlled and developed by the staff themselves. The project provided many instances of the use of *commodities* in a positive sense and as well as being used through veiled threats for those who wished to maintain the status quo. For example, the managing director preferred to use his charms and personality to control the situation rather than use his formal power. He had initiated the project as a means of appeasing his managers by bringing in someone from outside the company. Although he wished to change the ways of working he chose to do this without seemingly giving any one manager the power over the others and at the same time appearing to have no direct control over outcomes himself. But he could delay or speed up things by sheer force of personality whilst not appearing to be exercising his formal authority. (Stowell, 1988, pp 266-267). This approach was in contrast to that adopted by the Sales Director who attempted to maintain the status quo by making sure everyone knew of his contacts within headquarters staff (formal authority) and by nurturing his personal relationships with certain staff who opposed the changes (soft power) as a means of displaying the scope of the potential resistance to ideas that he did not share (Stowell, 1989, p 267).

Checkland and Poulter (2006) in *Learning for Action* (2006, pp35-37) provide an example of using the notion of commodity as part of Soft Systems Methodology Analysis 3 (Political). They describe a situation where two managers in a consultancy company were being interviewed together. The conversation between them became confrontational at which point one manager said, "...you are NKT, I'm KT". Company insiders would have recognised this shorthand and Knew Tom and Never Knew Tom. The significance being those who knew the recently deceased charismatic founder of the company, Tom, took on his reflected glory whereas the opinions of those that did not know him did not carry the same credibility. Clearly this distinction, highlighted by the notion of commodity, reveals an informal, but important, hierarchy. The example also serves to illustrate the way that commodities will change over time. In this case as the memory and influence of the founder fades so too will the "power" of those who chose to use "KT" as a commodity find its influence will also have faded.

In a research project Hart (2015) set up a series of workshops to help her research 'knowledge transfer'. The workshops were organised in an institution whose prime function was to provide advice to Government on a variety of issues surrounding defence. In this institution staff usually moved on to a different role every 2-3 years; consequently retaining knowledge within the organisation was operationally important. This meant that unless the 'knowledge' of the individuals was passed on to their replacement they had to effectively start again. As far as Hart's research was concerned this was an ideal situation in which to gain understanding about the importance and method of transferring knowledge. The workshops included various members (and grades) of staff who agreed to share their opinions to help gain understanding of what they thought about knowledge transfer. The participants were willing to participate and many gave up their free time to join the discussions.

It emerged in discussion that the participants were aware of the relationship between knowledge and its relationship to personal reputation. It was acknowledged that there was a desire by some individuals to maintain their reputation as experts that could act to frustrate knowledge sharing. Hart observed that the possession of tacit knowledge was used by some staff as a commodity of power contributing to an individual's reputation and offered the possibility of financial reward, e.g. a consultancy role. It seemed clear that this attitude would also affect wider sharing behaviour, but no one would admit to using this 'commodity' but acknowledged it existed. (Hart, 2014, p14). There seemed a consensus in the group that the hoarding of knowledge was a further expression of implicit knowledge; and every possibility that subject experts did 'guard' their knowledge, whether consciously or unconsciously

(Connelly et al., 2012). This strategy to 'protect' their knowledge, she commented, could be seen as an example of a commodity of power (Hart, 2014, p14).

Harts study coincided with a time of organizational change. At least one participant, a senior manager of the knowledge services department, did not attend all meetings because he was located in a different part of the country, but importantly, was due to transfer to this establishment within a few months. When present at the meetings he seemed threatened by the discussions (Hart, 2014, p165). It transpired that he had developed a knowledge resources 'system' and it is reasonable to assume that he expected to transfer the way he ran his present unit into the new location (Hart, 2014, p87, p132, p155). Hart observed that in each session, despite the meetings being set up to explore all aspects of knowledge transfer with no predefined outcome, he seemed to be at pains to ensure that the other participants were aware of his 'vast' experience. He saw his task to encourage staff in the operating department to provide his department with 'knowledge', a view that was at variance with the opinion of the operating department manager. It should be borne in mind that this research exercise was investigating the influences affecting knowledge sharing, yet this individual attempted to steer the discussion to his ends (Hart, 2014, p155).

Although her research was concerned with 'knowledge sharing', Hart was aware of the notion of commodity and was conscious of the way that individuals sought to apply their soft power. Thinking in this way gave context to the discussions and added insight and richness to her understanding of the situation of concern and its outcome.

EVOLUTION

We found that one difficulty for the practitioner and clients was how to introduce the notion of commodity in a real-world situation. For example, Cooray found that it was quite difficult for the clients to distinguish between formal power and informal power relations "…But with the incorporation of the notion of power within the mnemonic PEArL we found that it goes some way to address the difficulty and has proved to be a useful (Cooray, 2010, p144). In summary, PEArL stands for the following:

Participants

As the outcome cannot be known in advance, it is not possible to always include all those who may be affected by any proposed action. (That is potential customers and actors involved in any proposed intervention). At times it will be possible to make an intelligent guess, but we can never be certain. Also involving a wide range and number of participants often increases the cost and complexity of any investigation into a situation, so participation may be restricted. For this reason constrains the choice of participants. The criteria for inclusion is a matter that needs to be recorded. This sets a boundary, which may alter as the inquiry proceeds.

Engagement

Participants may take part in only certain aspects of the inquiry, or be a main advocate of the findings. The manner in which people were engaged in the learning process is also then a matter to record. The methods and tools employed are relevant, as techniques may favour some participants and not others. The time taken will also reflect the resources available to the inquiry process and provide insight into the environment in which the learning took place.

Authority

Any process of inquiry within a social organisational setting will be sanctioned by a person, or group concerned with the situation under investigation. This individual or group of people may be involved further with the actual inquiry process, or not; but the initial authorisation for the inquiry process is a matter to record such permission may involve the participants being supported through time allowed for investigation, or in financial support. This concept of authorisation for the inquiry is resonant with the 'Owners' category in CATWOE (SSM see Checkland, 1999). Recording who authorised the inquiry and for what purpose, is essential when considering if the inquiry has been pertinent to the situation of concern.

Relationships

The models produced during the inquiry process will reflect the intended relationships to be maintained within any proposed intervention. An examination of these relationships may be useful in questioning any undeclared assumptions and worldviews (Weltanschauungen) held by the participants. These relationships (and how they are employed) may provide insight into how individual power and control have been dealt with by participants during the inquiry and with the intended intervention. Reflecting on the intended relationships within the situation of focus may also help to identify potential areas of conflict.

Learning

Participants in the learning will have gained individually from being made aware of the perspective of others. The active process of engaging in discussion may engender an 'appreciation' of the possibilities and constraints within a situation. Any agreed intervention (or non-intervention) into the situation would reflect the transformation that has occurred due to the process of inquiry and so is a crucial element to record when considering authenticity. The practical outcome of the inquiry will include the success, or failure, of any intervention. Evaluating the intended and unintended consequences of intervention (or non-intervention) into the situation of concern enhances the shared 'Appreciation' of the participants and is crucial to making judgements concerning authenticity. These issues can be summarised using the mnemonic PEArL

Thinking of power within the mnemonic PEArL[1] seems to offer a practical way of teasing out 'soft' power.

PEArL enables the researcher and client to be explicit about who is included and excluded from the situation of interest. By using PEArL it allows the practitioner to ask about **A** and about the lower case **'r'** (which relates to soft or the commodity of power) and helps the researcher, and often the client, to explain the power relationships within the boundary of the situation of interest that are thought to exist (Stowell, 1989; Stowell & Welch, 2012 pp74-76; Checkland, 1999). Cooray, in her research into a city library reflected upon 'r' and writes that

The experiences during the group sessions at the Portsmouth City Central Library show a close affinity with the informal power relations discussed by Stowell (2000) and Checkland (Checkland and Poulter 2006) than the formal power relations discussed by critical theorists, since in several instance individuals used commodities in order to exert power and control (Cooray, 2010, pp130).

In two workshops facilitated by Bednar and Stowell, A. in 2011 they found the mnemonic a useful way of helping participants to appreciate the way in which individuals react when they meet for the first time. For example, Stowell. A., (July 9th2008 report on the Social Services workshop) was able to recognise the actions of some of the group and the way that they used their assumed status from their formal role in their organisation to try to exert influence (commodity). She noted that this strategy had little effect upon those not connected with the same organisation and was soon recognised and abandoned by the individual concerned. In one workshop she noted that "The group appeared at ease and willing to discuss conflicting views. They accommodated each other by either changing terminology to accommodate a group understanding or agreed to have two different subsystem*s* (Stowell, 2008, Facilitator's notes). In the reported workshops there was no evidence of attempts to exercise control over the other participants during group work, as Bednar reported (facilitators notes. July 8th 2008): 'Informal powers were low key and while some may have been used I believe that this was more personality trait than intention. I would argue that members went out of their way to try to avoid imposing their beliefs on others during this session.' (Interestingly Bednar's reflection about personality trait is a good example of a commodity of power as, in this instance, someone used their personality to good effect at times). On this point it is worth noting that participants might use other ways of trying to influence outcomes. For example, they may use their personal charm or, in a negative way, withdraw cooperation as suggested by the case study reported above (Stowell, 1989).

Experience shows that power need not be used as a force for suppression and coercion. It can be used in a supportive way for an individual or group. Jary and Jary observed, "Although power …is often seen in negative terms, as involving coercion and conflicts of interest...'power' can also be seen in more positive terms, as enabling." (Jary & Jary, 1995, p513). Power can be creative and inspirational. For example, we can recognise the power of love, the power of poetry and the power of laughter as positive uses of power. Whilst it is easy to accept that there are examples of individual strategies for dominance what we are discussing here is not such ambition, but the kind of situation that ordinary individuals might face in their working environment. In such cases the use of personal power is more about job security than malevolent plotting. For example, Raven's (1992) reworked 'Power Interaction Model', "…describes the agent as a rational decision maker who weighs various costs and benefits of the power bases available to him or her before invoking one of them to influence the target.'(Mundate et al., 2003, p7). The observation highlights a view that individuals, here they are referred to as agents, have various kinds of power at their disposal and they select one or more they think appropriate to influence their 'target'. It is recognition of the dynamic of a social situation in which the actors are in a kind of negotiation with each other. Patočka discussing the constitution of language raises a point relevant to this when he writes, "…when surrounded by mere things we are alone, whereas the presence of the other acts on us as a specific stimulus, we feel observed, at the mercy of the other's understanding." (Patočka, p.94). It is often the case that it is only in situations where the outcome of discussions has a real impact upon participants do we begin to 'find out' what makes us tick. Gadamer makes such a points where he says "A man who is disguised does not want to be recognised, but instead to appear as someone else…we merely pretend, act a part, and create an impression."(Gadamer, 2004, p111). Our task is to think in terms of the way participants use their commodity of power and help 'unmask' the players and reveal the play.

Addressing the Criticisms

Since it was first proposed the notion of Commodity as a metaphor for organisational power has had its critics. Foremost amongst these criticisms is the reported difficulty some have in understanding its purpose. Simply stated the intention is to help participants gain insight into the use of power and its effect upon the outcomes. Thinking in terms of a metaphor is a way of providing participants with the opportunity to explore the nature of the interactions taking place within a social group. For example, observations about the language used and the non-verbal cues that are part of the way we interact, which is important to uncover when the situation is likely to be contentious. Our intention with the noun 'commodity' is by using a familiar and non-threatening word we can tease out any underlying meaning in the way in which the discussions evolve. It is not the intention to humiliate or belittle anyone (to do so would surely cut off any free discussions), but provide the participants with a way to recognize how soft power might influence outcomes and impede open discussions.

Commodity provides a neutral way of revisiting aspects of the debate and the language used, which sometimes has hidden meanings embedded within. But we should be aware of what Vickers refers to as our 'readinesses' and Patočka (p 72), the expression of original tendencies. We take this to highlight our tendency to "...see, value and respond to situations in familiar ways which, while they last, exclude the power to see other possibilities." (Vickers, 1983, p.54); to overcome this tendency we should attempt to suspend judgement. Our experience of life provides us with examples where we have noticed a look or expression exchanged between participant that has signalled a change in direction in the discussions or a hitherto unexpected alignment of views. But as participants we seek to suspend judgment about the existent and simply observe and absorb the situation as it takes place. By thinking in terms of commodity we can attribute certain words or actions to the way in which they seem to influence the 'shape' of the discussions; what commodities were engaged (Checkland & Poulter, 2006, p37).

The notion of commodity is not an attempt to produce a model of the world or of social interaction, but a way of helping to make sense of a complex social situation; the intention is to gain understanding. In any situation amongst groups in which a significant number of participants are known to each other, there is a common social and linguistic code of behaviour. A micro culture exists in which the language and actions of the participants are understood in ways that outsiders cannot always recognise. Not everyone in such a group as this will have the same understanding. Certain statements or expressions might be recognised and absorbed by some members of a group and may influence their support or otherwise. It is not unusual for expressions with 'veiled' implications for some and imperceptible for others to determine outcomes. It is only in being part of that culture that the hidden meanings of what is being said can be understood. The difficulty is that not every participant will have understood what is being said in the same way. It is to this end the notion of commodity is proposed. As a minimum it can provide useful insights for the facilitator or researcher, but if used sensitively it can be of benefit to all participants. What we advocate is think of it as a way of highlighting hidden attitudes which manifest in different ways - as a sort of 'intellectual coat hanger'. The idea is to highlight without prejudice or pre-formed notions on why these commodities are being employed. It is an attempt to find a neutral ground from which to proceed; prejudging no solution.

The easiest way to think about 'commodity' is as a form of hermeneutic circle, where understanding a part modifies how we understand the whole. The task of the facilitator is to steer the clients and, where possible, the 'consumer', through an exercise "...in which constantly exercised understanding understands itself" (Gadamer, 2004, p268). It is an intellectual device to surface 'Soft power' that is described in language familiar to participants. We seek to understand its purpose and why it is being used.

What Are Commodities and How Can the Idea Be Used?

Providing a list of examples of 'commodities' is counterproductive as the way in which soft power is used is specific to a particular group and situation. Whilst we acknowledge, as Ryle remarks, "Understanding is still psychological divining..." (Ryle, 1990, p52), as individuals we call upon our experiences and understanding of the culture of which we are a part, to make sense of interactions between participants. The lack of examples are less an indication of the poverty of the idea, but more a reflection of the kind of thinking that needs to 'hypothesise' and make a clear connection between the example and the situation. There is a contradiction between this idea and the mindset that demands such a logical connectivity, in that this kind of thinking implies some form of proof. The idea underpinning 'Commodity' is the phenomenology of Husserl (e.g. 2002), and as such we should recognise it as 'a phenomenon', which '...is something uncovered, manifest, no less than the extent itself, but rather it contains something that has claim to recognition..."(Patočka, 2016, p39). The situation and those within it will determine what is powerful or not.

Perhaps Patočka, once more, encapsulates the essence of what we are trying to achieve when he says, " ... the works of man as an individual as well as those of history are given in the original - is a phenomenon whose originality must first be described and analysed and only then interpreted, in such a way that the phenomenon does not vanish with the interpretation." (Patočka, 2016, p127). Phenomenology, is a reflective philosophy (Patočka, p 123) which is in contrast with scientific reductionism that is concerned with hypothesis and proof about what makes things exist how can they be explained. Unlike phenomenology, implicit to scientific thinking is that the world exists within a set of structures with the implication that a failure to acknowledge this is to concede to mysticism. The 'certainty' of scientific discovery creates a reluctance to trust our experience of the 'real world' and our observations of life. Instead we fall back on the thesis of 'what is', developed from hypotheses of explaining the world. But adopting a phenomenological approach we become aware that what we experience is a function of our 'being in the world'; Dasein (Heidegger, pp.67-77, 1962). In a practical sense this proposes a 'hermeneutic phenomenology' "...which proceeds by means of interpretation... the goal of phenomenology is to bring out what is implicit in that understanding..." (Cerbone, 2008, p27-29)

The Relevance of Language to Commodity

Natural language philosophy provides us with some insights into the way we might set about observing group interaction. For example, Ryle (1990, pp 211-212) suggests that observations fall into two categories, there are observations where we are trying to find out something and the other where we have found something out. If we accept Ryle's contention then in the first case we are using our senses; looking, listening, feeling, to explore the situation and gain understanding, in the other we have found out by using some sort of 'model' on which to base our observation. It is to the former we link the notion of commodity as we try *absorb* what is taking place, unfettered by the need to fit our observations into a predefined framework. We are trying to gain insight into the way in which participants react to each other as the topic of concern is discussed. We seek to uncover the meaning behind the interactions. The interpretation of the interactions can only be uncovered by those with a vested interest in the situation. Gadamer, (1976) reminds us that interpretation is fundamental to understanding and that individuals interact with the world from a position of pre-understanding, or prejudice including assumptions implicit in the language used in a situation, (Gadamer, after Heidegger, called this philosophical Hermeneutics). What we can take from this is that each participant will have their own view of the situation and, in many cases, how it should be addressed. This is the antithesis of the way many researchers set about 'finding out. Instead they seek to find a causal relationship between what they observe and a predetermined 'model'. This creates a difficulty for a researcher because most methods of inquiry means that the inquirer learns about the situation through *their* interpretation of the situation and of the language used because we pay more attention to the method than the situation. What we seek is to enhance the participants understanding and understanding is gained through a descriptive analysis rather than ontic 'certainty'.

The way that that the participants react is contextual and the language they use is appropriate to their situation and their role within it, which may not be familiar to the inquirer. Language should not be thought of as a series of signs offering some sort of meaning. What is said and how it is said has meaning to the listener and in particular those who are a part of the social grouping in which the 'conversations' take place, we should "...think about it *in situ*, embedded in the lives of those who speak it" (McGinn, 1997). Wittgenstein (1987, p.28) reminds us " ... let's not forget that a word hasn't got a meaning given to it, by a power independent of us...A word has the meaning someone has given to it." As organisational inquirers our task is to seek a way of understanding the clients' 'world-in situation' (Verstehen, Weber, 2012/1930) and in terms of the language they use to describe the way the world is for them. We seek to find a way of enabling the participants to represent their world independent of the conceptual, political and linguistic predisposition of the inquirer.

CONCLUSION

Having grown up in a world which subscribes to the scientific paradigm many find it difficult to look outside a world described according to a causal model into one where reality is established from subjective experience of relationships within the human environment. Even within the Systems community some are more comfortable 'imposing' a framework as a means of helping them make sense of what they perceive. Whilst we are finite beings and part of a world surrounded by concrete phenomena we are also in a world of lived-in experience accessible to all to a varying extent. What we make of our world is shaped by a series of concrete phenomena and impressions formed from our (subjective) experience of it and through the communities to which we belong. The naive world in which we start our life, full of 'feelings', instinct, wonder, observations and things we do not understand, is gradually eroded by the seduction of the notion of objectivity. Subjectivity, we are soon taught, is close to mysticism and belongs to the past. Explanation is gained through the application of a set of structures which provide a well-trodden outcome. If this fails then we set about trying to find new paradigms or new ways to explain something which, frankly, cannot be wrapped up into a package to be used until the next time it does not answer our question.

We are afraid of accepting that our naive insights may be of value. To accept that a simple statement uttered in discussion might have a greater depth for the person uttering it may be difficult for those who view the world solely in concrete terms. But the manner in which individuals interact may ring alarm bells amongst the participants. To dismiss such phenomena is to deny things that we have all experienced but have found difficult to express. Thinking in terms of a 'Commodity' as a metaphor for power has synergy with Heidegger's notion of hermeneutic phenomenology, which "...proceeds by means of interpretation" (Cerbone, 2008, p29). By allowing participants 'in' on the way in which people seek to influence outcomes acts as a means of reminding us that we should stop and think 'what does this mean'; 'why are they reacting in this way?' The point of the idea is to develop participants' awareness of the potential impact that some interactions might have on shaping the outcome. We make sense of this primarily through the spoken word as everything we do and understand happens via language. We can gain insight into the way participants attempt to shape the outcome through their use of language.

The way that the language used by individuals can be used as an attempt to transform a situation that they feel unable to control into one where they feel able to control – by the use of 'soft 'or commodities of power. Power is an integral part of political behaviour and its currency is relative to the degree of dependency upon a given 'commodity'. Commodities are part of a transformation process affecting the position held by each participant in a given situation. Starting positions are transformed by the trading of commodities and by the strategies chosen by participants. By making 'soft' power explicit opens up the possibility of a fruitful outcome. But the language and the meanings embedded within the 'conversations' can only be fully understood by those who are a part.

We have found from the practice that embedding 'commodity' (r) within the mnemonic PEArL it becomes a visual and practical way in which to explore the implications of 'soft' power (Champion and Stowell, 2001; Champion and Stowell, 2003; Stowell, 2009; Hart, 2014, pp131-132). It can help with the understanding of the dynamics of the group where individuals seek to 'persuade' others to accept their point of view. The lower case 'r' is an attempt to clearly differentiate 'soft' power from formal 'hard power as represented by the upper case 'A'. By surfacing the use of Soft power we can generate an exchange of ideas as a means of gaining an *Appreciation* of the situation of interest. The discussion is intended reach 'a fusion of horizons' (Gadamer, 2004). That is to say there is an exchange of ideas as a means of gaining an *Appreciation* (Vickers, 1983) of the situation of interest. PEArL provides a framework through which participants can discuss, without rancour, any difference of opinion and make visible (although this may be unsaid) the strategies of power applied to influence outcomes. Opinion, Gadamer says, is the enemy of understanding, it '...has the ability to propagate itself and acts to suppress questions' (Gadamer, 2004, pp 359-361). Depending upon the group and the skills of the practitioner this understanding can be uncovered for all to reflect upon and used as part of the process of gaining understanding.

These experiences argue powerfully for the epistemological/process approach to organisations as research objects and as objects to be managed. The lessons provided from the 300 plus real-world action research projects at Lancaster, and those undertaken subsequently elsewhere, provide a compelling argument for approaching each situation from the perspective of finding out rather than trying to fit it into a model; in other words, ditch ontology. Each intervention should be seen as providing an opportunity to learn about the way *that* enterprise operates and of the various factions within. By encouraging a culture of participation and transparency the practitioner/researcher provides an opportunity for all participants to feel they have a measure of control over outcomes. Thinking about power in this way provides a means of articulating it in terms that are appropriate to a given organizational culture, which can be understood by those most affected. What we are seeking is a method of bringing into public debate a description of those "commodities" that are used to influence that situation. Commodity is not a means of trying to control those with 'power' but a way of disarming the way in which it is used.

FURTHER RESEARCH

It should be noted that idea of power expressed above is also used as part of a method of inquiry into complex situations. By embedding Commodity, represented by the lower case 'r', within PEArL, and this in turn used within the Appreciative Inquiry Method (AIM), has shown it to be helpful in gaining an appreciation of the situation of interest. Our current research involves investigating how these ideas can be used in situations where decisions are taken 'remotely' where there is little or no physical contact between participants. In such cases all discussions take place via ICT's. We believe this research to be important because the turbulent environment in which an organization exists is now compounded by the ubiquity of technology. We live in an age of 'instant communications', which undoubtedly affects business decision-making. Many organizations now engage the expertise of individuals and groups situated in different geographic locations, where they may not share the language or the culture of the 'parent' company, creating difficulties for managers. Often these groups are required to play an important part in the formulation of strategic decisions, but clearly the traditional of developing strategies is no longer possible. Strategy building exercises are prime examples of a social process that necessitates strategic thinkers to interact, share information and collaborate. Traditionally these interactions were carried out in co-located teams, but the evolution of the use of technology in many organizations means managers now communicate with their virtual teams, digitally. But recent research shows that virtual teams behave differently from co-located teams and are more conflictual in nature. The literature indicates that conflicts in virtual teams lead to dissatisfaction among participants (e.g. Sarker and Valacich). This is a new phenomenon and creates new challenges for researchers, especially those who employ methods of Action Research (AR). To investigate this challenge Stowell and Cooray have begun to research decision making in virtual teams. To do this they have employed the Appreciative Inquiry Method (Stowell, 2012) and PEArL (Champion and Stowell, 2001; Cooray, 2010) with the lower case 'r' referring to 'soft' power. Combining these ideas has provided a suitable framework for researching this complex area and experience of how AR might be adapted to this new area of inquiry. Results from their research (Stowell and

Cooray, 2017) suggests that by allowing members to define their perceptions of the situation, e.g. why they view it in that particular way gives them the opportunity to gain an appreciation of other participant's perceptions. This means that they gain a richer *systemic* understanding of the problem situation. Participants seemed to have gained an understanding of the meaning and the context of the various suggestions made by other team members. The research so far has shown some encouraging results. If this initial research proves to be helpful in other areas then the framework might provide a way to reduce conflict in virtual team decision making. Adopting this approach might help to overcome the disadvantages of not being physically together and help to mitigate the miscommunications arising from the lack of social presence (e.g. Non-verbal cues) and the lack of opportunity for relationships to grow 'naturally, through non-work conversations. Plans are now in place to continue to explore the way in which the above ideas can be investigated.

REFERENCES

Aristotle. (1981). *The Politics*. Penguin Classics.

Beer, S. (1979). *The Heart of the Enterprise*. Chichester, UK: John Wiley and Sons.

Capra, F., & Luisi, P. L. (2014). The Systems View of Life: A Unifying Vision. Cambridge University Press.

Cerbone. (2008). *Heidegger, A Guide for the Perplexed*. Continuum.

Champion, D., & Stowell, F. A. (2001). PEArl; A Systems Approach to Demonstrating Authenticity in Information System Design. *Journal of Information Technology, 16*(1), 3–12. doi:10.1080/02683960010028438

Champion, D., & Stowell, F. A. (2003, February). Validating Action Research Field Studies: PEArL. *Systemic Practice and Action Research, 16*(1), 21–36. doi:10.1023/A:1021928511690

Checkland, P. B. (1999). *Systems Thinking Systems Practice. Includes a 30 year retrospective*. Chichester, UK: John Wiley and Sons.

Checkland, P. B., & Holwell, S. E. (1998). *Information, Systems and Information Systems*. Chichester, UK: Wiley and Sons.

Checkland, P. B., & Scholes, J. (1990). *Soft Systems Methodology in Action*. Chichester, UK: Wiley and Sons.

Checkland, P. (1981). *Systems Thinking, Systems Practice*. Chichester, UK: Wiley.

Checkland, P. (1999). *Systems Thinking, Systems Practice, 30 year retrospective*. Chichester, UK: Wiley.

Checkland, P. B., & Holwell, S. (1998). *Information, Systems and Information Systems*. Chichester, UK: John Wiley and Sons.

Checkland, P. B., & Poulter, J. (2006). *Learning for Action*. Chichester, UK: John Wiley and Sons.

Churchman, C. W. (1971). *The Design of Inquiring Systems, Basic Concepts of Systems and Organizations*. New York: Basic Books.

Clayton, E. (2005). *International Encyclopaedia of Philosophy*. Academic Press.

Clegg, S. R. (1989). *Frameworks of Power*. London: Sage.

Connelly, C., Zweig, D., Webster, J., & Trougakos, J. (2012). Knowledge Hiding in Organisations. *Journal of Organizational Behavior*, *33*(1), 64–88. doi:10.1002/job.737

Cooray. (2010). *End User Driven Development of Information Systems –Revisiting Vickers Notion of Appreciation* (PhD Thesis). University of Portsmouth, Portsmouth, UK.

Giddens, A. (2001). *Sociology* (4th ed.). Cambridge, UK: Polity.

Gadamer, H. G. (2004). *Truth and Method* (2nd ed.). New York: Continuum.

Hardy, C. (1996). *Understanding Power: Bringing about Strategic Change*. Wiley.

Harris, J. (2009). *Tönnies*. Cambridge, UK: Cambridge University Press.

Hart, P. J. (2014). *Investigating Issues Influencing Knowledge Sharing in a Research Organisation Using AIM* (Thesis). University of Portsmouth, Portsmouth, UK.

Heidegger, M. (1962). *Being and Time*. Malden, MA: Blackwell Publishing.

Husserl, E. (2002). Pure Phenomenology, its Method and its Field of Investigation. In D. Moran & T. Mooney (Eds.), *The Phenomenology Reader* (pp. 124–133). London: Routledge.

Jackson, M. (1982). The Nature of Soft Systems Thinking: The Work of Churchman, Ackoff and Checkland. *Journal of Applied Systems Analysis*, *9*, 17–29.

Jary, D., & Jary, J. (1995). *Sociology, Collins Dictionary*. Glasgow, UK: Harper Collins.

Marabelli, M., & Galliers, R. D. (2016). A Reflection on Information Systems Strategizing: The Role of Power and Everyday Practices. *Information Systems Journal.*

Marabelli, M., & Newell, S. M. (2014). Knowing, Power and Materiality: A Critical Review and Reconceptualization of Absorptive Capacity. *International Journal of Management Reviews, 16*(4), 479–499. doi:10.1111/ijmr.12031

McGinn, M. (1997). *Wittgenstein, and the Philosophical Investigations*. London: Routledge.

Latour, B., (1986). Visualization and Cognition: Thinking with Eyes and Hands. *Knowledge and Society: Studies in the Sociology of Culture Past and Present, 6,* 1-40.

Mundate, L., & Bennebroek Gravenhorst, K. M. (2003). Power Dynamics and Organisational Change: *An Introduction, Applied Psychology. International Review (Steubenville, Ohio), 52*(1), 1–13.

Patočka, J. (2016). The Natural World as a Philosophical Problem. Northwestern University Press.

Raven, B. H. (1992). A Power/Interaction Model of Interpersonal influence: French and Raven Thirty years Later. *Journal of Social Behavior and Personality, 7,* 217–244.

Ryle, G. (1990). *The Concept of Mind.* London: Penguin Classics.

Sarker, S., & Valacich, J. (2010). An Alternative To Methodological Individualism: A Non-Reductionist Approach To Studying technology Adoption By Group. *Management Information Systems Quarterly, 34*(4), 779–808. doi:10.2307/25750705

Stowell, F. A. (1989). *Change, Organisational Power and the Metaphor Commodity* (PhD thesis). University of Lancaster, Lancaster, UK.

Stowell, F. A. (2009). Soft Systems and Research. *Kybernetes, 38*(6), 883–900. doi:10.1108/03684920910973144

Stowell, F. A. (2012). *Peter Checkland, interview.* Retrieved from https:// www.youtube.com/watch?v=Pq90qS5FvBgSchutz

Stowell, F. A. (2013). Peter Checkland Interview. International Journal of Information Technologies and Systems Approach, 6(2), 53 – 59. doi:10.4018/jitsa.2013070105

Stowell, F. A. (2012). The Appreciative Inquiry Method – A Suitable Framework for Action research? *Systems Research and Behavioral Science, 30*(1), 15–30. doi:10.1002res.2117

Stowell, F. A. (2014). Organisational Power and the Metaphor Commodity. *International Journal of Systems and Society, 1*(1), 12–20. doi:10.4018/ijss.2014010102

Stowell, F.A., & Cooray, S. (2017). *Policy, participation and the challenges of virtual teams- a "Soft" approach? An account of a research project into the use of Soft Action Research.* (submitted)

Stowell, F. A., & Welch, C. (2012). *The Managers Guide to Systems Practice. Wiley.* doi:10.1002/9781119208327

Stowell, F. A., & Stowell, A. F. (2011). *Appreciative Inquiry Method – Day 2 Facilitator's notes – Social Services. University of Northumbria.* Workshops.

Stowell, F. A., & Allen, G. A. (1988). Co-operation, Power and the Impact of Information Systems. *Systems Practice, 1*(2), 181–192. doi:10.1007/BF01059857

Stowell, A. F. (2011). *Organising 'e' Waste* (PhD thesis). University of Lancaster, Lancaster, UK.

Truzzi, M. (1971). *Sociology: The Classic Statements.* New York: Oxford University Press.

Vickers, G. (1983). *The Art of Judgment, A Study in Policy Making.* London: Harper and Rowe.

Weber, M. (2012). *The Protestant Ethic and the Spirit of Capitalism.* Rennaisance Classics.

Weber, M. (2001). *The Protestant Ethic and the Spirit of Capitalism.* New York, NY: Routledge.

Wittgenstein, L. (1978). *The Blue and Brown Books.* Oxford, UK: Basil Blackwell.

ENDNOTE

1 PEArL see Stowell and Welch, 2012; Champion and Stowell (2001).

Chapter 8
Enterprise Systems, Power and Improvisation:
Equipping Universities for Mass Production?

David W. Wainwright
Northumbria University, UK

Teresa S. Waring
Northumbria University, UK

ABSTRACT

Enterprise systems (ES) have been extensively procured in large organizations but much research fails to develop sociotechnically informed approaches that facilitate their implementation whilst understanding the impact of integrated technology on professional working practices within complex organizational environments. This chapter takes a critically informed sociotechnical approach to power and improvisation in ES implementation. The contribution is a combined "circuits of power-improvisation" (CPI) framework which can facilitate a better understanding of ES implementation, sociotechnical theory, and practice. Practical lessons learned from the study may potentially be used to avoid some of the problems experienced with the over-zealous and rapid introduction of digital technologies into university organizations where the risk is that they become a student mass production system. It highlights the important role of power and improvisation, enabled and afforded by new digital technologies, in what may be misrepresented as planned strategic and deliberate organizational change.

DOI: 10.4018/978-1-5225-5996-2.ch008

INTRODUCTION

In the last two decades, increasingly complex forms of organization have developed in response to the pressures from globalisation, economic uncertainty and market instability. In addition, rapid technological developments, especially in information and communications (now commonly termed digital) technologies, have provided opportunities for the adoption of new management practices - especially those based on principles of performance monitoring, reporting, staff accountability and control. This is particularly prominent in both the private and public sectors, where objectives focusing on efficiency, competitiveness, quality, and accountability have led to an emphasis on the development of new metrics underpinned by a culture of performance management, greater centralisation and hierarchical systems. In order to facilitate and reinforce these new management regimes, deliberate organizational strategies have often entailed the procurement, adoption and implementation of Enterprise Systems (ES). This is particularly evident in sectors such as Higher Education in the UK where, in the past, there have been more professional and collegial forms of work organization, professional autonomy and localised decision making (Fowler & Gilfillan, 2003; Pollock & Cornford, 2004; Oliver & Romm, 2009; Wagner et al., 2010). The rise of managerialism, underpinned by ES technology and more recently artificial intelligence and machine algorithms, has given rise to a new form of technocracy based on key drivers of machine efficiency and effectiveness that has begun to dominate the human and social aspects of work and knowledge based systems that have been developed over many years in organizations. This raises new issues regarding the maintenance of an appropriate balance between human, social and technical aspects of work thus giving rise to the need for more informed sociotechnical approaches and methods to be adopted in ES implementations that also now extends to digital transformation strategies.

Emerging research (Wagner et al., 2010; Dong et al., 2009) would also appear to indicate that many organizations, although having ambitious Information Technology (IT) strategic plans in place, find large scale enterprise systems overwhelming during implementation and may adopt organizational actions which respond to unfolding events resulting in unexpected and unintended consequences of the implementation as it progresses. One such consequence may be the ability to engage in opportunistic re-organization strategies afforded by the assimilation of 'new managerialism' (Clarke & Newman 1997) within the organization and, with it, a culture of command and control that is often incongruent with existing management cultures and organizational structures (Deem, 2004; Deem & Brehoney, 2005). The effects of ES adoption and its relationship to the redistribution of power and the alteration of political structures may lead to new policies, strategic opportunities and threats within organizations. This is an area that appears to be under researched. However, ES, once adopted, can also act as a catalyst for organizational and work based improvisation, thereby creating new and unanticipated power relations, managerial roles and structures.

The aim of this chapter is to contribute both to sociotechnical theory and practice, by examining IS improvisation practice through the lens of a power and improvisation model that explicitly recognises the concept of power and its role in IT enabled organizational change within the wider context of enterprise systems adoption and implementation. Unlike Orlikowski's (1996) 'open-ended' highly configurable groupware IT, and Elbanna's (2006) 'successful' manufacturing ERP implementation, the ES under consideration here is one implemented within a highly complex UK Higher Education (HE) organizational environment. This critically informed and sociotechnical approach adopts the 'circuits of power' framework (Clegg, 1989; Silva, 2007) and relates this to an improvisational model of change to surface issues and actions that bring with them unintended consequences for organizations. This chapter is an updated version of our previous research study (Waring, Wainwright & Skoumpopoulou, 2016) with an extended discussion section that demonstrates the rapid acceleration of digital technology adoption in a university and the accompanying transformational and strategic changes; subsequently resulting in work force changes, deskilling and reskilling. Hence the notion of universities being increasingly viewed as a metaphorical 'production machine' to enable the mass production of students (Morgan, 2006).

The next section discusses the theoretical concepts that underpin this work and examines the fundamental ideas around improvisation and power and how the two may be linked. The primary research involved a longitudinal case study of a UK HE institution which implemented an ES and this is presented in the third section. The case is then deconstructed and analysed by relating it to the improvisation and power concepts previously formulated. The next section provides an updated account of the digital transformation strategy and its impact on the organization that was in many ways facilitated by the ongoing ES adoption. The final section presents the implications of this research for future ES adoption and implementation, and reflects on the utility of developing improved theoretical models of improvisation to surface issues of power and political relations in organizations. It is proposed that this will add to the body of sociotechnical theory and lead to a better understanding of improvisation enabling opportunities and actions to be taken that result in more ethical practice and increased value for stakeholders who may be the potential users and benefactors of ES applications and digital transformation programmes.

IMPROVISATION AND ENTERPRISE SYSTEMS

Both the research and practitioner literature concerning the implementation of enterprise wide systems (ES) in large organizations are replete with normative models, methods and guidelines providing prescriptions for implementation success. One area of research which appears to have relevance to the ES implementation debate, but which has tended to be largely overlooked, is the important role of organizational actors and improvisation during the process. This lack of formal recognition is rather paradoxical however, as it appears that no matter how highly structured and well managed the organization is, or how experienced the project teams are in terms of IS/IT implementation, improvisation is often brought to bear in situations where rubrics, formally planned approaches and methods fail (Cibbora, 1996). Research on improvisation has received increased recognition within the contexts of organizational change (Ciborra, 1996; Moorman, 1998), punctuated sociotechnical change and information systems (Lyytinen and Newman, 2008), management (Mangham & Pye 1991; Weick,1998), bricolage (Lanzarra, 1999), tailorability (Morch & Mehandjiev, 2000) and information systems development and implementation (Orlikowski, 1996; Elbanna, 2006). The literature also defines its other foundational areas as sense making (Weick 1995), innovation (Weick 1999; Kamoche & Cunha, 2001), emergent change (Mintzberg, 1994) and adaptive structuration (Desanctis & Poole, 1994). Nevertheless, what is often missing from this literature is a formal consideration of the power and political environment within which these improvisations occur.

Before considering how improvisation can inform ES implementation theory it is essential to better understand the concept, and its historical research context.

Defining the Concept and the Language of Improvisation

The concept and hence the language of improvisation is not only restricted to the information systems field and there has been a great deal of research carried out in anthropology, arts and management studies generally and organizational change management more particularly. Early organizational researchers grounded their studies of improvisation in the arts, especially jazz music owing to improvisation and the centrality of 'bricolage' (Cunha et al., 1999; 2012). Researchers then moved the debate away from the arts into management and organizational settings where they attempted to develop definitions of organizational improvisation and show its relevance in competitive environments (Ciborra, 1996).

When one considers the totality of definitions available some common themes relevant to management can be identified. First, in these contexts, improvisation must be *organizational* because actions are carried out by organizational actors within teams, on their own or on behalf of the organization. Second improvisation can also be seen as the *conception of action as it unfolds* and finally improvisation implicitly suggests that it can only take place where there are *resources* and individuals able to make use of these resources. This is where the concepts of '*improvisation*' and '*bricolage*' are closely related. For example, improvisation implies the pre-existence of a set of resources, be it a 'plan of action', knowledge or a social structure, upon which variations can be built (Orlikowski & Hoffman, 1997). Also, if improvisation means to respond in real time then it follows that improvisers cannot wait for optimal resources to be deployed and have to tackle the issues at hand with what is currently available (Cunha et al., 1999). *Bricolage* on the other hand has been defined as "making things work by ingeniously using whatever is at hand, being unconcerned about the 'proper' tools and resources" (Thayer, 1998:239). These concepts build upon original notions identified in anthropology, cultural studies and the work of Claude Lévi-Strauss.

Improvisation Research and Its Context Within the IS Field

One of the most influential conceptual thinkers and writers within IS on improvisation and IS development is Claudio Ciborra. Working between the interstices of the management, information systems and organizational behaviour fields, Ciborra (2002, p.49) expands on the essential differences and nuances between the terms bricolage, improvisation and hacking. He defines this as:

Bricolage means tinkering through the combination of resources at hand. These resources become the tools and they define in-situ the heuristic to solve the problem ...With bricolage, the practices and the situations disclose new uses and applications of the technology and things.......Improvisation puts an emphasis on the suddenness, extemporaneity and unpredictability of the human intervention, though highly situated....hacking is devising and implementing a program to perform a certain useful function by making use of any technology in an original, unorthodox, and often playful way....where hacking stands to software engineering tenets as bricolage stands to organizational procedures...

Ciborra (2002) expresses a caveat that these expressions can signify or be interpreted as derogatory practices or deviant management behaviours. He calls for a *'Copernican revolution'* where improvisation and bricolage is seen as the norm in strategy and systems implementation representing a valid alternative to rational and deterministic planning and decision making. These practices are particularly relevant to the adoption of ES providing a cognitive and organizational capability to convert these systems, applications and data into practical, situated, and unique knowledge for action. Ciborra (2002) also stresses the importance of considering 'mood' as a presupposition for improvisation as situated action. Ciborra's work, although mostly conceptual, acts as a complement to interpret more empirical studies that investigate improvisational change and methods closely related to bricolage.

Recognising that there was a need to develop better theoretical insights into large integrated IT systems implementation, Orlikowski and Hofman (1997) conducted an empirical study of organizational improvisation due to the adoption of IT groupware. Their original aim was to describe how improvised change can take place in a series of deliberate, emergent or opportunistic phases. Here three different types of change are recognised consisting of: anticipated, emergent and opportunity-based change (Figure 1).

Anticipated changes are those that are planned ahead of time and occur as intended. Emergent changes arise spontaneously from local innovations and are not planned or intended. Opportunity based changes are those that are introduced purposefully and intentionally in response to an unexpected opportunity, event or breakdown. Orlikowski and Hofman (1997) view these types of change as building on each other iteratively over time. The use of the new technology typically involves a series of opportunity-based, emergent and further anticipated changes the order of which cannot be pre-determined. This is because of the nature and complexity of change and its interaction with events and the environment within which it is taking place. They also suggest that "there is a discrepancy between how people think about technological change and how they implement it....that contributes to the difficulties and challenges that organizations face as they attempt to introduce and effectively implement technology-based change" (Orlikowski & Hofman, 1997:11). However, in their description of the case company Zeta, Orlikowski and Hofman (1997) make no mention or explicit reference to power or political actions that may influence or affect the outcomes of these improvisation types either in the supporting case study analysis or theory reviewed. In their study of groupware (open ended) technology adoption, the anticipated strategy and any unanticipated outcomes are

Figure 1. An improvisational model of change management over time (Orlikowski & Hofman, 1997:13)

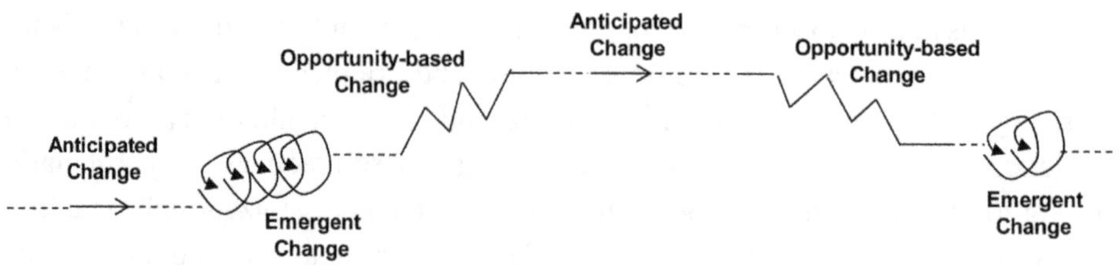

all viewed highly positively. Negative consequences related to actual and potential user resistance, due to power and political issues, are rather 'downplayed' resulting in a 'rose tinted spectacle' view of planned and emergent organizational change. Yet, the power and political implications do not form part of the improvisation framework in Figure 1.

POWER AND IFORMATION SYSTEMS

Although there are IS studies which address power and politics it is not the intention to explore them at any length in this chapter. The intention is to focus upon IS research that is based upon Clegg's (1989) 'circuits of power' (Silva, 2002; Silva & Backhouse, 2003; Silva, 2007). Silva (2007) argues that the 'circuits of power' provides a rich analytical framework for the study of power and IS and has been derived from a number of studies on power within sociology (e.g. Callon, 1986; Law, 1986; Latour, 1987) as well as insights from Foucault's (1977) disciplinary power. This framework (Figure 2) also has the added benefit in that it relates the concepts of power to organizations.

In Figure 2, *episodic power* relates to agency, and, within the context of IS, Silva (2007) suggest that the focus of study may be on 'A's' exercise of power over 'B' (A and B are different agencies) and 'B's' possible resistance to some aspect of an IS adoption. Law (1991) views this as 'power over' or causal power and suggests that individuals have a scope of actions and that specific scope of action is given by the structural characteristics of the individual's particular context. Clegg (1989) complements episodic power by drawing on the work of actor-network theory and specifically the concept of 'obligatory passage points'(OPP) and the principle of 'symmetry' (Callon, 1986; Latour, 1987). Silva and Backhouse (2003) suggest that OPPs are the results of four 'translational steps', problematization, interessement, enrolment and mobilization, after which B has no other choice but to accept the OPP. The movement between steps is called displacement and involves discursive practices. When displacement occurs, power is exercised. Silva (2007) argues that information systems can be viewed as OPPs.

The 'circuit of *social integration*' addresses dispositional power. This capacity is a characteristic that distinguishes between having and exercising power. Power in this context is a facility that entitles someone to exercise power - but it does not necessarily imply its exercise (Clegg, 1989). Clegg (1989) defines social integration in terms of the relation between rules of meaning and membership. The analysis of this circuit identifies the legitimate (formal rules) and illegitimate (informal rules) dimensions of power within the organization

Figure 2. Circuits of Power (Clegg, 1989:214)

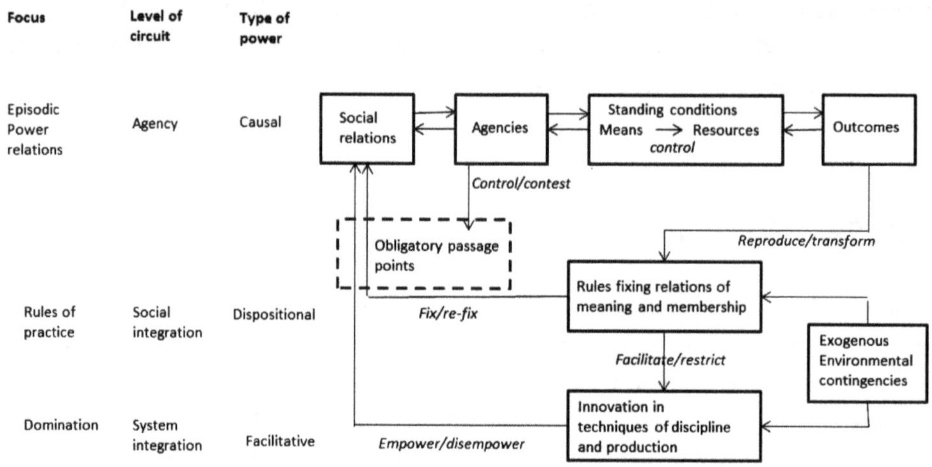

System integration within the context of power is viewed as facilitative and is understood in terms of its ability to produce and achieve collective goals. In the organizational context, system integration is the technological means of control over the material and social setting and the skills associated with these means (Lockwood, 1964). Besides the material means of production, Lockwood includes in systemic integration the material means of surveillance. In short, system integration is constituted by techniques of production and discipline. System integration deals with facilitative power because the material conditions of production might empower or disempower agencies in their productive activities. The circuit of system integration is the major source of change in the circuits of power framework, particularly when the material conditions of production are altered: hence its relevance to improvisation and our study. Changes in the circuit of systemic integration will entail new agencies, techniques and practices that the circuit of social integration might find difficult to resolve.

Improvisation, Power and Information Systems

Exploring the relationship between power and improvisation is challenging and, within IS, Silva (2002) has been one of the few researchers who have attempted this. Silva's (2002) case study, conducted in the context of a strategic outsourcing of information systems in two hospitals in Guatemala, took a narrow definition of improvisation as defined by Weick (1993) and through the lens of 'facilitative power' examined how the outsourcing was conducted. Although insightful it does not address more specific power issues within IS implementation. We propose that this work can be developed further by utilising Clegg's (1989) original 'circuits of power' framework and more recently Silva and Backhouse's (2003) adaption within the context of a system implementation where improvisation was an important activity at various stages of the project. Our intention is to demonstrate our approach by developing and applying a 'circuits of power – improvisation' (CPI) framework to an Enterprise System implementation within a university. We have identified in Table 1 where exploration of the data will occur.

IS improvisation does not take place without reference to organizational power and politics. Distinguishing the nature and circuits of power within an improvisation and the resultant changes has the potential to further our understanding of improvisation in an ES context. It is proposed that our understanding of the three types of improvisational changes (anticipated, emergent, opportunistic) identified by Orlikowski (1996) and Orliksowski and Hofman (1997) can be enhanced if examined against the combined circuits of power improvisation framework, Table 1. The three circuits of power, episodic, social integration and systemic integration may provide a richer account of the complex power and political relations within an organization type that contains both micro and macro-level improvisations and high levels of procedural memory. These typically encompass higher levels of routine knowledge

Table 1. The 'Circuits of Power-Improvisation' (CPI) framework and our interpretation

Circuit	Type of Power	Our Interpretation of the Framework in the Context of Improvisation
Episodic	***Causal Power***: When A makes B do something B otherwise would not do.	Who are the As and Bs when the implementation is initiated? Who are the systems champions and are there resisters? What improvisations occur and who are the As and Bs? What are their objectives? What strategies are employed within the improvisation are adopted by the As and Bs? What resources are mobilised by the As and Bs within the improvisation and what are the outcomes? Can opportunity-based outcomes be explained?
Social Integration	***Dispositional Power:*** Provides the conditions for As to exercise power. Rooted in rules of meaning and membership of the organization.	What are the organizational rules and norms prior to implementation that place the As and Bs within their respective position? How do the improvisations affect these rules and norms? Are there new rules and norms? What are the meanings ascribed to these rules and norms as sense making takes place?
Systemic Integration	***Facilitative Power:*** This power causes the organization to generate outcomes and is defined by techniques of production and discipline.	What techniques are employed by As to ensure Bs compliance during/after an improvisation? What are the work tasks that are affected and how have they changed? Have the improvisations led to the system as a resource to instil discipline? How has this occurred?

and well defined rules (Tjørnehøj & Mathiassen, 2010) and are typically resident within a professional and academic led university and HEI culture.

RESEARCH METHODOLOGY

The study of the ES, a university student administration system (USAS), implementation began in 2007 as an ethnographic study and lasted five years. Ethnography can be defined as a 'style of social science writing which draws upon the writer's close observation of and involvement with people in a particular social setting and relates the words spoken and the practices observed to the overall cultural framework within which they occurred' (Watson 2011:205). Watson (2011) argues that ethnography should involve participant observation, content analysis of documents, stories, myths, rituals, symbols and other artefacts. This may be supplemented and strengthened by interviews, statistical analysis and even small surveys. The research began with an in-depth critical analysis of the documentation leading up to the USAS implementation. This was followed during 2008 by 22 interviews averaging one hour each, with university staff who had been in the university for no less than five years. As in the case of Doolin (1998) the interviewees were interviewed more than once over the research time period.

Interviewees were taken from across the university, Modern (pseudonym). A non-directive interviewing technique was used which allowed respondents to express their own views about organizational life in their own words rather than force them into predetermined categories (Hirschheim & Newman, 1991). The interviews involved a discussion of issues surrounding the participants' prior experience of student information systems, the implementation of USAS, life in the organization and change during and after USAS went live. Interviews were audio taped with permission, transcribed and returned to the interviewees for verification. Anything that was felt by the participants to be problematic was removed from the transcript and after one interview a respondent decided to wholly withdraw her transcript.

Participant observation took place throughout the research study and was recorded using a diary. As a member of staff one of the researchers was able to participate in the activities which contributed to the academic role in providing student data. Working alongside other colleagues it was possible to observe the action of various individuals and interpret them in order to gain insight into the cultural manifestations of the organization (Bryman, 2004). Burgess (1984) also argues that participant observation can increase the richness of the research and Waddington (2004) suggests that being part of day to day activities or important events can provide valuable understanding of organizational practice which can become ritualised over time. In order to understand administrative working practices, the same researcher spent a number of periods of observation during peak times in the academic calendar: student enrolment in October, marks recording after assessment in February and examination board preparation time in June. The research team also recorded events and further progress using participative observation and document collection to subsequently update the research programme to record significant events in the associated digital transformation programme through to 2017.

Using a general inductive approach informed by grounded theory (Crabtree & Miller, 1999; King, 2004) the interviews, documentation and diary data were coded according to theoretical concepts suggested by the data rather than imposed by the researcher. The approach used involved a process of developing initial categories, grouping data, identifying patterns and then making comparisons to uncover shared elements and properties (Barley, 1990; Van Maanen, 1979). The documentation and transcripts were also read critically to identify statements which reflected values, beliefs and assumptions about USAS as well as for evidence of organizational stories, myths and rituals which may have arisen over the period of the research. The analysis of the case study is presented in a form of narrative taking into consideration the improvisations which took place and the elements of each circuit. Bearing these in mind the discussion explores what happened during the USAS implementation and the subsequent digital transformation programme. The intention is to identify those possible elements or factors that have resulted in the systems being adopted by Modern University in the manner it has. Our aim is not to arrive at a conclusion isolating certain factors or improvisations which have had maximum impact on the organization but to articulate an explanation covering a number of the power and political factors involved. The framework is a set of concepts that will support researchers of improvisation and power to collect, analyse and interpret data and practitioners to identify elements and processes that can hinder or support the introduction of a new ES and the associated digital transformations.

THE CASE OF MODERN UNIVERSITY: THE ADOPTION OF USAS

Modern University is a large city based university with over 3000 employees and at the beginning of this research was a decentralised organization consisting of nine academic faculties. In the mid-2000s Modern failed to deliver appropriate student information to central government and turned to an ES, called a University Student Administration System (USAS), solution to address their difficulties. However, where research and finance can be a major consideration in other higher education research studies, the focus at Modern was on the students, the courses offered and on the management of related data. A business case was accepted to make USAS the preferred alternative to upgrading an existing Oracle system. USAS is used by over 70% of the UK Higher Education (HE) market, by 25% of the Scottish Further Education (FE) market and has become the de facto standard for student information systems. Like many other integrated systems of its genre it has been built around an 'ideal' model of university administration consisting of modules to support admission of students to university, programmes to manage the curriculum through a standard three-year degree, a student module to manage enrolment, fees, progression throughout the degree and tools to 'enable' users to analyse, process and extract data. These enterprise systems are marketed as being highly flexible with the opportunity to customise to individual university needs. Thus, with other similar universities introducing it and at meetings with the vendors senior university managers were given assurances that the system could provide the solutions for which they were looking. Anticipated change from a senior university management perspective was centralisation of student data management and more efficient use of both academic and administrative staff resources.

Improvisation and Episodic Power

'Episodic power' focuses on the relationship between resources and outcomes and is intrinsically linked to the improvisations which occurred as organizational members drew upon available material, cognitive, affective and social resources (Cunha et al., 2012). Led by the Director of Finance, and supported by the university registrar and two other senior staff as well as representatives from the vendors, the plans for the implementation of USAS started in 2005. In order for USAS to be introduced into Modern University a budget was set up, a number of new groups or teams were established, and material resources released. The two groups that played a major role in the project were the USAS implementation team (technical staff employed from internal university resource and external agencies) and the USAS liaison group (senior administrative staff seconded from each of the departments) both of which were intended to be temporary structures to be dismantled after the system was up and running. There was little academic staff involvement. The USAS implementation team were seen as the technical experts, housed in their own suite of offices, trained by the USAS vendors, who would lead the technical aspects of the implementation. The USAS liaison group would support departments in getting the data into the system and extracting the requisite information. Examination of minutes of their meetings suggest the focus prior to 'going live' was on USAS technical issues and not on organizational change. This approach is not unusual within these types of information technology projects as the initial focus is to ensure that the physical resources are in place and being accessed by the appropriate staff (Coombs, Knights & Willmott, 1992; Parker, 2000). Nevertheless, the decisions on who has access to the new technology are political and subject to control of the dominant group (Doolin, 1998).

The USAS implementation team and the USAS liaison team worked closely together from 2005 until the system went live in 2007 and shared an excellent working relationship. One example of an area that they jointly worked on was student data migration from all the department systems to the new USAS system. USAS has a very rigid set of data structures which cannot be altered. It was the responsibility of the USAS implementation team to transfer the current university student data into the structures available in USAS. This was not easy as each department kept their student records according to their own needs. The USAS team were also faced with determining how the standardised approach to student data collection would be managed going forward. The Director of Finance and his advisors anticipated this would be a simple transfer of common data into structures congruent with university requirements and facilitated by the USAS vendors. However, Pollock and Cornford (2004) suggest that enterprise systems such as USAS overwhelm organizations so that they have little option but to implement 'default' settings which tend to emphasise similarities between organizations and not their differences. Thus, individual departmental needs are disregarded in the quest to be 'up and running'. The USAS liaison group was asked to assist and advise on new degree programme codes which they did without reference to the academic colleagues who actually managed the students. Nicolini (2007) sees this assumption of 'knowing what is best for the rest' as a feature of many ES implementations as a dominant group, in order to make sense of their identity, imposes a view of practice intended to enhance their credibility and competence. Nevertheless, this improvisation within USAS illustrates the use of affective resources where the two teams felt the need to support one another in overcoming a problem. In fact, the lack of academic knowledge was not seen to be an issue (Briefing paper from USAS Project Manager, 2005). What occurred was an unanticipated emergent change which only became visible when the system went live.

Although academic staff had not been consulted in the lead up to 'going live' with USAS it was the administrative staff who were feeling the stress in relation to the new role they were expected to play. It was anticipated that all administrative staff who dealt with student data would be trained on USAS in time for 'going live'. The USAS vendors had a specific approach to training which they passed on to the USAS implementation team. This approach ensured that everyone used the system in the same way. However, this training by the vendors was expensive and could not be rolled out across the university. Therefore, the USAS project manager decided that the USAS implementation team would provide training for all. The problem was there were insufficient trainers to cover every administrator who needed training on USAS. Hence the improvised training was done in large groups in information technology laboratories with little interaction. Staff complained about the content and quality of that training.

Added to this training issue the USAS implementation team would not allow cascade training *'in case the wrong message was given'* so tensions were high as August 2007 approached. The power to determine the nature of the training is an opportunity based improvisation and is indicative of the growing internal credibility of the USAS implementation team not just with the senior management of the university but also over the departmental staff taking part in the training. These staff may never have encountered the USAS team prior to training and this acts to transform and re-enforce their identity and position in the organization (Coombs, Knights, & Willmott, 1992).

Academic staff knew little about USAS until they returned from annual leave in September 2006. Student enrolment had taken place without their input and when term began there was chaos: staff were in the wrong rooms, students were on the wrong degree programmes, class lists were inaccurate. This was a result of a previous decision taken by the USAS implementation team and the USAS liaison team when developing degree programme codes for the system. When the Academic Heads of department approached the USAS implementation team they were told it was 'teething problems' or that they were the only department with that particular problem. The USAS liaison team were overworked trying to sort out problems within the departments; problems that had not been anticipated.

In the context of the USAS system, looking exclusively at the episodic circuit of power provides insight into the strong position of the Finance Director, based on numerous resources, and his ability to manage his expectations for USAS through his appointed teams. Nevertheless, by devolving responsibility, improvisations are made at points in the project which may not be congruent with the sentiment of what is ultimately the target outcome. Also, the result of an improvisation may not be felt for some time when it impacts on another aspect of organizational life. This is dealt with here through looking at the circuit of social integration.

Improvisation and Social Integration

This circuit deals with rules of meaning and membership. The USAS system influenced the way both academic and administrative staff interpreted the management style and priorities. The emphasis on the spending on USAS sent a strong message to all that USAS would be the only way that student data would be managed, that local discretion around student data was not to be tolerated and signalled the end of a more participatory approach to management. Here, we show how new social groups have become important within Modern University as well as the reactions and interactions of other staff with them. Tension between academics and administrative staff were high during 2007 and this was also the case within the administrative teams who were dealing with USAS on a day to day basis. Academic staff began to complain about their lack of information and inability to access the system. When they then tried to contact the USAS team for help they often found the phones 'off the hook' and emails not answered. Even going to their offices was impossible as the USAS team were in a suite of rooms protected by a digital lock. The eventual response of the USAS team was the development of 'rules of access'. Any enquiry or request for information had to be made through appropriate channels and only certain staff had access rights to the USAS team. Even though data was urgently needed to be entered onto USAS they enforced their training and access policy. Nevertheless, even those who had been trained still found USAS difficult to work with.

The improvisation of 'good housekeepers' as an informal group of USAS experts with advanced skills to support early system difficulties has proved to be a very powerful influence within Modern and this group has grown to over 100 strong in 2012. Although in some HE institutions academic staff are able to enter their marks into USAS, Modern University have chosen to only allow administrative staff to have direct access to the system. Licence costs have been raised as an issue - yet other more political aspects have been identified e.g. 'good house keepers' have some strong opinions about the ability of academics to deal with USAS if given access and it was not unusual to hear criticism of academics especially with regard to providing correct and timely information and inputting marks.

Improvisation and System Integration

This circuit concentrates on techniques of control, discipline and production and explores who has benefited or been empowered by the improvisation and who has been disempowered. From our research USAS is a system which is not easy to use and has caused a number of challenges for Modern University that were not anticipated at the start of the project. During the two years within which the system was introduced the USAS team built up a lot of technical knowledge and it became obvious that their skills could not easily be lost to the organization. They became a permanent group when departmental staff recognised they needed increasing amounts of student information not readily available without some degree of USAS programming. They play a major role in the university and even have their own newsletters which are sent out to departments when new changes to USAS are to be implemented. Academics have little knowledge of this. The improvised processes and control introduced in 2007 have also impacted upon academic staff. For example, at that time, academics no longer had access to the 'system' for recording of marks for examination boards and have found themselves much more regulated by the departmental USAS administrative requirements:

Examination boards are also conducted in a different manner. USAS uses algorithms to calculate marks for modules and then the overall year average:

Using USAS to calculate degree classifications has allowed management (both academic and administrative) to control academics behaviour at examination boards. Academics find it difficult to challenge the logic of the system and USAS as to do so would be to argue for paternalism and self-interest and deny students fairness (Clarke & Newman, 1997). Added to this are penalties for not abiding by the USAS procedures and rules. Administrators report academics to their managers when they fail to meet USAS marking deadlines and it can be seen from our research that academics are increasingly becoming subjugated to a system that challenges their sense of meaning, identity and reality through their participation in an increasing range of disciplinary practices (Coombs, Knights, & Willmott, 1992).

Within departments administrative staff who work regularly with USAS data have seen their status increase as managers require more information from the system. IT literacy is an essential skill for many administrative roles in Modern University and is reflected in salary levels. These new types of staff appear happier sitting at a computer than interacting with the student body and this is a cause for some concern with a number of academic staff. It is also divisive amongst administrative staff whose alternative skills are not recognised to the degree they once were.

Finally, it can be seen that management have used an opportunity based improvisation to replace academic staff roles with those of administrators. Prior to USAS, academics ran the student management systems and had access to all of the data. They also enrolled the students on their degree programmes and managed many of the processes around assessment periods. USAS and its centralising ability has facilitated the removal of the academic from many of these roles. To many academics this has been viewed negatively and some feel they have lost their status in the University. Even feeding back marks to students has been changed without academic staff consultation: students can now access their module marks from USAS (and latterly, post 2016, from the electronic Learning Platform (eLP) and electronic plagiarism and electronic marking assessment system (Turnitin)) almost as fast as the administrators are entering them.

Although it may be considered good practice to allow access to personal data it also involves communicating new processes to staff so that they can be prepared. Students are now more assertive than ever - and this is putting even more stress on university staff. This is not helped by the university documentation that continually refers to students as customers and, as is often the case, with that status comes new expectations.

Integration of USAS Beyond 2007: The New Digital Transformation Programme

USAS at Modern University was procured and implemented between 2005 and 2007 when it went fully live. This heralded a time of significant acceleration in terms of the adoption of new digital technologies, loosely coupled to a deliberate university 10-year corporate strategy and vision. This focused on increasing both efficiency and effectiveness in terms of streamlined administrative bureaucracy, enhancement of teaching and learning delivery and quality, as well as increased levels of research output and impact. External factors driving this strategy included: The Research Excellence Framework (REF) to assess and rank research quality, Border Control Agency (BCA) government immigration policy to monitor overseas student attendance, the National Student Survey (NSS) to rank universities on student experience, and more recently the Teaching Excellence Framework (TEF) to assess and rank teaching quality. USAS could therefore be viewed as being at the core of an ambitious digital transformational programme enabling the proliferation of new key performance indicator (KPIs) targets and measurement tools, computerised monitoring and business intelligence systems – all requiring aggressive organizational change initiatives. A plethora of new digital technologies and information systems were (opportunistic change) able to be spun off from the new ES and potentially integrated (emergent change) or used in a complementary manner (anticipated change) with the core system. Table 2 provides an illustration of these policy initiatives and the corresponding digital technologies/information systems that were adopted at Modern University.

Table 2. Digital Transformation Programme at Modern University

2007	2008	2012	2013	2014	2015	2016	2017
SITS (USAS) Blackboard e-Learning Platform (eLP)	Research Excellence Framework (REF) University League Tables (Times Higher) National Student Survey (NSS)	BCA Electronic Student Attendance Monitoring (SAM-1) barcode and smart card system and reporting pilot eLP now integrated with USAS	Student Attendance Monitoring – full roll out (SAM-2)	Postgraduate Researcher monitoring e-Vision system REF electronic monitoring of research performance outputs, strategy and impact	SAP self-service – MyHR. SAM-3 desktop – integrated with eLP for self-service reporting of student attendance	Self-serve Electronic travel and accommodation booking; Self-serve E-procurement Purchase to Procurement workflow system (P2P) link with SAP finance system Lecture video recording	Electronic Marking and Assessment (EMA) System integrated with eLP and also Turnitin plagiarism system Teaching Excellence Framework (TEF)

As can be seen in Table. 2, USAS has enabled a new generation of integrated systems to be developed in an opportunistic manner – whereby they have precipitated emergent changes in administrative procedures and work flow systems. This has rapidly changed the role and status of academics (centralised administration controlled procedures and policies resulting in less academic staff autonomy) in terms of programme/module development, teaching and learning management and pedagogic delivery. Key amongst this is the move towards making USAS and its additional HR and Finance modules self-service for both administrators and academics. This shifted the administrative work on to academics whereby an attendant effect was then to disestablish faculty administrative services and recentralise/ physically relocate them to fit within a 'One University' strategy. Many long and traditionally established roles then disappeared in what was in effect both a rapid and major business process transformation. This was mirrored by a new wave of performance monitoring and control systems aimed at academic staff. These new digital technologies included mandatory electronic Performance Development and Appraisal (PDA), Performance Review and Innovation Plans (PRIPs), Institutional Research Repository (to monitor staff research publication outputs) and a Research Grant monitoring system (to monitor staff grant proposals and bidding activity). More recent technologies also included electronic marking and assessment (EMA) for streamlining the assessment process and personal monitoring of staff timeliness and consistency in the performance of mark submissions and feedback, as well as video recording of main lectures and electronic student rating of programme modules. Many of these systems have been opportunistically integrated with USAS, a significantly more functional Electronic Learning Portal (eLP), and then accessed through a new Corporate website and intranet system.

DISCUSSION

In this section we explore how our work provides further insight into improvisation and how it relates to previous theoretical models. Orlikowski's (1996) work on improvisation is very insightful but it does not really address the very nuanced nature of improvisation and the difficult issue of power within it. Even the detailed work on power and institutionalization of IS discussed by Silva (2007) does not focus on the improvised actions which brought about the embedded IS. Thus, it has been our intention to surface these concerns and to explore them through the case study presented here. Figure 3 provides a timeline view of anticipated, emergent, problem and opportunity based change identified over the duration of the USAS (SITS technology) project.

Figure 3. A timeline showing improvisational change for the USAS (SITS) system

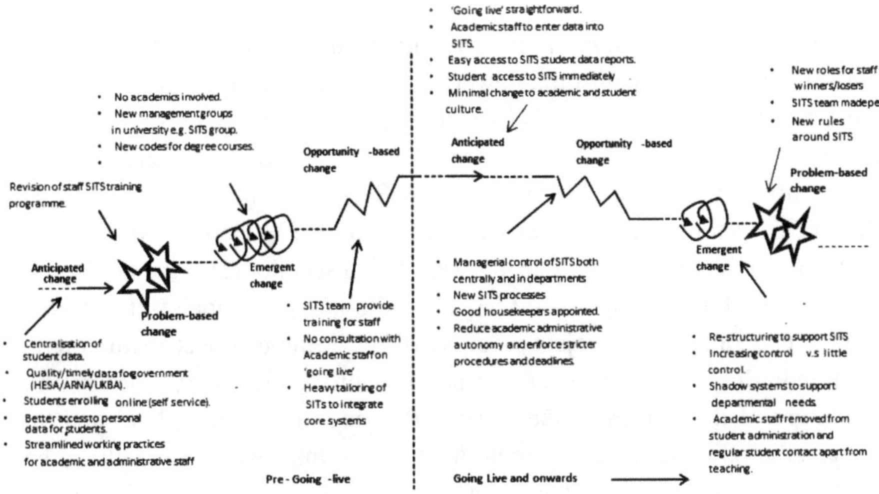

This represents a literal adaptation of Orlikowski's improvisational change model. Although useful in terms of providing an overview of the major events and changes affecting academics and administrators, it still does not provide sufficient means of analysing the deeper power relations and causal effects of agents (human, organizational and technology) that may determine outcomes and result in new embedded and institutionalized routines, practices and structures. The original model did not show problem based change as distinct from emergent or opportunity based changes. Problem based changes can be equally significant triggers for improvisation requiring rapid actions and use of available materials to hand. This is more characteristic of bricolage as opposed to anticipated changes where there is a timeframe for planned actions according to some pre-determined goals and plans. In our analysis we have therefore included a new category for problem based improvisational change as opposed to focusing on anticipated changes. The anticipated changes can be seen in Figure 3 alongside the identified sequence of emergent, opportunistic changes and problem- based changes over both the (pre) and (post) go live phases of the project. Added to this we have further deconstructed improvisational types related to emergent, opportunistic and problem based changes and mapped them to the various agencies and standing conditions as described by Clegg's (1989) circuits of power framework.

The Circuits of Power- Improvisation (CPI) Framework

It is evident from our research that improvisations are actions which do not occur without some reference to episodic power as actors adopt a bricolage approach to drawing on material, cognitive, affective and social resources (Cunha et al., 1999). At the start of the project, USAS was viewed as the OPP on to which actors from across the university were enrolled to join the USAS team and USAS liaison team. The Finance Director presented no alternative solution to Modern University's data issues (problematization). The two teams were socialised alongside the vendor team and became very close sharing a specific USAS suite and shared goals (interessement and enrolment). Thus, as the project progressed they became project champions and gained in credibility within the university (mobilisation) (Callon, 1986; Latour, 1987).

The nature of the improvisations during the lead up to 'going-live' and their link to episodic power can now be seen. Most of the improvisations outlined here were undertaken by agencies that drew on the financial support of the USAS project budget as well as the affective and social resources of being part of a team. They did not need at that time to refer to anyone outside the USAS project. Nevertheless, the teams were faced with issues that could not be addressed by financial resources e.g. codes for degree programmes or training of staff and at that point the improvisations called on other resources such as mental models and affective support of the other teams (Cunha et al.,1999). This is where episodic power comes to the fore as action is legitimised through standing conditions, resources and control (Figure 2). The mental models employed by the USAS teams for both the training and degree programme codes were limited as can be seen from the outcomes in our case study - yet they were still able to determine that action.

Training was a major issue in this implementation and even when the USAS team took control it was quickly realised that they had insufficient human resource to undertake the level of training that was required for all the administrators and academics across the university. Thus, new improvised rules emerged which targeted certain staff and not others. Now these rules have become embedded within the organization and have further emerged over time through the social and systemic integration process.

Although there are many examples of how certain actors called on a variety of resources to support them in their actions during the USAS implementation what is not apparent initially is the link between improvisation and social integration. For the circuit of social integration in which rules fix relations of meaning and membership, certain fixtures of meaning are privileged and certain membership categories are aligned with these meanings (Clegg, 1989). Although social integration began prior to going-live it is clear that once the USAS system was in place it became more evident. To illustrate, it is apparent from our study that prior to USAS being introduced into Modern University academic staff led on most student related activities. When the system was introduced into the faculties in August 2006 only administrative staff had had training and hence were the only ones with access to the data and student information. The difficulties at 'going live' led to faculty registrars and Deans exercising their dispositional power in order to 'get things sorted out'. Thus, new relationships emerge between the faculties and the USAS team which are determined by the dispositional power (Clegg, 1989) of the technologists. Yet within the faculties it is the administrators by virtue of their superior USAS knowledge which determines how USAS will be used, by whom and when access to it will be granted. The emergent nature of the Good Housekeepers improvisation has been useful to the faculties as they now have their own resource on which they can call for most USAS related business and therefore do not need to access the USAS team as frequently.

Clegg (1989: 224) views system integration as the circuit of power which is concerned with the empowerment and disempowerment of agencies' capacities, as they become "..more or less strategic as transformations occur which are incumbent upon changes in techniques of production and discipline" Improvisations which have taken place within Modern University have led to the empowerment and disempowerment of a number of staff. The centralised nature of the USAS system has led to a concentration of control at the centre of the university with little localised discretion in the faculties. Rules about all academic matters have ensured that conformity is the norm and intense monitoring is undertaken. Some of these rules are embedded within the USAS data structures and others have emerged through improvisations at key points in the academic year. Within the academic departments it is evident that there has been a power shift and in our interviews academic staff expressed concern about their perceived status and the increasing power of certain administrative staff.

The adoption and implementation of USAS was used to force through major centralised administrative and structural changes that then continued through later years through to 2017. Integration technologies building on enterprise application integration platforms have enabled ambitious automation of existing administrative workflows and business processes. This has been driven by opportunistic change to enable and consolidate the downsizing of core (lower skilled) administrative staff with a deliberate strategy for centralisation of services (with an increasing administrative management cadre) and the adoption of a new public management and performance monitoring (targets and league table driven) culture. Problem based change could also be identified, specifically related to the development of class based and online student attendance monitoring systems to comply with government immigration and BCA policy. Problem based change could also be seen as a driver for the rapid adoption of electronic marking and assessment (EMA) integrated with plagiarism detection software (Turnitin) in order to centralise the assessment administrative process and monitor staff performance with respect to achieving mandatory 21 working day feedback turnaround targets (to improve student feedback for NSS scores). Taken as a whole, the 10 year cycle from 2007 to 2017 might be seen as demonstrating emergent change driven by incremental integration of USAS, SAP and the eLP and then punctuated by rapid opportunistic and problem based changes reacting to government policy for immigration control and the competitive marketisation of the UK university sector.

The new digital transformation programme at Modern could also be seen to mirror significant changes in power relations – both between core academic staff and senior executive academic management and also academic staff and the rise of the new managerial administrative and centralised service functions. In terms of episodic power, resources were increasingly siphoned from faculty budgets who had to pay very high financial overhead costs/contributions to support the increasingly centralised university service departments. These new service departments expanded rapidly, especially functions such as Corporate Marketing, based on a deliberate change strategy that focused on raising the university standing in published university sector league tables. This follows a general international trend where administrative staff now outnumber academic staff at over two thirds of UK universities (Jump, 2015) with an increasing number of higher level managers brought in to manage the expanding administrative service departments! (Ginsberg, 2012). In Modern University, the deliberate strategy to expand student numbers and increase teaching and research quality ratings precipitated a culture of performance management, enabling greater hierarchical control paralleled by new academic reward and incentivisation regimes. Alvesson and Spicer (2016) also note the effects of coercive power and the embracement of neo-liberalist agendas. They identify other effects such as manipulation and mainstreaming (game playing to optimise institutions positions on the university ranking league tables with recruitment drives to develop 4* by 4* research factories), domination (reshaping of academic norms and values), and subjectification (incentivising the creation of new academic and managerial identities). Disturbingly, along with Martin (2016) they identify an increasing culture of defeatism (and willingness) amongst academics to go along with the increased managerialism and production factory ethos. Academics are seen to either delude themselves or justify this by an acquiescence to a 'game playing' culture based on self-preservation and personal career advancement (what they term subjectification).

At Modern, this 'academic gamification culture' could be seen to be accelerated due to the overwhelming roll-out of the new digital technologies enabling new levels of performance management and hierarchical control. The circuit of social integration was enabled by changed job descriptions, new centralised policies and procedures and structural organizational change both for academics and also administrative staff. Systems integration was also responsible for moving towards a new 'one university' culture and only made possible due to the enterprise application integration innovations. It was now 'one size fits all' heralding a loss of academic integrity, autonomy and control in traditional decision making involved with faculty strategy, programme development, management and assessment processes.

The CPI Framework and Sociotechnical Theory Development

A number of studies have researched ES from a sociotechnical perspective, specifically to better understand ERP implementation (Lyytinen et al., 2009; Kwahk & Ahn, 2010) as well as investigating the role of improvisation and change in Information Systems development and implementation (Lyytinen & Newman, 2008). Mumford (2003, 2006) stated that sociotechnical systems should emphasise quality of working life and humanism at work. This approach should facilitate new forms of organizational innovation, often as a result of the adoption of new technologies that remove elitism, promote flatter hierarchies, develop multiskilling and facilitate more democratic processes of decision making. However, existing methods for sociotechnical design, such as ETHICS (Mumford, 2003), PSIC (Lyytinen & Newman, 2008) or SPRINT (Wastell, 2011), treat technology as neutral and do not explicitly deal with complex organizational power relations, neither focus on the challenges of improvised strategies and working practices often enabled by rapid new technology adoption typical of enterprise systems projects. Stahl (2007) recognises and critiques some of these limitations and proposes that sociotechnical approaches can be enhanced through adopting principles and theory associated with critical research in IS and computer and information ethics. The CPI framework proposed in our research and illustrated in this paper is 'critically' informed and can be used to surface issues of power, domination and control whilst taking a pragmatic view of ES implementation and digital transformation projects. This is where planned and deliberate strategies are supplemented by periods of opportunistic, emergent and problem based organizational change. This also responds to recent critiques and identified challenges for sociotechnical theory, methods and approaches relating to its inherent idealism and inability to influence IS innovations during design, use and implementation whilst neglecting power dynamics and the inability to legitimize and sustain spontaneous innovation (Avgerou & McGrath, 2007; McGrath, 2005).

CONCLUSION

IS research in general, and sociotechnical research in particular, must be prepared to build on theoretical models and frameworks that have the potential to provide better insight into the complex area of digital technology adoption and implementation. This must involve consideration of the power and politics which envelop these types of projects and recognition that improvisation is a reality when implementing them. Clegg's (1989) theoretical exploration of power offers great explanatory potential for large scale information systems adoption and implementation. Its later adaptation and use as a model by Silva and Backhouse (2003) demonstrates how it can be used to provide much deeper insight and a rich narrative for the institutionalization of an IS. Similar parallels can be drawn with the case of Modern where there were very obvious exogeneous factors associated with government policy that could be seen as triggers for rapid change and also a source of system integration and facilitative powers. Digital transformation projects are characterised by the importance of improvised change as opposed to a reliance on more rational strategic and anticipated plans for the new developments. In the case of Modern, the administrative staff seized an opportunity to redevelop their skills and knowledge to operate and redefine the new systems whilst simultaneously utilizing dispositionary and facilitative powers for new forms of both social and systemic integration. This is seen to be increasingly the case in organizations characterised by minimal structures and low procedural memories (Tjørnehøj & Mathiassen, 2010) but also perhaps more surprisingly in professional and administrative led organizational cultures where there might be high procedural memory and large bureaucratic structures but also opportunities for radical shifts in roles and responsibilities; hence the growth of managerialism (Deem, 2004), new public management, marketisation and neo-liberalist agendas (Alvesson & Spicer, 2016).

Within the context of ES implementations, and even more recently digital transformation projects, there are relatively few sociotechnical, ethnographic and longitudinal studies that have been undertaken or published. Those that are published tend not to adopt a critically informed stance but focus on improvisation and bricolage as a means of effecting strategic change and innovation as demonstrated by the use of bricolage for IS and business strategy alignment (Ferneley and Bell, 2006), improvisation for health information systems success (Heeks, 2006), rapid adoption of groupware technologies (Orlikowski & Hofman, 1997), and the effective management of outsourcing (Silva, 2002). Our exploration of the use of an adapted CPI framework more closely follows the later work of Silva and Backhouse (2003) but places additional emphasis on the types of improvisation undertaken as the ES project unfolds and how they relate to the different circuits of power identified. Our study is also relevant for professional and administratively intensive organization cultures such as universities but may also be applicable to other types of organization that exhibit similar characteristics. Besides being a tool for academic researchers to analyse and understand ES adoption and implementation post hoc, the CPI framework might also be used by more critically aware or ethical practitioners. This may be used more in real time as events unfold and materials to hand are utilised.

Change can happen extremely rapidly resulting in social, technical and systemic integration. Understanding the consequences of actions earlier in the ES implementation might negate undesirable effects and the over hasty institutionalization of potentially inefficient and damaging work practices and reorganization. The case of ES adoption and implementation is still live and on-going at Modern university. The student administration system has now been effectively institutionalized where academics no longer question its existence, its extension or its role within the academic decision-making processes. It has now become the centre-piece or core of a highly ambitious digital transformation strategy and project. This heralds a new dawn for universities to become mass producers (production machines) of students with increasingly centralised bureaucratic processes and more worryingly, less academic freedom and autonomy. This may lead to professional deskilling resulting in heavily formulaic teaching and learning strategies and standardised content delivered by lower paid associate academic staff. New digital technologies have provided the embedded agency to make this possible and are reinforced by both the social and systemic integration circuits of power.

In the future we anticipate that the CPI framework will be evaluated for its potential to identify improvisation types as they occur and assess the potential changes needed for a more ethical exercise of episodic, social and systems integration circuits of power.

REFERENCES

Alvesson, M., & Spicer, A. (2016). Conditional Surrender? Why do Professionals Comply with Managerialism? *Journal of Organizational Change Management*, *29*(1), 29–45. doi:10.1108/JOCM-11-2015-0221

Avgerou, C., & Mcgrath, K. (2007). Power, Rationality, and the Art of Living through Socio-Technical Change. *Management Information Systems Quarterly*, *31*(2), 295–315. doi:10.2307/25148792

Barley, S. R. (1990). Images of imaging: Notes on Doing Longitudinal Fieldwork. *Organization Science*, *1*(3), 220–247. doi:10.1287/orsc.1.3.220

Bryman, A. (2004). *Social Research Methods*. Oxford, UK: Oxford University Press.

Burgess, R. G. (1984). *In the Field: An introduction to Field Research*. London: George Allen and Unwin. doi:10.4324/9780203418161

Callon, M. (1986). Some elements of a sociology of translation: domestication of the scallops and the fishermen of St Brieuc Bay. In J. Law (Ed.), *Power, Action and Belief* (pp. 196–233). London: Routledge and Keegan Paul.

Ciborra, C. (1996). Improvisation and Information Technology in Organizations. *Proceedings of the International Conference on Information Systems*, 369-380.

Ciborra, C. (2002). *The Labyrinths of Information: Challenging the Wisdom of Systems*. Oxford, UK: Oxford University Press.

Clarke, J., & Newman, J. (1997). *The managerial State: Power, Politics and Ideology in the Remaking of Social Welfare*. London: Sage.

Clegg, S. (1989). *Frameworks of Power*. London: Sage Publications. doi:10.4135/9781446279267

Coombs, R., Knights, D., & Willmott, H. C. (1992). Culture, Control and Competition; Towards a Conceptual Framework for the study of Information Technology in Organizations. *Organization Studies*, *13*(1), 51–72. doi:10.1177/017084069201300106

Crabtree, B. F., & Miller, W. L. (1999). Doing Qualitative Research. *Sage (Atlanta, Ga.)*, 1999.

Cunha, M. P., Clegg, S. R., & Kamoche, K. (2012). Improvisation as "real time foresight". *Futures*, *44*(3), 265–272. doi:10.1016/j.futures.2011.10.009

Cunha, M. P., Viera Da Cunha, J., & Kamoche, K. (1999). Organizational improvisation: What, when, how and why. *International Journal of Management Reviews*, *1*(3), 299–341. doi:10.1111/1468-2370.00017

Deem, R. (2004). The knowledge worker, the manager- academic and the contemporary UK university:new and old forms of public management. *Financial Accountability & Management*, *20*(2), 107–128. doi:10.1111/j.1468-0408.2004.00189.x

Deem, R., & Brehony, K. J. (2005). Management as ideology: The case of 'new managerialism' in higher eduction. *Oxford Review of Education*, *31*(2), 217–235. doi:10.1080/03054980500117827

Desanctis, G., & And Poole, M. S. (1994). Capturing the Complexity in Advanced Technology Use: Adaptive Structuration Theory. *Organization Science*, *5*(2), 121–147. doi:10.1287/orsc.5.2.121

Dong, L., Neufeld, D., & Higgins, C. (2009). Top management support of enterprise systems implementations. *Journal of Information Technology*, *24*(1), 55–80. doi:10.1057/jit.2008.21

Doolin, B. (1998). Information technology as disciplinary technology: Being critical in interpretive research on Information Systems. *Journal of Information Technology*, *13*(4), 301–311. doi:10.1057/jit.1998.8

Elbanna, A. R. (2006). The validity of the improvisation argument in the implementation of rigid technology. *Journal of Information Technology*, *21*(3), 165–175. doi:10.1057/palgrave.jit.2000069

Elbanna, A. R. (2007). Implementing an integrated system in a socially dis-integrated enterprise: A critical view of ERP enabled integration. *Information Technology & People*, *20*(2), 121–139. doi:10.1108/09593840710758040

Ferneley, E., & And Bell, F. (2006). Using bricolage to integrate business and information technology innovation in SMEs. *Technovation*, *26*(2), 232–241. doi:10.1016/j.technovation.2005.03.005

Foucault, M. (1977). *Discipline and Punishment: The birth of the prison*. London: Penguin.

Fowler, A., & And Gilfillan, M. (2003). A framework for stakeholder integration in higher education information systems projects. *Technology Analysis and Strategic Management*, *15*(4), 467–489. doi:10.1080/095373203000136051

Ginsberg, B. (2011). *The Fall of the Faculty: The Rise of the all administrative university and why it matters*. Oxford, UK: Oxford University Press.

Heeks, R. (2006). Health information systems: Failure, success and improvisation. *International Journal of Medical Informatics*, *75*(2), 125–137. doi:10.1016/j.ijmedinf.2005.07.024 PMID:16112893

Hirschheim, R., & And Newman, M. (1991). Symbolism and information systems development: Myth, metaphor and magic. *Information Systems Research*, *2*(1), 29–62. doi:10.1287/isre.2.1.29

Jump, P. (2015). Academics in the Minority at more than two thirds of UK Universities. *Times Higher Education*. Retrieved on July 11, 2017 from https://www.timeshighereducation.com/news/academics-minority-more-two-thirds-uk-universities#survey-answer

Kamoche, K., & And Cunha, M. P. (2001). Minimal Structures: From Jazz Improvisation to Product Innovation. *Organization Studies*, *22*(5), 733–764. doi:10.1177/0170840601225001

King, N. (2004). Using templates in the thematic analysis of text. In C. Cassell & G. Symon (Eds.), *Essential Guide to Qualitative Methods in Organisational Research* (pp. 256–270). London: Sage. doi:10.4135/9781446280119.n21

Kwahk, K. Y., & Ahn, H. (2010). Moderating Effects of Localization Differences on ERP use: A Socio-Technical Systems Perspective. *Computers in Human Behavior*, *26*(2), 186–198. doi:10.1016/j.chb.2009.10.006

Latour, B. (1987). *Science in Action*. Cambridge, MA: Harvard University Press.

Law, J. (1986). *Power, Action and Belief*. London: Routledge and Kegan Paul.

Law, J. (1991). Power, discretion and strategy. In A Sociology of monsters: Essays on power, technology and domination (pp. 165-191). Routledge.

Lockwood, D. (1964). Social and System Integration. In G. K. Zollschan & W. Hirsch (Eds.), *Exploration in Social Change* (pp. 244–257). London: Routledge and Kegan Paul.

Lyytinen, K., & Newman, M. (2008). Explaining Information Systems Change: A Punctuated Socio-technical Change Model. *European Journal of Information Systems*, *17*(6), 589–613. doi:10.1057/ejis.2008.50

Lyytinen, K., Newman, M., & Al-Muharfi, A. R. A. (2009). Institutionalising Enterprise Resource Planning in the Saudi Steel Industry: A Punctuated Socio-Technical Analysis. *Journal of Information Technology*, *24*(4), 286–304. doi:10.1057/jit.2009.14

Mangham, I. L., & Pye, A. (1991). *The Doing of Managing*. Oxford, UK: Basil Blackwell.

Martin, B. R. (2016). What's Happening with our Universities? *Prometheus*, *34*(1), 7–24. doi:10.1080/08109028.2016.1222123

Mcgrath, K. (2005). Doing Critical Research in Information Systems: A Case of Theory and Practice Not Informing Each Other. *Information Systems Journal*, *15*(2), 85–101. doi:10.1111/j.1365-2575.2005.00187.x

Mintzberg, H. (1994). The Fall and Rise of Strategic Planning. *Harvard Business Review*, (Jan-Feb): 107–114.

Moorman, C., & Miner, A. S. (1998). The Convergence of Planning and Execution: Improvisation in New Product Development. *Journal of Marketing*, *62*(July), 1–20. doi:10.2307/1251740

Morch, A. I., & Mehandjiev, N. D. (2000). Tailoring as collaboration: The mediating role of multiple representation and application units. *Computer Supported Cooperative Work*, *9*(1), 75.

Morgan, G. (2006). *Images of Organization* (2nd ed.). Beverly Hills, CA: Sage.

Mumford, E. (2003). *Redesigning Human Systems*. Hershey, PA: Information Science Publishing. doi:10.4018/978-1-59140-118-6

Mumford, E. (2006). The Story of Socio-Technical Design: Reflections on its Successes, Failures and Potential. *Information Systems Journal*, *16*(4), 317–342. doi:10.1111/j.1365-2575.2006.00221.x

Nicolini, D. (2007). Stretching out and expanding medical practices: The Case of Telemedicine. *Human Relations*, *60*(6), 889–920. doi:10.1177/0018726707080080

Oliver, D., & Romm, C. (2009). ERP adoption –What do they say about it? In C. Brooke (Ed.), *Critical Management Perspectives on Information Systems* (pp. 103–131). Oxford, UK: Butterworth-Heinmann.

Orlikowski, W. J. (1996). Improvising Organizational Transformation over Time: A Situated Change Perspective. *Information Systems Research*, *7*(1), 63–92. doi:10.1287/isre.7.1.63

Orlikowski, W. J., & Hofman, J. B. (1997, Winter). An Improvisational Model for Change Management: The Case of Groupware Technology. *Sloan Management Review*, 11–21.

Parker, M. (2000). *Organizational Culture and Identity*. London: Sage.

Pollock, N., & And Cornford, J. (2004). ERP systems and the university as a "unique" organisation. *Information Technology & People*, *17*(1), 31–52. doi:10.1108/09593840410522161

Silva, L. (2002). Outsourcing as an Improvisation: A Case Study in Latin America. *The Information Society*, *18*(2), 129–138. doi:10.1080/01972240290075057

Silva, L. (2007). Epistemological and theoretical challenges for studying power and politics in information systems. *Information Systems Journal*, *17*(2), 165–183. doi:10.1111/j.1365-2575.2007.00232.x

Silva, L., & Backhouse, J. (2003). The circuits of power framework for the study of power in the institutionalisation of information systems. *Journal of the Association for Information Systems*, *4*(6), 294–336. doi:10.17705/1jais.00041

Stahl, B. C. (2007). Ethics, Morality and Critique: An Essay on Enid Mumford's Socio-Technical Approach. *Journal of the Association for Information Systems*, *8*(9), 479–490. doi:10.17705/1jais.00138

Thayer, L. (1998). Leadership/ communication: a critical review and a modest proposal. In G. M. Goldhaber & G. A. Barnett (Eds.), *Handbook of Organizational Communication* (pp. 231–263). Norwood, NJ: Alex.

Tjørnehøj, G., & And Mathiassen, L. (2010). Improvisation during process-technology adoption: A longitudinal study of a software firm. *Journal of Information Technology*, *25*(1), 20–34. doi:10.1057/jit.2009.20

Van Maanen, J. (1979). The Fact of Fiction in Organizational Ethnography. *Administrative Science Quarterly*, *24*(4), 539–550. doi:10.2307/2392360

Waddington, D. (2004). Participant Observation. In C. Cassells & G. Symon (Eds.), *Essential guide to qualitative methods in organizational research* (pp. 154–164). London: Sage. doi:10.4135/9781446280119.n13

Wagner, E. L., Newell, S., & Piccoli, G. (2010). Understanding Project Survival in an ESS Environment: A Sociomaterial Practice Perspective. *Journal of the Association for Information Systems*, *11*(5), 276–297.

Waring, T., Wainwright, D., & Skoumpopoulou, D. (2016). Enterprise Systems Adoption: A Power and Improvisation Perspective. *International Journal of Systems and Society*, *3*(1), 49–66. doi:10.4018/IJSS.2016010104

Wastell, D. (2011). *Managers as Designers in the Public Services – Beyond Technomagic*. Devon, UK: Triarchy Press.

Watson, T. J. (2011). Ethnography, Reality, and Truth: The Vital Need for Studies of 'How Things Work' in Organization and Management. *Journal of Management Studies*, *48*(1), 202–217. doi:10.1111/j.1467-6486.2010.00979.x

Weick, K. E. (1993). The collapse of sensemaking in organizations: The Man Gulch disaster. *American Quarterly*, *38*, 628–652.

Weick, K. E. (1995). *Sense Making in Organizations*. Sage.

Weick, K. E. (1998). Introductory Essay: Improvisation as a Mindset for Organizational Analysis. *Organization Science*, *9*(5), 543–555. doi:10.1287/orsc.9.5.543

Weick, K. E. (1999). Theory Construction as Disciplined Reflexivity: Trade Offs in the 90s. *Academy of Management Review*, *24*(4), 797–806.

Chapter 9
Factors Influencing Knowledge as Practice in the Knowledge Workplace

Penny Hart
University of Portsmouth, UK

ABSTRACT

This chapter examines issues around sharing subjective knowledge as practice in an organization, exploring what knowledge means to those working with it, how knowledge sharing happens in practice, and what influences its effectiveness. It is suggested that knowledge as practice as internalized by individuals can be difficult to communicate for a range of reasons, many relating to the effects of organizational subcultures on employee self-efficacy. A case study is used to illustrate this, based on fieldwork carried out in a research organization using the appreciative inquiry method to look at influences on the individual practice of knowledge sharing. An environment that promotes cognitive and affective trust amongst an organization's employees may go some way to address these issues.

INTRODUCTION

In organisations, different types of knowledge can be recognised. For those working in an organisation whose business is knowledge development and deployment, knowledge as practice is a crucial attribute and complements the explicit knowledge available in an organisation. In knowledge as practice, workers make use of their subjective understanding and expertise in exercising judgement on how explicit knowledge can be used. This knowledge as practice complements explicit knowledge but is

DOI: 10.4018/978-1-5225-5996-2.ch009

hard to pin down, and harder to communicate or share. In this chapter we explore what knowledge as practice means to those working with it, how knowledge sharing happens in practice and what influences its effectiveness. A case study is used to illustrate this, based on fieldwork carried out in a research organisation using the Appreciative Inquiry Method (e.g. Stowell & Welch, 2013) to look at influences on the individual practice of knowledge sharing. It is suggested that knowledge as practice as internalised by individuals can be difficult to communicate for a range of reasons, many relating to the effects of organisational culture on employee self-efficacy.

Organisational Knowledge

The effective sharing of knowledge is a critical success factor for organisations, but there are ontological questions about what is meant by 'knowledge'. Knowledge management as a discipline over the past two decades has privileged particular views of what organisational knowledge is, treating it as an asset or 'content', which can be created, manipulated and stored. Knowledge appears to be used as a synonym for information, this trend emphasised by the move to data analytics (Pauleen & Wang, 2017). On the other hand, Spender's (2009) concept of knowledge as practice allows for inclusion of the human element - subjectivity of experience, exchange of tacit aspects, and how this is expressed in social networks (Dixon, 2010) and used to inform expertise. This interpretivist view of knowledge sees it as socially-constructed, context-sensitive and dependent on history (von Krogh, Roos & Slocum, 1994).

The different concepts of knowledge, and the interaction between them, can be a source of innovation for the organisation (Cook & Brown, 1999). Tsoukas (1996) suggests that explicit, objective knowledge can only be made use of once it is appreciated subjectively, taking into account context, tacit knowledge, motivation, and the other aspects relating to human subjectivity. In this way it is 'reattached to and embedded in the ongoing processes of the organisation' (Spender, 1996): knowledge as practice. Individuals will attribute different meanings to the same phenomena and may change their minds about meaning over time. The decisions they make are not solely on a rational basis, but are influenced by their individual and subjective perspectives. An interpretive approach to investigating individual ideas about knowledge sharing, for both explicit knowledge and knowledge as practice, is therefore worthwhile.

Factors in Knowledge Sharing

Knowledge sharing in organisations enables members to benefit from each other's knowledge and expertise, allowing them to contribute ideas and experience (Ipe, 2003;

Hansen, Moore & Lovas, 2005). Organisational culture and structure, management support and incentives have all been found to influence the likelihood of employees to engage in knowledge sharing. Examples are given of breaking down silos within the organisation, encouraging inter-team communication (Tsai, 2002), providing time and resources for knowledge sharing, and correctly structuring tangible and intangible reward mechanisms (Dyer & Nobeoka, 2002). Motivations for sharing knowledge can be examined in terms of impression management and power, social costs, the experience of learning and evaluation apprehension (Wang & Noe, 2010).

Knowledge sharing literature exploring the individual perspective focuses on the social and psychological drivers which influence knowledge sharing. Sociological theories are invoked to provide explanations of factors intrinsic to the individual: self-efficacy, enjoyment in helping others, concern for one's reputation, and trust, both that others will reciprocate in the sharing practice, and in the organisation's procedural fairness. They also include interpersonal factors as they affect the individual: the social norms and networks in which the individual and knowledge sharing practice are situated, including theories about the operation of communities of practice (Jeon et al., 2011; Su et al., 2012) the perceived costs and benefits, and apparent effectiveness of sharing (Dyer & Nobeoka, 2002; Cabrera & Cabrera, 2002; Chiu, Hsu & Wang, 2006; Collins & Smith, 2006). Social Capital Theory has been cited in work on knowledge sharing motivation (Huysman & Wulf, 2006), contribution of dimensions of social capital to knowledge sharing in organisations (Widen-Wulff & Ginman, 2004) and the pressures on frequently consulted knowledge workers (Oldroyd & Shad, 2012). Social Exchange Theory, assuming that in attempting to maximise their likelihood of achieving their aims in social exchanges, humans are rational and make rational decisions, has been used to inform ideas about reciprocity in knowledge sharing (Bock & Young-Gul, 2002; Hall in Widen-Wulff & Ginman, 2004; Liao, 2008). The concept of self-efficacy, the belief in one's own abilities, has been adopted from Social Cognitive Theory and used in models of motivation (Bock, Zmud & Lee., 2005; Lin, 2007; Tsai & Cheng, 2010). Other sources comment on the need for organisational support in informal knowledge sharing (Marsick, 2009) and the need for a common language in the use of story-telling for the transmission of tacit knowledge (Whyte & Classen, 2001).

Action research studies have shown the importance of contextual factors when sharing tacit knowledge, and the influence of internal and external boundaries (Fernie, Green & Weller, 2003; Dawes, Creswell & Pardo, 2009; Kimble, Grenier & Goglio-Primard, 2010; Peet, 2012). An action research approach has advantages for enquiring into individual's perception about knowledge sharing and the influences on it when considering phenomena as integral to the individual as tacit knowledge or motivation and intention to share. In the case study, Appreciative Inquiry Method (AIM) was used for these reasons: it allows for enquiry into the factors influencing knowledge sharing, and crucially, it permits the participants in the study to do this for themselves, to reflect on their perceptions of knowledge sharing, with minimal influence from the researcher.

Appreciative Inquiry Method

AIM was developed from the Soft Systems Methodology research of Checkland et al. (1981, 1991, 1999), to enquire into systems which are complex or difficult for participants to discuss. Making use of the ideas from Geoffrey Vickers's work, AIM draws on his concept of Appreciation (Vickers, 1965), discovering the 'shifting perceptions, judgements and structures of the world of culture' which influence 'how human beings and groups deliberate and act'. It involves cycles of enquiry that help participants arrive at a shared appreciation of the situation as it is' (reality judgement) and as it 'ought to be' (value judgement) and to consider the feasibility of actions to improve it, according to organisational norms and values (Stowell, 2013). Participants are helped to think about their often hard to describe situations ideas in a transparent and unrestrained way, taking into account 'difficult to measure' aspects of human behaviour and action'. They explore the shape of their knowledge, according to their language, using connections that make sense to them (Stowell, 2013).

Discussion of differences of opinion and clarifying issues help increase the participants' collective understanding; the debate throughout the Method's stages moves the learning forward (Champion & Stowell, 2003). The importance and potential difficulties caused by power relationships are acknowledged, and so the cultural and political feasibility of action becomes part of the conversation. In practice, AIM provides insight into participants' assumptions and perceptions, and a framework within which accommodations of different views can be reached and standards and norms in the situation modified, in a way which is 'agnostic to the enquiry' (Checkland, 1985; Stowell, 2013)). Reflection on the participants' understanding is carried out through the lens of Vickers's 'Appreciative System': their readiness to notice aspects of the situation, influenced by their mutually related judgements of reality and value.

The credibility of an AIM study is established using the notion of recoverability – the use of a 'framework of ideas' (in this case Interpretivism) to underpin the methodology (Checkland & Holwell, 1998), and by Champion and Stowell's work on authenticity through the PEArL framework (Champion & Stowell, 2003). The framework was formulated as a methodological tool, rooted in systems ideas, to support claims to authenticity, with respect to Participants, their Engagement, the Authority and relationships in the research exercise and the Learning experienced. The 'relationship' aspect of PEArL in particular enables reflection on the soft power (commodities of power (Stowell, 2000)) which can influence how participants discuss and accommodate their perspectives (Champion & Stowell, 2003; Stowell, 2013). The study documentation, in the form of system maps, root definitions and models, supports recoverability and transferability of the findings, and claims to authenticity are supported by the reflections on the process of enquiry.

CASE STUDY

The study was carried out in a research organisation amongst participants were interested in promulgating good knowledge sharing practice. The remit of the organisation's management was to maintain the science and technology knowledge base for the benefit of its clients, to develop its expertise in new areas and to act as an interface between the client departments, industry and academia, in work on client projects. The management was also required to exploit the organisation's intellectual property by developing new products, by publishing in peer-reviewed literature and by increasing its customer base. The organisation's departments were supported by a central Knowledge Service function which developed strategies and procedures to take advantage not only of explicit knowledge but also to attempt to capture and promulgate what is 'known and used'. The importance to the organisation of explicit scientific and technical knowledge was reflected in the organisational structure, where the Knowledge Services department occupied a position alongside the operating departments. This was believed to work well, but could occasionally result in competing claims to knowledge and expertise during project work and the lack of clarity on responsibility for items of knowledge and their use. The organisation operated in a challenging funding environment, including changes to procurement procedures by its clients, with a resulting impact on its revenue from projects and competition for qualified staff. External fund holders' lack of appreciation of the organisation's knowledge capability led to poorly informed investment and disvestment decisions, and lack of investment in knowledge and information management capability. The study allowed investigation of the effects of this on individuals' knowledge sharing practices.

Initial discussions for the study revealed that there was concern within the organisation with issues which influence how knowledge was defined and used. While strategies and procedures have been created to take advantage of explicit scientific and technical knowledge it was recognised that knowledge as practice was equally valuable. The means of capturing this more subjective and tacit knowledge was influenced, however, by the view of knowledge as an asset. This appeared to influence mechanisms set up to share knowledge as practice: subject experts were considered to be 'resources' from whom this knowledge was to be extracted. Apparent issues were how to make knowledge management devices more effective, and how far capturing tacit knowledge was possible given the constraints of the organisation.

Participants in the study represented a range from senior to more junior staff, selected for their involvement in knowledge sharing practice. They included representatives from both the operating departments and from Knowledge Services. The participants were directly engaged in knowledge sharing practice at an operational level, or were using their experience of knowledge sharing in determining strategy. The boundary for the study was dictated by the responsibilities and formal influence of the participants. The researcher's initial position was that knowledge could be most usefully defined as knowledge as practice and was interested the issues affecting this in the organisation. The Appreciative Inquiry Method was used to explore the question: What are the issues around knowledge sharing in [the organisation]?

A composite map (Figure 1), showing subsystems of interest, was agreed by the participants. Root definitions were developed for each subsystem and were used as an agenda for discussion about what 'ought' to be. This stage was crucial in exposing participants' norms and values and developing their awareness of each other's perspectives. The most root definitions the participants thought most relevant are given in Table 1.

The activity models produced from the revised root definitions were used as an agenda for continued discussion. This meeting was the final activity of the study, where the participants were closely observed to assess their shared Appreciation of knowledge sharing in the organisation. The study concluded once this cycle of

Figure 1. Composite map (AIM)

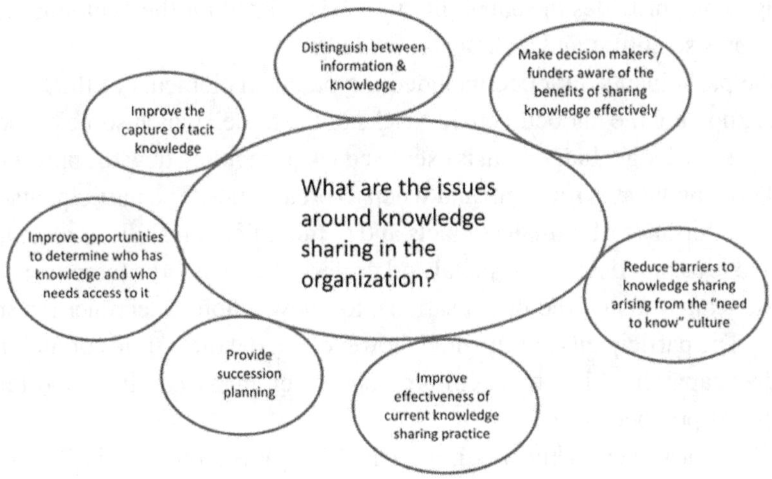

Table 1. Subsystems and their root definitions

Improve opportunities to determine who has knowledge and who needs it	Improve existing "mechanisms" to increase shared awareness within the organization of who has knowledge which can be applied on specific "projects" as a means of improving delivered outcomes.
Make decision makers/funders aware of the benefits of sharing knowledge effectively	Ensure the cost and value of maintaining the organization's science and technology research expertise, including all published work, is understood by decision makers at every stage of the project life cycle.

enquiry was complete. The PEArL framework (Stowell & Champion, 2003) was used to reflect on the study process throughout.

CASE STUDY FINDINGS: ASPECTS OF KNOWLEDGE

Over the course of the study the participants revealed their views of knowledge as being 'in-head' knowing, having experience and expertise in practice requiring the internalization of information, and the development of a personal capability in a scientific and technical area. Sharing this capability, for example among the organisation's subject experts, can generate new knowledge (both explicit and 'knowing') (Tsoukas, 2011; Polanyi & Prosch, 1975; Spender, 1996). Objective, explicit knowledge can only be 'known' by the knower appreciating it subjectively. The participants' views accorded with Orlikowski's (2002) definition of knowledge, as being at any given time, what practice has made it. Knowledge as practice, combined with judgement, provides the capability which is needed for the exploitation of the organisation's scientific and technical knowledge.

For the participants, practice included the practical elements of the knowledge domains, knowing the undocumented 'tricks of the trade', which some participants called tacit knowledge. Judgement is exercised when deciding how to apply what one knows, knowing what is relevant, and when to break rules. The participants agreed about the importance of human contacts and communication skills as a component of knowledge as practice: to use social and professional networks (amongst experts, within the team, outside the organisation), to know whom to contact for specific problems. The participants commented, however, on the difficulties of maintaining knowledge capability due to a culture that de-emphasized the importance of knowledge as practice.

The different aspects of knowledge used in the organisation could be illustrated by considering what was needed in a client-funded project (the usual mode of working):

explicit knowledge derived from published work in a literature review, knowing what is available from the Knowledge Services repositories and departmental resources, finding people with the requisite expertise (deploying social and professional networks) and the capability to make use of this knowledge and network (practice). The participants emphasized the importance of knowledge as practice both to show value for their clients and to work most effectively. Not having time or resources to practice caused problems for maintenance of capability.

Tacit Knowledge

An important aspect of knowledge as practice is the tacit understanding which underpins its use and the participants reflected on the different forms of tacit knowledge at some length. Their initial view was that it was problematic for knowledge sharing: it is 'that which is hard to communicate, which can't be written down, causing practical difficulties'. The participants commented on its use in knowledge as practice: its role in 'in-headedness', the ability to make judgements, to use experience, to deploy the 'tricks of the trade'; its importance and its difficult-to-grasp nature. There were difficulties in realizing what tacit knowledge the organisation had, for example that possessed by subject experts or embedded in accepted working practices. Another view was that some knowledge was 'deliberately tacit' ('implicit' in Simard and Jourdeuil (2012), Li and Gao (2003)). According to one participant, risk-aversion and a propensity to blame in parts of the organisation led to non-documentation of failure, the consequence being that valuable lessons were lost. The hoarding of knowledge was a further expression of implicit knowledge; a possibility that subject experts 'guard' their knowledge, whether consciously or unconsciously (Connelly, Zweig, Webster & Trougakos, 2011), could be seen as a commodity of power (Stowell, 2000). Yet another perspective was that of interpersonal skills: 'how to handle personalities' in the words of another participant– how an individual's self-efficacy or belief in one's own capabilities in interpersonal and communication skills could help to remove or avoid personal barriers to knowledge sharing. Deploying interpersonal skills was also required when developing one's social and professional network, to be called upon when engaging on project work.

A final comment was that 'real' tacit knowledge could not be known, or individuals did not realise what they knew, and therefore it would be unable to be articulated. Transfer of this tacit knowledge was only effectively done by working with the owner of the knowledge ('apprenticeship'). Succession planning could therefore be a problem: where the matching of a departing member of staff with his replacement in a suitable project could not be done effectively without awareness of what needed to be transferred. Logistical difficulties stood in the way of capturing those categories of 'tacit knowledge' that were amenable to capture. Time and resources needed to overcome these were not always forthcoming; a cost-benefit analysis was conducted when deciding whether difficult–to-capture knowledge was important enough to record.

Effects of Knowledge Sharing Infrastructure

The many mechanisms used for knowledge sharing shows the need for staff to deploy another type of knowledge as practice: being aware of what mechanisms are available, and how best to use them. When discussing how these mechanisms were used to share the scientific and technical knowledge, distinctions were drawn between knowledge's day-to-day use and maintenance, its transfer as part of succession when staff leave a role, and how to re-establish capability once lost.

A full catalogue of explicit scientific and technical knowledge repositories was available to the participants, together with externally sourced consultation with professional and scientific bodies, with contractors and by maintaining links with expert staff who had retired or changed their role. Efforts were made in the organisation, via training, registers of staff knowledge, staff portfolios and contacts, social media tools and bulletins, to determine what staff knew and to make sure they knew how to deploy their knowledge. These facilities however were poorly used. Factors militating against sharing were possible misconceptions about using sharing mechanisms, or a lack of motivation for doing so. More community management of this resource was felt to be needed, to encourage staff engagement.

Effective use of scientific and technical knowledge in the organisation was promoted by providing formal opportunities for sharing: subject expert seminars, training and workshops, a mentoring system. The Knowledge Services department embedded staff in each operating department to provide services and to identify resources that could be added to the main repository. The participants agreed that responsibility for knowledge as practice meant not only developing and maintaining knowledge, but also sharing it when leaving a role. There were problems maintaining knowledge as practice developed in the course of a project, in that once a project was completed, maintaining capability could not depend on there being further projects to help with this (Kakabadse, Kakabadse & Kouzmin, 2002). This was discussed in the context of succession planning. Taking responsibility for knowledge sharing was affected by the mismatch between timescales for projects which would maintain knowledge capability and the requirements for staff rotation. Other factors included the dispersed responsibilities which were a consequence of matrix management for projects. Although collective responsibility for knowledge sharing was encouraged (Ellis, 2001 as cited in Riege, 2005; Zack, 2002; Zhang, Voegl & Zhou, 2012), strategies were needed to overcome the logistical difficulties.

Effect of Value of Knowledge to the Organisation and Its Clients

The most important issue for the participants was that contribution of knowledge as practice was undervalued by the clients and fund-holders in the organisation who were results driven and did not appreciate the need to maintain knowledge capability for future work. Pressure from clients for quick, cheap, results-only reports, and their unwillingness to fund knowledge activities caused problems in maintaining capability and in making space for knowledge sharing initiatives. This was felt to lead to increased cost and inefficiencies where processes had to be repeated to recover lost expertise when needed, and to the reduced capability of the organisation to compete for future work.

External influences on the fund holders were also an issue. The organisation's staff both collaborated and competed with other organisations on a European scale, and the remit had been moved away from being a knowledge generator towards being a knowledge aggregator and exploiter. Therefore it was less easy to defend requests for budgets to maintain capability. The way those in positions of political and financial power perceived the value of knowledge appeared to influence attitudes and practice in the organisation. This negative agency (von Krogh, as citen in Easterby-Smith & Lyles, 2011) was expressed in the resources they made available and in their willingness to engage with capability maintenance.

Impact of Knowledge Sharing Cultures in the Organisation

It became apparent that sharing and use of knowledge of either type was affected by the organisational cultures and subcultures operating on the individual. These can shape assumptions about knowledge, set the context for social interaction to control its use, and help define who generates, shares and uses it (De Long & Fahey, 2000). A number of informal subcultures existed alongside the formal organisational culture of the organisation, driven by the different values placed on the knowledge offering. Each influenced an individual's perceptions about the possibility of, and motivation for, knowledge sharing, about how self-efficacy was developed and practised.

Subcultures influenced by the organisation's professionalism agenda were:

- The mandated culture of sharing, supported by the knowledge sharing infrastructure. The belief was that, to maintain knowledge capability, scientific and technical knowledge, the knowledge as practice and business intelligence needed to exploit them should be shared, and that staff should engage with the structures and mechanisms provided.

- The 'you should know' culture, related to expectations about professionalism and self-efficacy, but which could manifest itself either in risk avoidance activities, or in refusal to impart information.

Subcultures influenced by the value the client places on knowledge were:

- The 'need to know' culture, relating to past working practices espoused by longer-serving staff in operating departments, and still in operation in parts of the organisation. This suggested that certain scientific and technical knowledge should be restricted in distribution to those who needed to know it to carry out their work. There were situations where this is the case, but 'need to know' could be used as a 'reason' for not sharing knowledge.
- The risk-averse culture, where staff perceived that there was a tendency to assign blame rather than to accept and learn from errors. It was associated with a reluctance to take responsibility for maintaining knowledge. This was a response to the client's demand for quick answers and lack of tolerance for the nuanced knowledge as practice approach, and similarities could be seen with the 'you should know' culture.
- The change of culture also being imposed by the external requirement to integrate existing knowledge from internal and external sources rather than to generate it. This emphasis on exploitation rather than exploration (Kakabadse et al., 2003) was driven by external cost considerations.

Statements made by the participants, such as 'People should just use the mechanisms' and 'people's perceptions are their problem', showed that in imposing the sharing culture, there was a lack of acknowledgement as to why staff believed as they did. It was possible, therefore, to see how individuals came to perceive a conflict between the official, mandated requirement to share knowledge and those internal subcultures which acted to decrease willingness to share – to expose oneself to challenge. 'You should know' and 'need to know' may have been used to conceal a practitioner's lack of knowledge or competence, enabling him/her to avoid taking responsibility for ensuring that knowledge is shared. The conflict for individuals between the mandated culture and the messages from clients had an effect on motivation, self-efficacy and willingness to take responsibility for outcomes.

Motivation for Knowledge Sharing

The expectation for the mandated culture was that processes were in place to develop an individual's capability, as part of the professionalism agenda. Extrinsic motivation to share, such as reward for reuse of knowledge as well as for innovating, was discussed, but Knowledge Services focused on influencing the behaviour of the staff, persuading them to internalise sharing attitudes. The participants showed their understanding of the role of intrinsic motivation as they considered tacit knowledge and succession planning, and speculated on reasons for not sharing. Aside from the use of the 'need to know' culture, and logistical barriers, fear of being challenged was an issue.

Self-efficacy was encouraged by the organisation under study as part of the professionalism agenda (Swales, Wright & Oxenham, 2011) and as a factor in the mandated culture of sharing. An individual who had self-efficacy was in a position to generate and exploit knowledge for the benefit of project clients, colleagues and the organisation. In the 'you should know' subculture, it was assumed that employees possess self-efficacy. The informal subcultures may subvert the main culture of sharing in an organisation, and their effects on motivation and self-efficacy (Figure 2). In one participant's view, employees who had underdeveloped self-efficacy could deploy a range of avoiding strategies which affect knowledge sharing behaviour (for example, claims of difficulties with technology when the real problem is the employee's lack of ability to communicate effectively). However, in a risk-averse subculture, the self-efficacious person concerned about his/her reputation may also realise the issues with a situation, but remain silent about them or create defensive reports. Li and Gao (2003, p.8) comment on the unwillingness to disclose knowledge for 'organisational reasons'. The knowledge involved becomes implicit and concealed rather than shared. The mandated knowledge sharing culture may have addressed the 'you should know' and 'need to know' subcultures, but may not have been able to influence the risk-averse subculture because of external influence and attitude of clients, fund holders/decision makers at different levels of projects. Concern about reputation could also act to prevent knowledge sharing. The example was cited of a subject expert who retired, to be re-employed shortly afterwards as a consultant. Here, the possession of tacit knowledge was a commodity of power (Stowell, 2000), contributing to the individual's reputation but also affecting his wider sharing behaviour.

A self-efficacious person may conceal politically inconvenient knowledge or knowledge which might result in a loss of individual or organisational reputation (as in project failure) despite the fact that reasons for failure are valuable knowledge in their own right. S/he may also hoard knowledge to maintain power or reputation, as when subject experts move into consultancy roles. A person who lacks self-efficacy may attempt to avoid challenge by invoking the 'need to know', or attributing problems to technology rather than lack of communication skills. The 'you should know' culture may inhibit that person from seeking the knowledge s/he is supposed to have already.

Figure 2. Effects of organisational subculture on self-efficacy in knowledge sharing

	Mandated culture	**Sub-cultures**
Individual possesses	Professionalism: use of knowledge for benefit of clients	Concealing knowledge
	Self-efficacy	
Individual doesn't possess	Processes to develop (Swales et al., 2011)	Avoidance strategies

Summary of Case Study

A main outcome of the study is that it suggests a link between how knowledge as practice is valued externally and organisational subcultures, and provides support for the explanation that the characteristics of these cultures affect knowledge sharing practice in individuals. Participants believed that what constituted tacit knowledge, how it could be used or shared and the success with which this can be done was influenced by factors not necessarily under their control. The preference of clients for explicit knowledge was at the expense of the knowledge as practice that supports it.

On a practical level, clients' perception of the value of knowledge as practice influenced the availability of project funding, which affected whether staff can be allocated to a project (for knowledge sharing, training of new staff, transfer of knowledge as part of succession planning), and hence the organisation's ability to create and exploit knowledge. Clients' requirements for 'the answer' and their intolerance of failure contributed to 'you should know', 'risk-averse' and 'blame' subcultures. These informal subcultures acted to subvert the sharing of knowledge, as self-efficacious individuals concealed politically inexpedient knowledge, or hoarded it to maintain their reputation, and others engaged in distracting or avoiding behaviours, showing a general reluctance to take responsibility for the ownership or sharing of knowledge.

DISCUSSION

The findings of the study show how the value placed on different aspects of knowledge (and how this is expressed in organisational cultures) can influence knowledge sharing practices in individuals. Participants appreciated the need to internalise scientific and technical knowledge to develop personal capacity, and to use judgement to exercise that capacity effectively. This suggests that on an individual level, explicit knowledge and knowledge as practice can be seen as two necessary aspects of capability.

Providing knowledge sharing infrastructure of various sorts can facilitate sharing but does not guarantee it. An interesting question is why better use was not made of it in the study organisation, considering that participants valued both explicit knowledge and knowledge as practice, and it could be assumed that aids to internalising and using what they know would be welcomed. Knowledge sharing is something which all participants could agree on and is an imperative for the organisation's success, but it is complicated by organisational politics and cultures and by logistical barriers. Answers can be suggested by examining the role of tacit knowledge in knowledge as practice, and of trust in the sharing of that knowledge.

The contributions of tacit knowledge to knowledge as practice can take several forms, as illustrated in the case study. There was a disconnect between an appreciation of tacit knowledge's essential 'unknowability' and the view that it could be realised and exploited if the right mechanisms were found. For example, tacit knowledge held in the organisation itself was an issue for succession planning; if not known what it is or where it resided, matching successors with departing experts was seen to be difficult. There was also comment on intentionally tacit, or implicit, knowledge, which implies that participants can articulate their knowledge but keep it to themselves. These aspects move away from purely individual perspectives, introducing the supra-individual influences discussed below.

A third aspect of tacit knowledge is to do with knowing how to use interpersonal skills to build trust, from where knowledge sharing becomes more likely. Numerous authors have commented on the role of trust in knowledge sharing (Holste & Fields, 2010; Kucharska & Kowalczyk, 2016; Werner, Blass-Franken & Martin, 2016). Both cognitive-based trust (trust in the efficacy of another) and affective trust (interpersonal trust, based on perception of another's values and motivations) are involved (Lewis & Weigert, 1985). If dialogue and talk are to be added as essential components of knowledge management as Crane (2013) suggests, the building of trustful relationships is important. In the case of the organisation, the task of building this trust was often given over to the Knowledge Services staff embedded in operational area, who could be conceived of as 'other'. Overall, these insights into tacit knowledge provide more nuance than is allowed for in Nonaka's (1994) schema (Crane, 2013).

The interplay between cognitive-based and affective trust are illustrated by the need for succession planning and re-establishing capability in a knowledge area once it has been lost. Staff are being asked to 'give up' their knowledge, their contacts and the way that they think: it is illuminating that the case study participants from Knowledge Services talked about 'exploiting' knowledge. Seeing staff as a resource to be mined can have unpredictable effects on morale and self-efficacy. The issues this causes at the personal and intersubjective level this became apparent when the participants discussed the difficulty of obtaining tacit knowledge from subject experts who were retiring or changing role. From the participants' perspective, subject experts experienced the social dilemma of whether to share knowledge (Dyer & Nobeoka, 2002). This was described by Hsu, Hsu, Yen and Chang (2007) as a balance between the positive effects of reputation maintenance, satisfaction and self-efficacy on the one hand, and the negative effects of loss of power, a particular point to note if the expert hoped to return as a contractor. Motivations to share were examined using Social Capital Theory in studies by Huysman and Wulf (2006) and Cabrera and Cabrera (2005): acting as 'rational beings', the subject experts should be convinced by the argument that knowledge sharing was needed for the success and survival of the organisation. The contingency of the subject experts' behaviour in this organisation (Shoghi et al., 2013) suggested that social dilemma and affective trust considerations were a better explanation of failure to share their knowledge.

The ways in which an individual's self-efficacy motivated them to share knowledge was explored. Developing self-efficacy in gaining and using knowledge was encouraged by the professionalism agenda, which was generally espoused by the participants. Showing cognitive-based trust, they respected the capabilities of subject experts and valued them as a resource - the reputational aspects of self-efficacy were pronounced. A mandated culture encouraging knowledge sharing conflicted however with informal subcultures which variously promoted a controlling view of knowledge capacity, and punished those who deployed it in a way inconvenient to internal politics. To allow individuals' professionalism and self-efficacy to contribute effectively to knowledge sharing, it is clear that subversive subcultures should be addressed (Cameron, Quinn & DeGaff, 2005) and that practical steps to establish trustful relationships could go some way to alleviating these problems.

CONCLUSION

Individuals engaged in knowledge work draw on their experience when describing what knowledge means to them, and these descriptions can be nuanced and insightful. So their knowledge about knowledge can be a useful resource for researchers. The way their views are shaped and how they can deploy this knowledge depends on the specific context of the organisation however. The case study established the importance of knowledge as practice, both to the individuals and to the organisation to maintain capability, but also the structural problems in maintaining it. The inter-connected organisational influences such as subcultures, professionalism agendas and the funding environment combine to influence individuals in their practice. An emergent theme, trustful professional relationships, indicates where an organisation's managers should turn to resolve some of the knowledge sharing issues discussed in this chapter.

REFERENCES

Bock, G., Zmud, R., & Lee, J. (2005). Behavioral Intention Formation in Knowledge Sharing: Examining the Roles of Extrinsic Motivators, Social-Psychological Forces, and organisational Climate. *Management Information Systems Quarterly*, *29*(1), 87–111. doi:10.2307/25148669

Cabrera, A., & Cabrera, E. (2002). Knowledge-sharing Dilemmas [Electronic version]. *organisation. Studies*, *23*(5), 687–710.

Cameron, K., Quinn, R., & DeGaff, J. (2005). *The competing values framework: creating value through purpose, practices, and people*. Ann Arbor, MI: Addison-Wesley.

Champion, D., & Stowell, F. (2003). Validating Action Research Field Studies: PEArL. *Systemic Practice and Action Research*, *16*(1), 21–36. doi:10.1023/A:1021928511690

Checkland, P. (1985). From optimizing to learning: A development of systems thinking for the 1990s. *The Journal of the Operational Research Society*, *36*(9), 757–767. doi:10.1057/jors.1985.141

Checkland, P. (1991). From framework through experience to learning: the essential nature of action research. *Information systems research: Contemporary approaches and emergent traditions*, 397-403.

Checkland, P. (1999). *Systems Thinking, Systems Practice*. Chichester, UK: John Wiley & Sons.

Checkland, P., & Holwell, S. (1998). *Information, systems and information systems: making sense of the field*. Chichester, UK: John Wiley & Sons.

Checkland, P. B. (1981). *Systems theory, systems practice*. New York: Wiley.

Chen, S., Chuang, Y., & Chen, P. (2012). Behavioral intention formation in knowledge sharing: Examining the roles of KMS quality, KMS self-efficacy, and organisational climate. *Knowledge-Based Systems*, *31*, 106–118. doi:10.1016/j.knosys.2012.02.001

Chiu, C.-M., Hsu, M.-H., & Wang, E. (2006). Understanding knowledge sharing in virtual communities: An integration of social capital and social cognitive theories. *Decision Support Systems*, *42*(3), 1872–1888. doi:10.1016/j.dss.2006.04.001

Collins, C., & Smith, K. (2006). Knowledge exchange and combination: The role of human resource practices in the performance of high-technology firms. *Academy of Management Journal*, *49*(3), 544–560. doi:10.5465/AMJ.2006.21794671

Connelly, C., Zweig, D., Webster, J., & Trougakos, J. (2012). Knowledge hiding in organisations. *Journal of Organizational Behavior*, *33*(1), 64–88. doi:10.1002/job.737

Cook, S., & Brown, J. (1999). Bridging Epistemologies: The Generative Dance Between organisational Knowledge and organisational Knowing. *Organisation Science*, *10*(4), 381–400.

Crane, L. (2013). A New Taxonomy Of Knowledge Management Theory: The Turn To Knowledge As Constituted In Social Action. *Journal of Knowledge Management Practice*, *14*(1). Retrieved from http://www.tlainc.com/articl332.htm

Dawes, S., Cresswell, A., & Pardo, T. (2009). From 'Need to Know' to 'Need to Share': Tangled Problems, Information Boundaries, and the Building of Public Sector Knowledge Networks. *Public Administration Review*, *69*, 392–402. doi:10.1111/j.1540-6210.2009.01987_2.x

De Long, D., & Fahey, L. (2000). Diagnosing cultural barriers to knowledge management. *The Academy of Management Executive*, *14*(4), 113–127.

Dixon, N. (2010). *Common Knowledge: How Companies Thrive by Sharing What They Know*. Cambridge, MA: Harvard Business School Press.

Dyer, J., & Nobeoka, K. (2002). Creating and managing a high-performance knowledge-sharing network: The Toyota case. *Strategic Management Journal*, *21*(3), 345–367. doi:

Easterby-Smith, M., & Lyles, M. (Eds.). (2011). *Handbook of organisational Learning and Knowledge Management* (2nd ed.). Chichester, UK: John Wiley & Sons.

Fernie, S., Green, S. D., & Weller, S. J. (2003). Dilettantes, discipline and discourse: Requirements management for construction. *Engineering, Construction, and Architectural Management, 10*(5), 354–367. doi:10.1108/09699980310502964

Hansen, M., More, M., & Lovas, B. (2005). Knowledge sharing in organisations: Multiple networks, multiple phases. *Academy of Management Journal, 48*(5), 776–793. doi:10.5465/AMJ.2005.18803922

Holste, S., & Fields, D. (2010). Trust and tacit knowledge sharing and use. *Journal of Knowledge Management, 14*(1), 128–140. doi:10.1108/13673271011015615

Hsu, C., Hsu, C., Ju, T., Yen, C., & Chang, C. (2007). Knowledge sharing behavior in virtual communities: The relationship between trust, self-efficacy, and outcome expectations. *International Journal of Human-Computer Studies, 65*(2), 153–169. doi:10.1016/j.ijhcs.2006.09.003

Huysman, M., & Wulf, V. (2006). IT to support knowledge sharing in communities, towards a social capital analysis. *Journal of Information Technology, 21*(1), 40–51. doi:10.1057/palgrave.jit.2000053

Ipe, M. (2003). Knowledge Sharing in organisations: A Conceptual Framework. *Human Resource Development Review, 2*(4), 337–359. doi:10.1177/1534484303257985

Kakabadse, N., Kakabadse, A., & Kouzmin, A. (2003). Reviewing the knowledge management literature: Towards a taxonomy. *Journal of Knowledge Management, 7*(4), 75–91. doi:10.1108/13673270310492967

Kimble, C., Grenier, C., & Goglio-Primard, K. (2010). Innovation and knowledge sharing across professional boundaries: Political interplay between boundary objects and brokers. *International Journal of Information Management, 30*(5), 437–444. doi:10.1016/j.ijinfomgt.2010.02.002

Kucharska, W., & Kowalczyk, R. (2016). Trust, Collaborative Culture and Tacit Knowledge Sharing in Project Management – a Relationship Model. In *Proceedings of the 13th International Conference on Intellectual Capital, Knowledge Management & Organisational Learning: ICICKM 2016* (p. 159-166). Humanistic Management Network, Research Paper Series No. 49/16.

Lewis, J., & Weigert, A. (1985). Trust as a Social Reality. *Social Forces, 63*(4), 967–985. doi:10.1093f/63.4.967

Li, M., & Gao, F. (2003). Why Nonaka highlights tacit knowledge: A critical review. *Journal of Knowledge Management, 7*(4), 6–14. doi:10.1108/13673270310492903

Liao, L.-F. (2008). Knowledge-sharing in R&D departments: A social power and social exchange theory perspective. *International Journal of Human Resource Management, 19*(10), 1881–1895. doi:10.1080/09585190802324072

Lin, H. (2007). Effects of extrinsic and intrinsic motivation on employee knowledge sharing intentions. *Journal of Information Science, 33*(2), 135–149. doi:10.1177/0165551506068174

Marsick, V. (2009). Toward a unifying framework to support informal learning theory, research and practice. *Journal of Workplace Learning, 21*(4), 265–275. doi:10.1108/13665620910954184

Nonaka, I. (1994). A Dynamic Theory of organisational Knowledge Creation [Electronic version]. *organisation. Science, 5*(1), 14–37.

Oldroyd, J., & Shad, S. (2012). Catching Falling Stars: A Human Resource Response to Social Capital's Detrimental Effect of Information Overload on Star Employees. *Academy of Management Review, 37*(3), 396–418. doi:10.5465/amr.2010.0403

Orlikowski, W. (2002). Knowing in Practice: Enacting a Collective Capability in Distributed Organizing. *Organisation Science, 13*(3), 249–273.

Pauleen, D., & Wang, W. (2017). Does big data mean big knowledge? KM perspectives on big data and analytics. *Journal of Knowledge Management, 21*(1), 1–6. doi:10.1108/JKM-08-2016-0339

Peet, M. (2012). Leadership transitions, tacit knowledge sharing and organisational generativity. *Journal of Knowledge Management, 16*(1), 45–60. doi:10.1108/13673271211198936

Polanyi, M., & Prosch, H. (1975). *Meaning*. Chicago: University of Chicago Press.

Riege, A. (2005). Three-dozen knowledge-sharing barriers managers must consider. *Journal of Knowledge Management, 9*(3), 18–35. doi:10.1108/13673270510602746

Ryle, G. (1949). *The Concept of Mind*. London: Hutchinson.

Simard, A., & Jourdeuil, P. (2012). Knowledge Manageability: A New Paradigm. In Y. Al-Bastaki & A. Shajera (Eds.), *Building a Competitive Public Sector with Knowledge Management Strategy*. Idea Group.

Spender, J.-C. (1996). Organisational Knowledge, Learning and Memory: Three Concepts in Search of a Theory. *Journal of Organizational Change Management, 9*(1), 63–78. doi:10.1108/09534819610156813

Spender, J.- C. (2009). Organisational Capital: Concept, Measure or Heuristic? *Organisational Capital: Modelling, Measuring and Contextualising*, 5-23.

Stowell, F. (2000). Organisational Power. *Systemist, 22*(2), 106–127.

Stowell, F. (2013, February). The Appreciative Inquiry Method – A Suitable Candidate for Action Research? *Systems Research and Behavioral Science, 30*(1), 15–30. doi:10.1002res.2117

Swales, S., Wright, J., & Oxenham, D. (2011). *Systems Skills Development – The Yellow Brick Road Approach.* UKSS.

Tsai, M. T., & Cheng, N. C. (2010). Programmer perceptions of knowledge-sharing behavior under social cognitive theory. *Expert Systems with Applications, 37*(12), 8479–8485. doi:10.1016/j.eswa.2010.05.029

Tsai, W. (2002). Social structure of 'coopetition' within a multiunit organisation: Coordination, competition, and intraorganisational knowledge sharing. *Organisational Science, 13*(2), 179–190.

Tsoukas, H. (1996). The Firm as a Distributed Knowledge System: A Constructionist Approach. *Strategic Management Journal, 17*(S2), 11–25. doi:10.1002mj.4250171104

Tsoukas, H. (2011). *How should we understand tacit knowledge? A phenomenological view. In Handbook of organisational Learning and Knowledge Management* (2nd ed.). Chichester, UK: Wiley.

Vickers, G. (1965). *The Art of Judgement.* London: Chapman and Hall.

Von Krogh, G., Roos, J., & Slocum, K. (1994). An essay on corporate epistemology. *Strategic Management Journal, 15*(S2), 53–71. doi:10.1002mj.4250151005

Wang, S., & Noe, R. (2010). Knowledge sharing: A review and directions for future research. *Human Resource Management Review, 20*(2), 115–131. doi:10.1016/j.hrmr.2009.10.001

Werner, R., Blaas-Franken, J., & Harry, M. (2016). The impact of (low) trust on knowledge sharing. *Journal of Knowledge Management, 2*(20), 199–214. doi:10.1108/JKM-10-2015-0391

Whyte, G., & Classen, S. (2012). Using storytelling to elicit tacit knowledge from SMEs. *Journal of Knowledge Management, 16*(6), 950–962. doi:10.1108/13673271211276218

Widen-Wulff, G., & Ginman, M. (2004). Explaining knowledge sharing in organizations through the dimensions of social capital. *Journal of Information Science, 30*(5), 448–458. doi:10.1177/0165551504046997

Zack, M. (2002, April). A strategic pretext for knowledge management. *Proceedings of the Third European Conference on organizational Knowledge, Learning and Capabilities.*

Zhang, X., Voegl, D., & Zhou, Z. (2012). Effects of information technologies, department characteristics and individual roles on improving knowledge sharing visibility: A qualitative case study. *Behaviour & Information Technology, 31*(11), 1117–1131. doi:10.1080/0144929X.2012.687770

KEY TERMS AND DEFINITIONS

Action Research: Research consisting of collective enquiry amongst participants in social situations, in order to address a problem or situation of interest. The method privileges the capabilities of the participants themselves, the view being that they have the experience in the situation, and are capable, through cycles of reflection, of developing insights and solutions.

Affective Trust: Interpersonal trust based on one's judgements about another's motivations, values, and intentions to behave. Affective trust develops from interactions between individuals, and between individuals and organizations. An example of affective trust in a knowledge-intensive organization is sharing knowledge which involves a transfer of perceived power from the sharer to the receiver.

Cognitive-Based Trust: Trust based on one's judgements about another's skills, competencies, and reliability, based on perceived facts. An example of cognitive-based trust in a knowledge-intensive organization is one's belief that another has expertise in a particular knowledge domain.

Knowledge as Practice: Internalized, subjective understanding of explicit knowledge, such that it can be used to inform judgement and actions. Knowledge as practice develops from individuals' personal experiences, deploys their tacit knowledge, and is affected by personal attributes such as self-efficacy.

Knowledge Sharing: The activities by which knowledge is exchanged between individuals and groups within an organization. The knowledge may be explicit or tacit, and sharing is important to the organization's goals and effective working.

Motivation: The desire to perform an action. Motivation may be extrinsic, influenced by factors in an individual's environment, such as reward or sanction; or it may be intrinsic, to do with an individual's self-perception, such as pride in work, altruistic beliefs, concern for reputation.

Organizational Subculture: A set of beliefs and values held by a group within an organization that extends or differs from the dominant (often mandated) culture. It may be a local phenomenon (such as in a team) or one held by individuals with specific roles or histories within the organization.

Self-Efficacy: One's belief in one's own competence to achieve a task or succeed in a situation. In the context of a knowledge-intensive organization, self-efficacy involves an individual's confidence that they have expertise in knowing and making use of knowledge.

Tacit Knowledge: Experience-based knowledge about a domain ("know-how"), which can be used to inform knowledge as practice, but which is difficult to put into words, and therefore difficult to share or transfer.

Chapter 10
Understanding the Whole From the Parts:
A Systemic Analysis of Classical Dressage

Daune West
University of the West of Scotland, UK

ABSTRACT

This chapter reports on an application of systems theory to a complex area of human endeavor, classical dressage. The area is well represented in a rich literature dating back to the time of Xenophon (c.380BC) and has many practitioners worldwide today. The author uses her interpretive systems perspective to explore classical dressage theory and practice and, throughout, uses examples from the classical equitation literature to support and illustrate the analysis presented. The chapter offers a description of classical dressage from considering its component parts and suggests that it concerns not only the "correct" and ethical training and riding of horses but, fundamentally, the personal development of the trainer/rider. The chapter concludes by (1) inspecting the relationship between classical and competition dressage and suggests that they contain important elements that are mutually exclusive and (2) considering the potential contribution of "systems" to the study of the human-horse relationship as complementary to the increasingly popular approach of equitation science.

DOI: 10.4018/978-1-5225-5996-2.ch010

INTRODUCTION

Classical Dressage is an approach to the training and riding of horses that, according to its commentators, has its origins in the writings of the Greek general Xenophon (c. 380BC) and a practice that may stem back even further into the Hittite civilization (Pereira, 2001). Classical Dressage has a very rich literature and is practiced worldwide but has rarely been subject to a critical evaluation from an academic perspective although it has been referred to as 'academic equitation' in the past because of its advocates' interest in the studied relationship between theory and practice. Riders from this tradition have dedicated their lives to trying to understand the biomechanics of the horse and to explore how they can augment the natural movement and beauty of the horse when ridden. In the past, the reasons for following the Classical path have been varied but today it is frequently considered to be an art form by its proponents. This way of thinking has meant there has been some divide between the 'classical' proponents and those who are involved in the modern sport of competitive dressage, where focus appears to be upon the breeding and training of horses capable of executing more exaggerated movements as opposed to 'correct' movements (as laid down by the 'rules' of the governing body of sporting international dressage, which were developed out of 'classical principles').

Over the last thirty years or so, the training methods of some successful international dressage riders have been severely criticized (e.g. Heuschmann, 2006; 2012) since they appear, to some, not only to breach ethical boundaries but also to be practicing equitation that has moved a long way away from what some argue to be the 'correct', tried and trusted 'classical' methods developed over the last millennium. Arguments about the ethics of one particular training method, rollkur and its associated LDR (low, deep, round), adopted by some competitors, are rife today online (e.g. Shand, 2017; https://epona.tv/farewell-falsterbo). Calls to return to the 'classical' traditions represented and intended when competitive dressage was first established has kindled increased interest in the approach over the last few years. However, what exactly constitutes 'classical' is not easily recognized or agreed: there are many examples of individual trainers and riders who claim a classical approach, some who adhere strictly to the teachings of one or another classical 'master, some who 'mix and match' influences, and some who seem to use the classical label as a marketing tool but with little classical content in their work. Unsurprisingly, individual's and groups' defense of the 'classicism' of their approach have been vigorous, even aggressive at times.

The author of this paper has been a student of Classical Dressage for more than fifteen years: originally, it was the relief in finding a way of riding that was not so rule-based, that addressed nuances and the complexities of the human horse relationship, that offered thoughtful analysis, and a rich supply of approaches to problem solving that attracted me to this school of thought. In trying to follow the advice of many 'Classical' commentators who state the importance of combining theory with practice, I have immersed myself in the writings of both old and modern 'masters', travelled to take instruction from selected trainers, and I attempt to practice classical principles in the riding and training of my horses. The primary influence on my approach has been from Portugal and France and, more recently, from the Austro-Hungarian tradition. The more I studied, the clearer the 'mess' that is Classical Dressage, described above, emerged. When asked what 'Classical Dressage" was, I found it difficult to give a precise answer: I could give many examples but not a clear definition that satisfied my understanding and intuitive 'feel' for the approach, gained from reading the literature and taking part in the practice.

In trying to make sense of this 'mess' I started to apply my 'systems' training (in a professional capacity I have been involved in studying, teaching and practicing 'systems' inquiry for almost thirty years). I undertook a systems study with my Portuguese Classical Dressage trainer (of more than 10 years) to try and understand what Classical Dressage was to him. I found the results fascinating (West & de Bragança, 2012). This led me to undertake a similar study with another Classical Dressage trainer, but one with a Swedish/German background: the results were different although a similar core idea was present (but not one that appears in the usual attempts to describe classical dressage) (West, 2012). It seemed likely that each person I spoke to would give me a different answer to the question: "What is Classical Dressage?"

Consequently, rather than try to develop a single systems description of Classical Dressage by considering its overall 'purpose', it seemed it might be more appropriate to first try and understand it from a consideration of its parts, as represented in its literature and practice. In this way I could allow a description of Classical Dressage to emerge, as a whole, from the study of its parts. This chapter, then, describes my use of a 'systems lens' to explore what appear to be the component parts of this complex human activity and to identify and present what I believe to be key principles of Classical Dressage. A result of this analysis puts me in conflict with some Classical Dressage advocates in terms of the relationship between classical and competitive dressage: my analysis suggests that they do not make easy bed-fellows since important features for each derive from incompatible paradigms. The chapter concludes by arguing the important contribution of systemic inquiry to complement the increasingly popular approach of Equitation Science in the development of our understanding of the horse-human relationship.

USING A SYSTEMS LENS TO EXPLORE CLASSICAL DRESSAGE

The remainder of this chapter is structured under six 'systems themes' derived from an overall examination of the Classical Dressage literature (reported in detail in West, 2015). Many quotations from the literature are provided to illustrate and explain the issues discussed under these headings with the aim of allowing the words of the classical 'Masters' and commentators to present their own descriptions of Classical Dressage: it is hoped that their narratives, with their own language, associated context and examples, will enable the reader to enter into their 'worlds'. Of course, the selection of these narratives depends purely on the author's perception, selection and interpretation and, as such, I am aware of the highly subjective nature of the work presented.

Systems Theme 1: The Role of History, Culture and Tradition

A concept fundamental to interpretive systems is that of 'context', especially in relation to past events and ideas (Vickers, 1965). Looking at its 'context' gives us a useful starting point and insight into the nature of Classical Dressage because of its long history and various contributing factors. Consequently, this section is presented in two parts: the first deals with the way that history and culture has helped to shape the 'Classical Dressage' of today whilst the second part considers how tradition and the lessons from the past influences and directs the classical trainer and rider.

The Role of History and Culture

Classical Dressage has a long and colourful history. The form we see today has evolved out of an interesting mix of historical, geographical and cultural influences. Destriers, medieval war horses, were often heavy, cold-blooded breeds, strong and explosive but not particularly nimble. However, as warfare developed, and the benefits of the maneuverability of the smaller horses native to some regions became clear, horses were valued and trained for more than just their ability to produce power in straight lines (Hyland, 1998). The new war vehicle still needed to be strong and courageous but also easily controlled and quick to react. In addition to being merely a vehicle it could be trained to offer offensive and defensive battle movements and some have suggested that perhaps we see the legacy of such movements in the high school 'airs above the ground' practiced today by, for example, the Spanish Riding School and the Cadre Noir (bayareabert, 2009; Husky005reloaded, 2010). In southern Europe, using the horse as a working farm animal also required certain physical and mental characteristics of the horse and we can see this style of 'working' riding continued today in the Camargue, the Iberian peninsula (e.g. lanchonman, 2006), north and south America on ranches and in the sport of Doma Vaqueira/Working Equitation (e.g. Arenalilla, 2007; HorseOsteopath, 2010). Linked to this working style of riding comes the development of the use of the horse in bull fighting where the bravery, skill and athleticism of the horse is tested to the full (e.g. Rpgdressage, 2012). Moving away from the practical utility we come to the use of the horse for 'display'. In the Middle Ages, when not used in battle, valuable, highly trained horses were kept 'tuned' and used in displays such as jousts and tournaments (e.g. Duarte, 1438). Later, the education of any young nobleman included horsemanship and schools, or academies, that explored, developed and taught this skill flourished all over Europe from the 16th to the 19th century (Loch, 1990). Innovative horsemen from these schools provide us with much of the writing we have on 'academic' equitation – the work of the 'old masters' (e.g. Grisone, 1558; de Pluvinel, 1626; de la Guérinière, 1733; Kerbrech, 1891). The Grand Écuyer or Master of the Horse was one of the most highly regarded positions at court and the careers of horsemen and their schools were made by royal or noble patronage and were often linked to the military academies (e.g. Bourgelet, 1754; Pembroke, 1773). Later on, the display of the skills of the teachers and students in these schools to the public in the 'circus' (e.g Baucher (1919), Fillis (1911), Steinbrecht (1885) in the late 19th / early 20th century and, more recently, Nuno Oliveira (Équitation - Le hibou vert, 2012)) continued the tradition of 'display'. Four national 'Schools' remain today: the Spanish Riding School in Vienna, the Escola Portuguesa d'Arte Equestre in Lisbon, Le Cadre Noir de Saumur, and the Real Escuela Andaluza de Arte Ecuestre in Jerez, (e.g. Laurioux & Henry, 2009) and continue the tradition of display. At the beginning of the 20th century, military riders, trained out of various European schools, were the successful participants at the first Olympic 'Dressage' competitions (the first 'military tests' were included in the 1912 Stockholm games). Today, the guardians of the heritage of the highly schooled horse for display (or 'art') teach and preserve the centuries old traditions (e.g. horseherovideo, 2010). A full discussion of the history of Classical Riding can be found in Loch (1990) but the key influences are illustrated in Figure 1.

An additional but important influence on the development of the classical tradition is the popularity over the centuries of a particular breed of horse, the Iberian/Barb, with their unique physical characteristics and temperament that make them eminently suitable for the work required (Loch, 1986). An example of the high status, desirability and value of these horses can be found in the published letters of the Earl of Pembroke (one of our classical riding practitioners and commentators from the 18th century), who, writing to his son making a Grand Tour of several years, constantly questions him about the purchase, health and travel arrangements of an expensive Iberian/Barb horse, the Aley Bey, he had sourced in Italy. He tells

Figure 1. Multiple-cause diagram showing the influences leading to the training of the horse in the 'classical' tradition

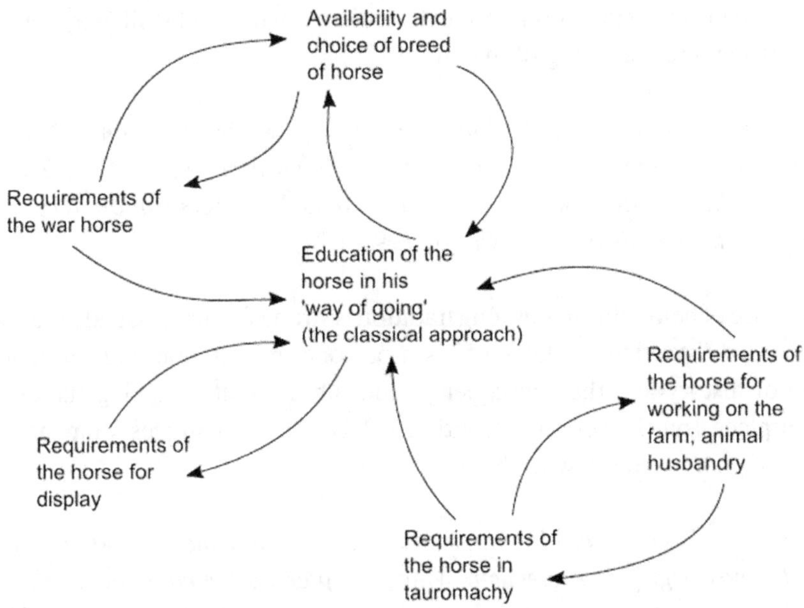

us that bringing the horse back to England "cost £81.16.0" (Herbert, 1939, p326) (about £14,500 in today's money).

The Importance of Tradition

Anyone studying the Classical Dressage literature is likely to gain a clear sense of belonging to a long tradition of exploration and learning about the training and riding of horses. As de Kunffy (1993) explains:

Horsemanship is an ancient art. Its body of knowledge is pragmatic and time tested; it cannot be re-invented. No one can relive the experience of millions of people who rode for millions of hours with millions of thoughts and feelings about it all. Those experiences have been sifted and synthesized for us, into accumulated knowledge that ought to be acquired by aspiring practitioners. All art remains only as good as those who practice it. (p19)

For this reason, Classical Dressage does not belong to any modern rider/trainer – there are no 'new', innovative approaches but long-time, tried-and-tested methods. As one learns more one takes on the responsibility of protecting the understanding and lessons of the past; one becomes a 'keeper' or 'guardian' of the body of knowledge and part of that on-going tradition. The classical rider does not invent new methods for self-promotion but protects and, if extraordinary in terms of skill, may contribute to the existing understanding of the art:

Instead, great masters learned all that the past offered and went on to build on it and perhaps to humbly add to it. Their contribution was civilized, because it was in the context of an artistic tradition, rather than a barbaric reinvention of that which already had its own history. (de Kunffy, 1993, p38)

The sense of belonging to this long tradition is strong for the classical practitioner. In a recorded interview, Belasik offers a personal recognition and reflection on the role of history and the consequent responsibility that following the classical tradition places on the serious practitioner. His comment suggests far more than just a physical connection with the past:

I think there's also another element, and that's ...the spiritual element and this is, to me, the most significant element. When you pick up the equipment...the longe line and the whip, you are picking up pieces of equipment that have changed hardly at all since historical times. You are almost picking up the same equipment that ... the Duke of Newcastle picked up, de la Guérinière, de Pluvinel. When you pick up this equipment and when you begin to use this equipment I think it's important that you recognize the tradition that comes down to you, and that you use this equipment with a sense of reverence to the process, a reverence to the horse.....

His recognition that being part of this tradition "...takes you out of yourself and makes you part of something that is much larger" (Belasik, 1993, side B, 1.38-4.24) seems to be an important part of the classical approach: it is a place where there is no room for the ego, a point which takes us nicely on to the next section which considers Classical Dressage as purposeful human activity.

Systems Theme 2: Classical Dressage as Purposeful Human Activity

From the description in the section above we see that in the past there were various reasons for engaging in a considered approach to the training and riding of horses: there was a practical need. The practice of developed horsemanship skills and the need to maintain expensive horses led to the display of the art form. Today, apart from tauromachy, the riders and trainers of Classical Dressage tend to practice their art, often privately, for their own pleasure and fulfillment. In this section the purpose of classical dressage as art and as practiced today is explored. First we look at the primary task of training the horse according to the classical approach, then look a little more closely at the reason why people engage in the activity and, finally, consider what appears to be an important emergent property of this purposeful human activity.

Classical Dressage: The Primary Task

There are many descriptions of classical dressage in the literature and common themes focus on the relationship between horse and rider with the ideals of attaining harmony and an augmentation of the horse's natural beauty and movement under its rider. An underlying provision is that all training should be without force and commentators refer back to the argument of Xenophon (c.380BC), perhaps the earliest recorded Classical horseman, who explained that:

For what the horse does under compulsion, as Simon also observes, is done without understanding; and there is no beauty in it either, any more than if one should whip and spur a dancer. There would be a great deal more ungracefulness than beauty in either a horse or a man that was so treated. No, he should show off all his finest and most brilliant performances willingly and at a mere sign. (Xenophon, p62-3)

Whilst classical dressage has a clearly recognizable 'face' to those following and engaging in it, it is not easy to find a clear description of its fundamental parts without resorting to examples, by trying to explain fairly subjective concepts or aligning oneself to a particular proponent or 'school', which frequently results in conflict between members or proponents when they meet (often in print, blogs, discussion boards, etc.). A good example of an informed debate which recognizes the difference in interpretation of the classical approach can be found in a lively discussion between Christoph Hess and Philippe Karl (Classical vs. Classique, 2007). In this way, for example, we might be able to recognize a French 'school' (e.g. Oliveira, 1965; de Bragança, 1975; Henriquet, 1991; Karl, 2008; Racinet, 2009), or a German 'school' (Steinbrecht, 1885; Meusler, 1937; Seunig, 1961; von Neindorff, 2005) and influences that seem to arise from practical issues such as the involvement in competitive dressage and the demands of particular horse breeds. The four European National institutions - the Spanish Riding School in Vienna, the Escola Portuguesa d'Arte Equestre in Lisbon, Le Cadre Noir de Saumur, and the Real Escuela Andaluza de Arte Ecuestre in Jerez, Spain, (e.g. Laurioux & Henry, 2009) claim to uphold the principles of Classical Dressage but financial pressures have potentially changed and threatening the existence (and purity) of these institutions (e.g. Levett, 2009).

It is usual to consider the writings of two classical horsemen, William Cavendish, Duke of Newcastle ("La Methode Nouvelle et Invention Extraordinaire de Dresser Les Chevaux" originally published in French in 1658 and in its new form in English in 1743) and François Robichon de la Guérinière ("École de Cavalerie" published in 1733), as the foundations of Classical Dressage (most particularly for their experiments in the training of the horse so as to achieve the 'lightness' and maneuverability of the horse so desired by the enlightened rider). Indeed, the Spanish Riding School (SRS) have always maintained that they implement the training, methods and philosophy of de la Guérinière and so are seen as a direct descendant of the teachings of this great Classical 'Master'. They also explain that the SRS "is the only institution in the world which has practiced for more than 450 years and continues to cultivate classical equitation in the Renaissance tradition of the Haute Ecole" (The Spanish Riding School, n.d.).

Loch (1990) offers a number of definitions of Classical Dressage from the literature, illustrating the different emphases and interests, and concludes with a clear if somewhat general description:

Taking all this into consideration, we may now define the Art of Classical Riding as the ability to teach the horse through the use of kindness, logic and exercises based on nature's laws of balance and harmony, to submit himself happily and proudly to the will of the rider without in any way upsetting his natural way of moving. (p18)

However, whilst this definition tells us what classical dressage is about, it does not really tell us why so many people are motivated to engage in this activity. By comparison, competitive dressage has a clear end result – a measurement of success demonstrated by the award of 'marks', rosettes, and prestige. So what is the purpose of Classical Dressage today?

Classical Dressage: Reasons for Its Continued Practice

In an attempt to understand classical dressage as a purposeful human activity system, the author undertook a 'soft' systems study to explore the reasons why different recognized classical riders and trainers continue to practice the classical approach. The results from this study indicated a range of different purposes. To illustrate this point, two descriptions are shown here. They are both provided in the form of a Root Definition (a modelling convention from Soft Systems Methodology (e.g. Checkland & Poulter, 2006)) that was produced by the rider/trainer and the author in collaboration. The first definition belongs to a Portuguese classical rider/trainer who comes from a family with a long classical equestrian heritage. He breeds Iberian horses and teaches horses and riders at his small farm in Portugal. He has been taught and influenced by riders/trainers from the French school of classical dressage and from Portuguese bullfighting (a full description of this particular study can be found in West & de Bragança, 2012). His explanation of what classical dressage is for him is as follows:

A rider owned, instigated and directed system for the purpose of experiencing brilliance by bringing about perfect connections in the horse-rider circular relationship through the practice of classical dressage. The experience of 'brilliance' and attainment of harmony brings about enjoyment to the rider to the extent that the result is like a special, personal secret between two parties. The outcome of the system is sought within the context of the current possible philosophy of the rider and the character and potential resistances on the part of both horse and rider. The system's activities must not in any way destroy the horse's natural movements.

The second example is from a Danish rider and trainer of both horses and riders. He has competed in dressage at an international level and he has been taught and influenced by riders/trainers from the German school of classical dressage (West, 2012). His explanation of what classical dressage means for him is expressed in the following Root Definition:

A team-owned and manned system to produce a healthy horse in such a way that its needs are met so that it can enjoy a good quality of life in whatever activity it is involved in with its rider. This activity will then be enjoyable for the rider. This process is constrained by the rider's mind.

These two examples appear to represent two very different personal reasons for engaging in the classical approach. The second description focuses very much on the production of the horse for the purpose of the rider and has more of a 'production' feel to it, whereas the first description is very much about the motivation to explore the magical, personal relationship between rider/trainer and horse. However, despite these differences, there is a clear common theme: the mind, attitude, and character of the rider.

The Rider's Journey: An Emergent Property?

A close inspection of the literature suggests that the role of the rider's mind may be a fundamental aspect of the classical approach which helps to differentiate between it and other approaches to equestrianism. For example, de Kunffy (1993) offers a clear description of the role of one's self in classical riding and explains:

The life of an equestrian involves one's personal inner life and values, and one's ethics and character traits. There is no question that horsemanship is a science of many branches (p18)

This idea of the training of the self seems to go back a long way. For example, King Duarte of Portugal, writing in the fifteenth century explained how horsemanship was a singular ability through which one could learn skills for life: to master one's fear, anger, frustration, and the ego and to develop judgment, tranquility, and wisdom (Duarte, 1438). He explains the relationship between riding and life itself:

… and the riders who have the same objective – to stay erect and firm on the beast – are able to do everything much better than the others; and practicing these principles, they would acquire the experience they need to succeed in everything they do. (p40)

Several centuries later de Pluvinel (chief écuyer to Louis XIII) argued that horsemanship offered something more than other arts and sciences since its consideration did not take place in the quiet of the study but in constant movement:

Thus Your Majesty can see quite clearly how useful this beautiful exercise [horsemanship] is to the mind, since it instructs and accustoms it to perform with clarity and order all these functions [scientific knowledge, generosity of spirit, soundness of judgment] amid noise, worry, agitation, and the fear of constant danger… (de Pluvinel, 1626, px)

This idea is just as familiar to modern 'masters'. For example, Podhajsky (1967), dressage Olympic medalist and one-time Director of the Spanish Riding School explains:

Equestrian art, perhaps more than any other, is closely related to the wisdom of life. Many of the same principles may be applied as a line of conduct to follow. The horse teaches us self-control, constancy, and the ability to understand what goes on in the mind and the feelings of another creature, qualities that are important throughout our lives. (p20)

Loch also explains how the classical approach requires a significant shift in the way we might think about the horse-rider relationship. She suggests:

Once the student can shrug off any idea of the horse as a vehicle for their own sport and achievement and begin, instead, to regard the horse as an intelligent living creature with his own needs and place in the scheme of things, perception will grow. As the horse is as super-sensitive and uncomplicated in his mental make-up as, indeed, we are ourselves, to understand him better will require very real and consistent development of our own mental faculties and powers. Indeed, when this mental aspect is lacking, the physical will suffer. (1997, p9)

De Kunffy (1993) suggests that equestrianism can be considered usefully as an apprenticeship; the education in its arts "should address the entire person, and thus develop the equestrian as a personality" (p41). He argues:

Beyond the necessary physical skills, which constitute the sport in riding, we need to address the rider's mind, that is, the science of riding, and the rider's spirit (de Kunffy, 1993, p1).

And finally, Henriquet suggests that the classical approach represents an example of the difference between training and education, the latter being a far more satisfying approach:

Equestrian art is distinguished from the usual current riding practice in that it educates, rather than merely trains. That is to say, it opens out the horizons of both horse and rider, rather than passing them along a production line. In this respect, it justifies itself as 'art' in a way similar to music and dance, which do not have rigid structures or boundaries. Furthermore, as with other art forms, it is self-fulfilling: the essence of equestrian art is that it is art. The sensation, for the rider, and the sight, for the onlooker, are complete in themselves (Henriquet, 1991, p1).

Taking all the above comments into consideration we might be able to recognise an important emerging property of the classical form of equestrianism. The primary task of classical equitation (as discussed above) might be to teach the horse to move under saddle as if at liberty using a kind, logical, natural and ethical approach. However, the reason for this activity would appear to be greater than the mere training of the horse. It is also about the personal development of the rider, the individual journey of discovery that the dedicated rider finds themselves travelling. The experience of brilliance, the virtue and the nobility that the schooled horse shares with his rider are undoubtedly strong motivating factors to encourage the classical rider on this journey.

Systems Theme 3: Measurement of Success – Relationship-Maintaining vs. Goal-Seeking

The measurement and validation of success in classical dressage is an interesting issue. If, as suggested above, Classical Dressage practitioners are involved in a personal journey with a non-human sentient being in an attempt to develop an art form that depends so much upon the development of 'feel', how can the success of this activity be judged? Vickers' work on 'appreciation' (1965) seems to offer us a way of thinking about this.

Vickers put forward the concept of an appreciative system as a way of explaining how we might consider the way in which individuals, or groups, make decisions about how to act. He explains that the appreciative system relates to:

...the inner coherence of that system of interests, expectations and standards of judgement which orders our lives, guiding action, mediating communication and making experience meaningful (1972, p155)

He offered the concept of relationship-maintenance as a way of describing the type of regulation that such a collection of parts relies upon. Vickers considered the all-pervasive concept of goal-seeking as being somewhat poverty stricken when used as a model to underpin human decision-making and action, but the maintaining of suitable and desirable relationships between the parts of the whole was a far more appropriate way to help us consider the management of complex human activity systems (Vickers, 1965). The discussion of the relevance of relationship-maintenance to classical dressage will be presented in two parts; the first dealing with the relationship between technique and process and the second with the measurement of success or the validation of process.

Technique and Process: Ends and Means

The training techniques of classical dressage, as we shall see below, are clearly identified. Many exercises, movements and 'scales' of practice are documented and explained: there is order and method in their arrangement. However, the application of the techniques in themselves do not constitute or guarantee either good practice or an output that is excellent. Belasik offers us an insightful example of the relative importance of technique and process. He explains how, in search of dressage excellence, he focused on perfecting his technique according to the 'rules' and practice laid down in the literature and taught by trainers he studied with. Some years down the line he realized that technique, although fundamental, is not sufficient on its own: riding as an art is stifled by imposing the order of technique. He begins to view the role of technique in a different way:

I begin to see technique as only a tool, a discipline to help you remain stable when facing the unconscious, the creative, the fear of the unknown, and the new. I found some breathing room to allow myself to experiment, to experience without constant stifling comparison and measurement. I was beginning to gain a certain confidence in my work. (1990, p88)

In this way he shows us how he begins to understand the importance of living the process rather than aiming constantly for the goal of perfection. This way of thinking is entirely consistent with the idea of a personal journey discussed above. De Kunffy pulls together the two ideas of 'process' and 'journey' when he explains:

A process knows no end, has no terminus and is never finished by arrival. Once the process 'arrives', it is eliminated. There is no joy in deadline. The desire to be something is ego-oriented and useless to the horse. Strive to do something instead! Do not seek arrival but enjoy the process of getting there (de Kunffy, 1993, p7).

The link to Vickers' concept of relationship-maintenance is clear – in classical dressage there is no sensible end point, there is no once-and-for-all goal since perfection is unattainable. The thrill and challenge is to manage the 'now' – the 'dynamic and ever changing situation': the rider's skill and understanding, the horse's strength and ability to perform the exercises. The rider must be judicious, making decisions in an instant about what to request, what to enforce and what to allow. Classical dressage is the continuous management of a complex set of activities and perceptions according to such classical ideals as harmony, equilibrium and lightness.

Measurement and Validation: External vs. Internal

Measurement is an underlying principle of dressage as competition which is very much goal driven. Tests are set to assess the performance demonstrated and to award rosettes and 'points' to those receiving the highest scores. Points accumulate over time and denote the level at which the horse is capable of performing. The tests themselves comprise a list of consecutive movements which are scored from 1 to 10. There are marks for 'collectives' (such as the overall obedience, suppleness, and compliance of the horse and the skill of the rider) but as these are considered at the end of the test it is likely that they become a judgment on the sum of the parts. The idea that those scoring the highest marks in competition are the best riders has been strongly argued against by De Kunffy (1993) who states:

The idea now spreading that successful competition on any level, local or international, should determine the ranking of riders is dangerous and faulty. (p27)

Such validation of the training process as judged by competition would seem to focus on output alone. This is in direct opposition to the classical principle that states that the 'end does not justify the means'. We have seen above how at the centre of classical dressage is the development of not only the horse according to certain ethical and biomechanical principles, but also of the rider. Focus on output is detrimental to the focus placed on the importance of process; it does not value the experience of brilliance, or of being 'at one' with the horse. Unsurprisingly, many classical riders avoid competition since the concept of external measurement to judge success makes little sense given the philosophy of their practice. Interestingly, in classical dressage, the measurement that is valued, the reference level, seems to be largely generated from within, especially because so much of classical riding is about 'feel' and, consequently, only the rider is in a position to experience this. This internal development of standards, or a reference level, is consistent with Vickers' view of this process in relationship-maintenance (West, 2005, p273). Mairinger (1983) helps to explain the responsibility of the rider to develop this 'feel' and to be able to judge personal success:

When you ride a horse do you really prepare yourself? Do you know exactly what you should know? Do you think about it? I will do my best to explain what you should do, how you should do it, and why you should do it. But I will not be able to tell you how the different movements will feel when you carry them out, and that is of the greatest importance – feeling when you are right and when you are wrong, feeling when the horse is moving properly, when his rhythm is right and his impulsion is right. No book can explain this satisfactorily. (p16)

And we might add, no judging onlooker may be in a position to understand what is being achieved exactly. Obviously, 'feel' needs to be educated. Throughout the literature we are told that measurements that are worthy of the classical rider involve asking questions about whether the rider is considered to be 'just', 'graceful', and able to demonstrate appropriate 'judgment'. It seems difficult if not impossible to equate numbers with such concepts in a sensible manner.

The need for judges in competitive dressage to reduce the assessment of a horse's training to the appraisal of a series of movements encourages a lack of systemic judgment; the appreciation of the whole (the harmony between horse and rider) are not of prime interest. Belasik picks up on this point, although he is referring to the process where trainers concentrate on the parts (ie movements and exercises) and forget the whole (interestingly, many judges are also trainers):

I think this points to a paradox that the western mind does not like to entertain: namely, that observations about riding and riding are two very distinct and separate worlds.

Although we need to use words to learn and to teach, they still are very limited. Breaking up the dynamic movements of riding a horse into 'successive entities' in order to talk about it is very artificial. The flow of riding is too continuous to keep stopping it, chopping it up with analysis of particular movements separated from the whole (Belasik, 1990, p108-9).

A final point, to differentiate between the process encouraged by competition and the process undertaken by the classical rider, is that whilst the latter breaks down their training into the parts of the 'movements', these 'movements' are practiced as a way of strengthening, straightening, and lightening the horse in the journey towards 'brilliance'. These 'movements' used in careful combination to fit the training requirements of the horse are the means to the end. 'Movements' in competition, on the other hand, can be seen as encouraging the execution of these movements for their own sake (i.e. to demonstrate competence in these movements).

Systems Theme 4: Communication, Control and the Connection Between the Parts

In the horse rider relationship, the rider teaches the horse to respond according to given signals. The classical tradition prides itself on these signals being as gentle and discrete as possible so as to be unseen by the spectator (e.g. Loch, 2003) These signals are dependent upon the rider being able to sit so as to move with the horse with as little disruption as possible so that the horse's movement is undisturbed. A quiet, balanced, tactful rider is able to 'whisper' the aids. To achieve this involves the management of a complex whole – the horse and rider in motion. Interestingly, while the horse's movement influences the rider's movement, so too the rider's movement influences that of the horse: a classic example of positive feedback. To discuss communication and connection between the parts in more detail this section is divided into 3 parts. The first deals with the mechanism by which the rider/trainer communicates with the horse by means of 'aids', the second focuses on the importance of the rider's 'seat' as the major 'control mechanism' and the means by which the communication is directed. The third part explains how, when the communication and control are appropriate, the horse and rider work in harmony as if they were a single entity.

Communication Through a System of 'Aids'

The way in which the rider/trainer communicates with the horse is through what are referred to as the 'aids'. Aids are the signals or instructions given by the rider to the horse using the hands, seat, legs, voice, spurs, and whip. Many equitation texts provide a thorough description of the aids, their use and their subtleties but they are often presented to the reader as a collection of individual movements or signals. However, in the classical tradition they can be seen as a complex set of signals that are used in combination, in greater sophistication as both the level of the schooling of the horse and the experience of the rider progress. Indeed, Stevens (1994) refers to the use of such signals as "an *aiding system*" [his emphasis] (p24). A useful way of thinking of the aids is as a language which is established by the rider in consideration of the horse's natural inclinations in response to stimulus. Mairinger (1983) explains the importance of the interplay between aids:

We should not really talk about the aids with your weight, legs, and hands because they don't exist separately – only by co-ordination of all three together. You cannot use only one. The use of seat, legs and hands can be passive or active. It is like a three-piece orchestra playing a waltz, rock-and-roll, and a Russian lullaby. All must be in tune, in the same time – then it is pleasant to listen to and pleasant to see. Your aids have to be tuned together. You can do it if you really learn body control. (p8)

The Rider's 'Seat' as Governor: Control of the Parts and Their Relationship to Each Other

Fundamental to the rider's ability to communicate with the horse via the aids is the need for what is called a good 'seat' – one that is quiet and in balance with the horse in motion. Quietness in motion is necessary if the communication of the aids, the language, is to be clear. The rider's seat (considered to begin at the rider's chest and end at the knee) is the subject of much debate, and lies at the heart of the classical approach such is its importance for the clear and steady communication of the aids (Loch, 1988). An early recognition of the central importance and systemic nature of the 'seat' is given by Bourgelet, writing in 1754:

I have no other Design in this Chapter, than to give an Idea of the correspondence that there is between all the Parts of the Body, because it is only by a just Knowledge of this mutual Relation of all the different Parts, that we can be enabled to prescribe Rules for giving that true and natural Seat, which is not only the Principle of Justness, but likewise the Foundation of all Grace in the Horseman. (p9)

We can see from this quote that the seat is the mechanism from which the desired characteristics of the classical approach derive. It controls, balances, energises and forms the movement of the horse. The other aids, such as those given by hands and legs supplement and complement the seat. To illustrate the power of the seat the power of the seat we can give an example provided by Mairinger. As a young rider in Vienna, Mairinger had difficulty controlling one of the stallions he was exercising outside. A train went past and the stallion bolted in fear. The Old Instructor, whose stallion it was, yelled at the young rider telling him "You are holding him too much." Mairinger reports he was "holding on for dear life". The next day the same thing happened. The Old Instructor demonstrated the control of the 'seat':

The Old Instructor said, 'Here, stay here, check him'. He mounted up and you could really see that weight sink into that saddle. The stallion was bouncing on the spot, his nostrils flaring, and looked about to explode at any minute. To rub it in the Old Instructor said, 'I told you. You held him too much'. He threw the reins away and the stallion could not do a thing. He couldn't do a thing because he had to balance himself under that weight (Mairinger, 1983, p45).

The classical rider/trainer soon begins to understand how discretion and judgment is required if the power of the seat is not to be exploited to the detriment of the horse where 'control' and 'governance' become brute force and domination (Belasik, 2001, p48).

The idea of the seat being a complex interplay between parts of the rider's body is explained further by Bourgelet, who instructs on the importance of the connectivity of the parts of the rider's body and explains how small movements in any part may change the whole. For example, in what must be one of the earliest 'systems' descriptions he explains how a necessary but difficult position (a good seat) may be achieved easily by the rider if he understands how the parts fit together and function as a whole:

The Horseman should present or advance his Breast, by this his whole Figure opens and displays itself. He should have a small Hollow in his Loins, and should push his Waist forward to the Pommel of the Saddle, because this Position corresponds and unites him to all the Motions of the Horse. Now, only throwing the Shoulders back produces all these Effects, and gives them exactly in the Degree that is requisite, whereas, if we were to look for the particular Position of each Part separately, and by itself, without examining the Connection that there is between the Motions of one Part with those of another, there would be such a Bending in the Loins, that the Horseman would be, if I may so say, hollow-back'd; and as from that he would force his Breast forward, and his Waist towards the Pommel of the Saddle, he would be flung back, and must sit upon the Rump of the Horse. (p7)

For the classical rider the seat is the mechanism by which the rider channels and manages the power of the moving horse.

Communication and Control to Achieve Harmony: The Horse and Rider as One

When a high degree of communication between rider and horse is achieved then the harmony, balance and lightness so desired by the classical rider may be experienced. Henriquet, (1991) explains:

Equestrian art requires that the two participants, a horse and rider, fuse into one entity. This fusion becomes wonderful when it is so complete that physical communication between two creatures seems replaced by the harmonious movement of [a] single being. (p1)

This description not only explains the whole that emerges from complete communication but also illustrates the way in which the brilliance sought by the Portuguese rider/trainer, as described above, might be achieved. Whilst we might recognize this output of the experience of 'brilliance' as being of a spiritual nature there is also a practical output of complete communication and control, namely 'collection'. Collection can be considered to be the greatest proof of a well-schooled horse: it is the culmination of long training and demands a supple, powerful and obedient horse. True collection allows 'lightness' and elegance, and represents schooling to the highest level. In his essay on 'collection' Belasik (2009) tries to explain how at lower levels of schooling the rider works the horse from outside the 'system' in that they influence the horse's way of going from outside. However, collection has a very particular effect on the horse/rider relationship:

As a horse collects a curious phenomenon begins to occur. The horse becomes more manoeuvrable but is less stable. In the levade, the horse is quite beautifully balanced but quite unstable – easy to affect, easy to upset with either good or bad riding. The rider's influences and forces change from external suggestions outside the system of the horse to internal forces within the new system of horse and rider. The rider and horse become one system; one body. The clichés about becoming one with your horse actually have physical merit as you enter into the world of collection. (p56)

Systems Theme 5: Principles of Practice and Equifinality

The method by which a horse is schooled according to the classical tradition is well documented in the literature (e.g. d'Endrody, 1959; Seunig, 1961; Karl, 2008; Kottas-Heldenberg, 2010). In this section we look first at the popular idea of a scale of training which attempts to offer a clear step-by-step process for the development of a well-schooled horse. In the second section we consider how the setting out and determination of the 'method' by which classical training can be applied may lead to conflict of interpretation and the need to understand that classical 'principles' need to be just that – guiding principles.

Establishment of Method: The Training Scales

A description of requirements of the horse's development can be found in references to the 'Training Scales'. The Training Scales are often associated with German riders and trainers of the twentieth century given their development out of the work of Steinbrecht (Zettle, 1998, p12) and their acceptance and implementation by the very successful German equitation teams in this and the previous century (German National Equestrian Federation (GNEF), 1997, p136-141). The GNEF handbook explains that the training of a horse is to be considered as:

... a programme of systematic physical education, or a 'gymnasticising' process, aimed at developing to the full the horse's natural physical and mental attitudes and making it into an obedient riding horse with a broadly based training, and one which it is a pleasure to ride. (p136)

The influence of this sentiment is clearly visible in the Root Definition belonging to the Danish (German trained) classical rider/trainer described above. The term 'systematic' seems to refer to an ordered method of training and may be adopted in this methodical way by those who are not firmly embedded in the classical approach. Although the term 'systemic' is not mentioned, the interdependency of the different levels is recognized as the GNEF (1997) state:

None of the... qualities can be considered in isolation – they are all interdependent. They must be developed in accordance with a systematic plan, though not singly and in a rigid order. (p137)

The training scales (illustrated in Figure 2) show progressive steps of outcomes to be sought in the training of the well-schooled horse. There are sometimes arguments as to the meaning of some of the terms (e.g. 'schwung' and Durchlässigkeit) as meaning is lost in the translation from German to English. (Exploration of the meaning of such terms, however, can be so important in developing personal understanding: language limits our ability to describe tacit experience, such as 'feel', and hence to pass on our understanding.) However, regardless of interpretation, the hierarchy of elements is clear and Zettle (1998) explains the underlying systemic nature of the training scales:

These elements can be thought of in terms of the three pairs (rhythm-relaxation, impulsion-contact, straightness- suppleness) leading to collection. Rhythm and relaxation go together throughout the horse's training. Schwung is the 'go' for which contact provides the control. Suppleness is the softness and flexibility which is needed for straightness. Collection is the embodiment of them all. (p14)

This last point helps to link the practicalities of the systematic training scales to the discussion of the purposeful human activity that is Classical Dressage. Zettle explains that when attaining the highest level of the scale, collection: "while the rider will have feelings of riding more and more uphill, the end result will be both horse and rider truly in heaven." (ibid. p14).

Equifinality: The Many Roads to Rome

Whilst the Training Scales discussed above represent an approach to explaining the systematic process of training, the way in which these accepted principles are achieved seem to generate conflict (e.g. Classic vs. Classique, 2007). Even the idea of the contents and order of a training scale may be disputed. For example, Karl, whose riding and teaching is influenced by the French classical rider/trainer Baucher (1919) offers an alternative Training scale (Karl, 2008, pp156-158) which he teaches in his School of 'Légèreté" which offers a classical training programme. Given such conflict, the challenge is to understand how particular exercises and techniques work, to be able to 'read' the horse, and to know what is required at any point in time. The following of method without reflection is not part of the classical approach. Belasik (1990) offers an insight into his personal moment of realization of this aspect of the classical approach. In his search for an understanding of and excellence in the practice of classical dressage, he explains his growing frustration and almost destructive anger at the complexity and conflict he witnesses as he travels to train with respected classical trainers around the world. At a point where

Figure 2. The Training Scales (adapted from German National Equestrian Federation (1997) p136 and Zettle (1998) Figure 1: The scales of training)

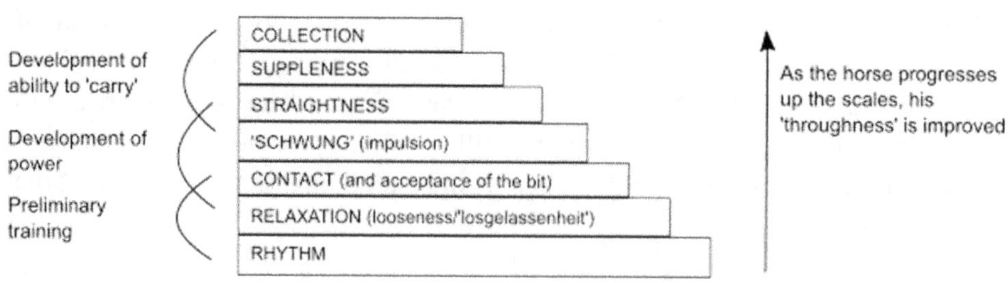

he is almost in despair he understands, suddenly, there may be more than one way to achieve an end result, and 'rules' are to *support* learning:

Then, almost at dusk, in all its ridiculous simplicity, I feel the answer. All the tension is gone. .... These masters are not all wrong. They are all right! There are great piaffes in Portugal, but there are great piaffes in France, Russia, England, Spain, Austria and Germany; and there are great piaffes in America. How could five, ten or fifteen horses be trained differently and independently, yet all be doing the exact same movement correctly? It is now clear to me that different paths can lead to the same result. In western education we are so steeped in our precious scientific method that it seems inconceivable that there could be more than one correct answer for a question or problem. In reality there are all kinds of correct answers simultaneously existing side by side. You can take your pick. They will all work if correctly applied. There seems to be no such thing as only one correct answer, but there is only one correct way of answering. (p76)

Systems Theme 6: Environmental Factors and the Parts Making the Whole

Before attempting to draw all the parts discussed above into a whole, we need to consider additional environmental influences that help to shape the theory and practice of Classical Dressage. These factors are presented in the literature by commentators from the earliest times (e.g. Xenophon, c380BC; Cavendish, 1743 (originally published in 1658); de la Guérinière, 1733; Loch, 1997; Belasik, 2012) and include issues such as the horse's breeding (and therefore its physiological and psychological characteristics), the horse management system in place which includes the horse's general care, its feed, stabling and turn-out practices, and the work of its veterinarian, dentist and farrier. Zettle (2007) adds other influencing factors for those horses in competition such as owners, grooms and judges.

Figure 3 represents an attempt to describe the many parts of Classical Dressage, discussed above, and show how they appear to interact as a whole.

The representation above suggests that at the heart of this system is the horseman's developing appreciation of their riding. This appreciation is acquired as a result of their knowledge, both theoretical (from reading) and practical (from riding), and out of this knowledge comes reflection and judgment which feeds back into the cyclical process of appreciating (or learning). The influence of an instructor or trainer may be considerable. If the instructor/trainer does not share the Classical principles of the rider then conflict is inevitable at some point. In the lower part of the diagram we see the influences of and on the horse. Given the complexity of this set of interacting parts it is not surprising that it takes many years of dedicated

Figure 3. Multiple-cause diagram showing complex interacting factors of an 'academic' or 'classical' approach to equitation (Classical Dressage)

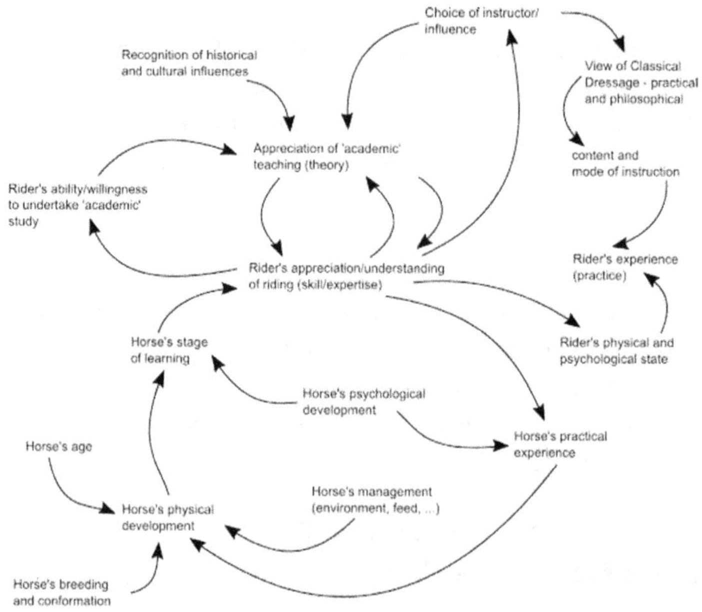

study for both rider and horse to attain the high ideals of the classical approach: it is a long journey with no 'quick fixes' for either party.

SUMMARY

In the various sections above we have looked at the role of history, culture and tradition and how these things have evolved into what we call 'classical' dressage: The past is very important to the present. We have explored the purpose of an activity that seems to exist for no real practical reason other than for the art form itself. The motivation for continuing this activity seems to be often for the pure joy of witnessing the 'brilliance' that can be experienced when horse and rider become 'one'. The horse offers us something that we treasure very highly. We have considered what would appear to be an important emergent property of this activity which is not just the training of the horse but also the training of the rider. Learning to ride well in the classical style is a lifetime's work and the journey will force the rider to confront their own fears, worries, anger, confusion as well as the problems of the 'ego'. Linked to this we have explored the idea of the importance of process over technique in the classical 'journey' and the way in which external measurement makes little sense in an activity where so much is dependent upon 'feel' and what might be referred to as 'internal' validation. This has been an important point in illustrating one of the differences between classical and 'competitive' dressage where external assessment is the way in which success is measured. The objective measurement of an art form is difficult and yet this is, largely, the approach in dressage competition.

In the above discussion we have considered the 'aids' as the vehicle of communication and the rider's 'seat' as the governor of action which is used judiciously in the work towards 'collection' which is the result of long training and perfect communication and control between horse and rider. True collection is the foundation of brilliance and harmony; the two (horse and rider) acting as 'one'. We have seen how the classical tradition has many exercises and movements that are used to train the horse progressively through various defined 'scales', or 'levels'. But we have also seen that there is a need to understand that there may be many different ways to implement a set of principles.

The aim in this paper has been to offer a systemic analysis of classical dressage. The work presented provides an overview of the many parts which make up the classical approach and explores the way in which they might be considered to fit together as a whole. The whole is, without doubt, complex and requires a reflective rider capable of weaving together theory and practice within general principles. A procedural, step-by-step, method is not a suitable model for the classical approach. For me, there is a deep feeling that all these parts fit together as a sensible and reasonable whole and I hope to continue the exploration of these parts and the whole for many years to come.

CONCLUSION

As a result of the above investigation, two particular areas seem worth revisiting in this final section: (i) the relationship between classical and competition dressage, and (ii) the importance of developing a complementary relationship between systems and science in helping us to understand complex situations.

Classical vs. Competition Dressage

For close on thirty years now, commentators have been criticizing those involved in competitive dressage (riders, trainers, judges, the FEI) for moving away from classical 'ideals'. The analysis offered above has suggested that classical dressage involves: the personal journey of the rider in terms of their own character development (including the suppression of the ego), the development of the horse's natural movement, the development of the rider's 'feel' (internal) for the correctness of the horse's movement and the search for harmony, balance, being-at-one, and brilliance experienced in the human-horse relationship – altogether a rather inward focusing activity that exists for its own sake. By comparison, competitive dressage involves ribbons, prestige, and at the higher levels, large financial gain, for those who *win*. Most importantly, it involves external, numerical measurement for clearly identified patterns executed within a set time, based upon judges' observations (according to an interpretation of the 'rule' book). Personally, these two points seem to represent an incompatibility that cannot be 'fixed'; they represent two diametrically opposed ways of thinking about the world. We might suggest that either we consider the two formats separately (e.g. Welz, 2010), or that the only way forward is to ensure that competition dressage is trained, ridden and judged exactly according to classical principles (e.g. Hector, 2017). Either way is likely to be impractical and unacceptable. It is suggested here that only when science can protect the treatment of horses in competition, by a thorough understanding of horse locomotion and biomechanics under saddle, that horse welfare can be enforced, by regulations and by law, to ensure ethical use of horses in sport. Such a change would place even more burden on judges and, hopefully, extend the responsibility of riders, trainers, breeders and sponsors. Such science will also support, or dismiss, the handed-down philosophies and directives of the classical 'masters'. Either way, the result must benefit the horse. The role of science becomes fundamental, but will its application be sufficient?

Science and Systems Working in Complement

The story presented above is the result of the author using her understanding of interpretive, or 'soft', systems theory to help make sense of a very old tradition for the training and riding of horses. There are other attempts on-going to help us learn more about the special relationship between humans and horses much of which is coming out of the relatively new field of "Equitation Science" (International Society for Equitation Science, 2018). Such research relevant to riding or dressage has tended to focus on issues that have a welfare factor, such as the position of the horse's head and neck (e.g. van Dierendonck *et al.*, 2012) and tension on the reins (e.g. Clayton *et al.*, 2003; Glißmann & König von Borstel, 2012). The results have been interesting but such studies seem to encourage more specific and detailed analysis of individual parts of very complex processes. The danger of such an approach is a lack of any context in which to judge and use the outputs of the science. For example when judging rein tension, the results do not tell us whether the tension comes from the rider using their hands, their back or whether it is a result of the horse pulling. The implications for this lack of understanding are serious in terms of erroneous interpretation as Belasik, reflecting upon research he participated in at the McPhail Center, illustrates: he and his horse were fitted with rein tension sensors so rein pressure could be recorded electronically when he was riding. The resulting pressure graphs of the signals looked as if the rider was implementing repetitive half-halts or corrections through the reins. When he saw this data he knew something was wrong with the initial researchers' interpretation regarding the weight and tension he was creating in the reins:

If you had looked at the graph after I finished, you would have seen the recording starting by spiking pretty hard and gradually getting lighter. .... The horse started out heavy on the forehand; each time he landed he tried to fall forward. I blocked this with a passive hand and supported my hand with my back, seat and thighs. My hand was still. The spike was occurring form the horse hitting the bit and reins. ....I was not pulling on the reins, but by me setting limits the horse corrected himself. The graph needed analysis. Another case of what you see might not be what is actually happening. (Belasik, 2009, pp74-5).

Interestingly, some scientific researchers now appear to be recognising the limitations of science to fully explain the human-horse relationship (e.g. Egenvall et al. 2016; Leemans, 2016). Belasik's example illustrates how the need to try to understand the nature of the horse/rider relationship necessitates more than a scientific approach: to truly understand the implications of the small parts of the whole depends upon us understanding the wider picture, or the whole system. This means we need to have a clear idea of the system we are exploring and to understand its boundaries. To complement science thoroughly is a serious and important challenge for systems science.

ACKNOWLEDGMENT

The author would like to acknowledge the patience, generosity and input of a number of people who, thereby, made the original paper possible: Mike Watson, D. Francisco de Bragança, Markus Bauer, and Malcolm Bronte-Stewart.

REFERENCES

Arenalilla. (2007). *Doma Vaquera*. Retrieved on November 23, 2017 from: http://www.youtube.com/watch?v=nk_n1yQlmT0

Baucher, F. (1919). *The Principles of Horsemanship*. London: Vinton & Co.

Bayareabert. (2009). *Spanish Riding School Vienna Philharmonic Beliebte Annen Polka Johann Strauss Elisen Polka Op 151*. Retrieved on November 23, 2017 from http://www.youtube.com/watch?v=Pn6Jmqs9IHQ

Belasik, P. (1990). *Riding towards the light: An apprenticeship in the art of dressage riding*. London: J.A. Allen.

Belasik, P. (1993). *An Overview of the Classical Principles behind the Training of Horse and Rider: An interview with Paul Belasik, Part One: Breaking the Young Horse* [Audio cassette]. Telford, PA: Base 10 Studios.

Belasik, P. (2001). *Dressage for the 21st Century*. London: J.A. Allen.

Belasik, P. (2009). *A Search for Collection: Science and Art in Riding*. London: J.A. Allen.

Belasik, P. (2012). *Nature, Nurture and Horses: A Journal of Four Dressage Horses – From Birth through the First Year of Training*. Vermont: Trafalga Square Books.

Bourgelat, C. (1754). A New System of Horsemanship: from the French of Monsieur Bourgelat by Richard Berenger. LaVergne, TN: Ecco Print Editions. doi:10.5962/bhl.title.33714

de Bragança, D. (1975). *L'Equitation de Tradition Francaise*. Paris: Belin.

Cavendish, W. (2000). *Duke of Newcastle, (1743). A General System of Horsemanship*. London: J.A. Allen.

Checkland, P. B., & Poulter, J. (2006). *Learning for Action: A Short Definitive Account of Soft Systems Methodology and its use for Practitioners, Teachers and Students*. Chichester, UK: John Wiley and Sons, Ltd.

Classical versus classique: with Christoph Hess and Philippe Karl. (2007). [DVD]. Thomas Vogel, pferdia tv.

Clayton, H. M., Singleton, W. H., Lanovaz, J. L., & Cloud, G. L. (2003). Measurement of rein tension during horseback riding using strain gage transducers. *Experimental Techniques*, *27*(3), 34–36. doi:10.1111/j.1747-1567.2003.tb00112.x

D'Endrody, A. L., & Col, Lt. (1959). Give your horse a chance: the classic work on the training of horse and rider for all disciplines. London: J.A. Allen.

De Kunffy, C. (1993). *The Ethics and Passions of Dressage*. Middletown, MD: Half-Halt Press.

Duarte, King of Portugal (1438) The Royal Book of Horsemanship, Jousting and Knightly Combat (Livro Da Ensinanca De Bem Cavalgar Toda sela). (2005). (A. F. Preto & L. Preto, Trans.). The Chivalry Bookshelf.

Egenvall, A., Rhodin, M., Roepstorff, L., Eisersiö, M., & Clayton, H. M. (2016). Rein tension peaks within canter (poster 39). *Proceedings 12th International Equitation Science Conference: Understanding horses to improve training and performance.* Retrieved on November 23, 2017 from: http://equitationscience.com/previous-conferences/2016-12th-international-conference

Epona, T. V. (2017). Retrieved on November 23, 2017 from: https://epona.tv/farewell-falsterbo

Équitation - Le hibou vert. (2012). *Nuno Oliveira resumen.mpg.mp4*. Retrieved on November 23, 2017 from: https://www.youtube.com/watch?v=o53YxuOCHuY

Fillis, J. (1911). Breaking and Riding: With Military Commentaries. Lincoln, NE: University of Nebraska Press.

Glißmann, C., & König von Borstel, U. (2012). Alternatives to conventional evaluation of rideability: rein tension and behavior from performance test dressage training. 8th International Equitation Science Conference, Edinburgh, UK.

Grisone, F. (1558). *Ordini di Cavalcare*. Retrieved on November 23, 2017 from: http://www2.vetagro-sup.fr/bib/fondsancien/ouvonline/1355-2/1355-2.htm

International Society for Equitation Science. (2018). Retrieved on November 23, 2017 from: (http://equitationscience.com/

de la Guérinière, F. R. (1733). School of Horsemanship (T. Boucher, Trans.). London: J.A. Allen.

German National Equestrian Federation. (1997). *The Principles of Riding: the official instruction handbook of the German National Equestrian Federation – Book I* (C. Belton, Trans.). Buckingham, UK: Kenilworth Press.

Hector, C. (2017). *Guarding the Classical Principles*. Retrieved on November 23, 2017 from: http://www.horsemagazine.com/thm/2017/06/guarding-the-classical-principles/

Henriquet, M. (1991). Henriquet on Dressage (H. Nelson, Trans.). London: J.A. Allen.

Herbert, L. (Ed.). (1939). *Henry, Elizabeth and George: Letters and Diaries of Henry, Tenth Earl of Pembroke and his Circle* (Vol. 1). London: Jonathan Cape.

Heuschmann, G. (2006). *Tug of War: Classical versus ""Modern"" Dressage - Why Classical Training Works and How Incorrect "Modern" Riding Negatively Affects Horses' Health.* Trafalga Square, VT: Reina Abelshauser.

Heuschmann, G. (2012). *Balancing Act: The Horse in Sport – an Irreconcilable Conflict?* London: J.A. Allan.

Horseherovideo. (2010). *Advanced work in hand with Francisco de Bragança.* Retrieved on November 23, 2017 from: https://www.youtube.com/watch?v=2DnQ1s-Fj6E

HorseOsteopath. (2010). *Pedro Torres and Oxidado Style Section Working Equitation European Championships 2009 - Medium.m4v*. Retrieved on November 23, 2017 from: http://www.youtube.com/watch?v=DzvopwrzjIo

The Spanish Riding School. (n.d.) Retrieved on November 23, 2017 from http://www.srs.at/en_US/start-en/

Husky005reloaded. (2010). *Cadre Noir de Saumur @ Paris – Bercy*. Retrieved on November 23, 2017 from: http://www.youtube.com/watch?v=1WlcXhTjqi0

Hyland, A. (1998). *The Warhorse: 1250-1600.* Stroud: Sutton Publishing Ltd.

Karl, P. (2008). *Twisted truths of modern dressage: A search for a classical alternative.* Brunsbek: Cadmos Verlag.

Kerbrech, General Faverot de. (1891). *Dressage Méthodique du cheval de selle: d'après les dernieres enseignements de F. Baucher*. LaVergne: Nabu Public Domain Reprints.

Kottas-Heldenberg, A. (2010). *Kottas on Dressage*. Shrewsbury, UK: Kenilworth Press.

lanchonman. (2006). *Garrocha*. Retrieved on November 23, 2013 from: http://www.youtube.com/watch?v=H9Y1F_f87F0

Laurioux, A., & Henry, G. (2009). *The Great European Schools of Classical Dressage: Vienna, Saumur, Jerez, Lisbon.* Brunsbeck: Cadmos.

Leemans, I., Willemsen, M. E., Douwes, B., Steenbergen, M. Y., van Iwaarden, S., & Rettig, A. (2016). Influence of rider's actiontype profile on rein tension, Poster 41. *Proceedings 12th International Equitation Science Conference: Understanding horses to improve training and performance*. Retrieved on November 23, 2017 from: http://equitationscience.com/previous-conferences/2016-12th-international-conference

Levett, S. (2009). *Tradition and Change at the Spanish Riding School*. Retrieved on November 23, 2017 from: http://www.viennareview.net/news/ideas-and-trends/tradition-and-change-at-the-spanish-riding-school

Loch, S. (1986). *The Royal Horse of Europe: The Story of the Andalusian and Lusitano*. London: J.A. Allen.

Loch, S. (1988). *The Classical Seat*. London: Unwin Hyman.

Loch, S. (1990). *Dressage: The Art of Classical Riding*. London: The Sportsman's Press.

Loch, S. (1997). *The Classical Rider: Being at one with your horse*. London: J.A. Allen.

Loch, S. (2003). *Invisible riding: The secret of balance for you and your horse*. Surry, UK: Horse & Rider Magazine.

Mairinger, F. (1983). *Horses are made to be horses: A personal philosophy of horsemanship*. Adelaide: Rigby Publishers.

Müseler, W. (1937). *Riding Logic*. London: Eyre Methuen.

Oliveira, N. (1965). Reflections on Equestrian Art. London: J.A. Allen.

Pembroke, H. (1978). *Earl of (1773). Military equitation or a method of breaking horses and teaching soldiers to ride*. London: J.A. Allen.

Pereira, C. H. (2001). *Étude du premier traité d'équitation portugais: Livro da ensinça de bem cavalgar toda sela du roi Dom Duarte*. Paris: L'Harmattan.

Podhajsky, A. (1967). The Complete Training of Horse and Rider. London: The Sportsman's Press.

de Pluvinel, A. (1626). The Maneige Royal. London: J.A. Allen.

Racinet, J-C. (2009). *Falling for Fallacies: Misleading Commonplace Notions of Dressage Riding*. Cadmos: Brunsbeck.

Rpgdressage. (2012). *Paulo Caetano Bullfight 2012*. Retrieved on November 23, 2017 from: http://www.youtube.com/watch?v=Vz14t8Wuek&playnext=1&list=PL246A51242E32977A&feature=results_main0

Seunig, W. (1961). The Essence of Horsemanship. London: J.A. Allen.

Shand, M. (2017). *The Concordia Connection – putting the horse first*. Retrieved October 27, 2017 from: https://www.facebook.com/groups/Concordiaequestrians

Steinbrecht, G. (1885). The Gymnasium of the Horse. Franktown, VA: Xenophon Press.

Stevens, M. J. (1994). *A Classical Riding Notebook*. The Kenilworth Press.

Vickers, G. (1965). *The Art of Judgment: A Study of Policy Making*. London: Chapman and Hall.

Vickers, G. (1972). *Freedom in a rocking boat: changing values in an unstable society*. Harmondsworth, UK: Penguin Books Ltd.

van Dierendonck, M., van Dalum, M., Beekmans, M., & Christensen, J. W. (2012). Acute stress responses of dressage horses ridden in three different head and neck positions. *8th International Equitation Science Conference*. Retrieved on November 23, 2017 from: http://equitationscience.com/previous-conferences/2012-8th-international-conference

von Neindorff, E. (2005). *The Art of Classical Horsemanship*. Brunsbek: Cadmos.

Welz, Y. (2010). *What has happened to Dressage?* Retrieved on November 23, 2017 from: https://www.thehorseshoof.com/HLC_dressage.html

West, D. (2005). Vickers' concept of 'Relationship-maintenance' as an alternative to 'Goal-seeking' models of organization: A difference in the notion of control. *Systemic Practice and Action Research*, *18*(3), 261–274. doi:10.100711213-005-4814-7

West, D. (2012). An application of 'systems science' to classical dressage. *8th International Equitation Science Conference*. Retrieved on November 23, 2017 from: http://equitationscience.com/previous-conferences/2012-8th-international-conference

West, D., & de Bragança, F. (2012). A systemic approach to eliciting and gathering the expertise of a 'knowledge guardian': An application of the appreciative inquiry method to the study of classical dressage. *Systemic Practice and Action Research*, *25*(3), 241–260. doi:10.100711213-011-9223-7

West, D. (2015). Classical Dressage: A Systemic Analysis. *International Journal of Systems and Society*, *2*(1), 23–43. doi:10.4018/ijss.2015010102

Xenophon. (2006). *The Art of Horsemanship* (M. H. Morgan Reprint, Trans.). New York: Dover Publications.

Zettle, W. (1998). *Dressage in Harmony: From Basic to Grand Prix*. Boonsboro, MD: Half Halt Press.

Zettle, W. (2007). *The Circle of Trust: Reflections of the Essence of Horses and Horsemanship*. Boonsboro, MD: Half Halt Press.

Chapter 11
Radicalization and Recruitment:
A Systems Approach to Understanding Violent Extremism – New Developments Through FRAM

Anthony Masys
University of Leicester, UK

ABSTRACT

As described in the GTI, ISIL's transnational tactics in combination with lone actor attacks inspired by the group drove an increase in terrorism to its highest levels ever across Europe and many OECD countries (upwards of a 650 percent increase in deaths since 2014). The attacks by ISIL in Paris, Brussels, and in Turkey's capital Ankara, were amongst the most devastating in the history of these countries and reflect a disturbing return of the transnational group-based terrorism. Actor network theory (ANT) was applied as a systems lens to open the "blackbox" of terrorism. The systems view facilitated by ANT highlighted how dynamic networked actors shape radicalization through the actor network process of translation. This chapter applies functional resonance accident model (FRAM) methodology. The FRAM method was used to analyze how radicalization activities (as described through ANT) take place and where and how intervention strategies can be designed to interfere with the radicalization process.

DOI: 10.4018/978-1-5225-5996-2.ch011

INTRODUCTION

Recent events such as the terrorist attack in Algeria (January 2013), the Boston Marathon Bombing (April 2013), attacks in Paris (2016), Belgium (2016), Manchester (2017), UK (2017) highlight the requirement for greater understanding regarding the radicalization and recruitment of terrorists. As detailed in the US Department of State Report (2011), over 10,000 terrorist attacks occurred in 2011, affecting nearly 45,000 victims in 70 countries and resulting in over 12,500 deaths. With a focus on the outcomes and results of terrorist activities, terrorism itself often becomes a 'blackbox' concept that does not capture the essence of the radicalization process nor the mechanisms of recruitment. This *conceptual paper* introduces Actor Network Theory (ANT) as a systems lens to open the 'blackbox' of terrorism. This systems lens '…is a discipline for seeing wholes. It is a framework for seeing interrelationships rather than things, for seeing patterns of change rather than static snapshots' (Senge, 1990,p. 68). The systems view facilitated by ANT is supported and informed by methods of network analysis and conceptual modelling that highlight how dynamic networked actors shape the radicalization process through the actor network process of *translation*.

Terrorism is often labeled a global threat. The recent terrorist events highlight issues pertaining to recruitment, radicalization and supporting structures that enable terrorism worldwide. These events and others challenge our understanding regarding the impact of ideology and local context on the motivations and recruitment for terrorism. How do we collectively deal with such complex issues as terrorism, often termed 'wicked problems'? The 'fixes that fail' archetype detailed by Senge (1990, 2006) and Sterman (2000) describe how action taken to rectify a situation, often only results in short term gains without addressing the foundational issues of the problem space. It forces one to revisit how we conceptualize 'work as imagined' in terms of terrorist activities and the actual 'work as done' (Hollnagel, 2012).

What is required in such wicked problems is greater understanding of the dependencies and interconnectivity that arises from a systems view of the problem space. This conceptual paper highlights actor network theory (ANT) as a systems lens to begin to reveal and unravel the complexities associated with terrorism and radicalization. In particular, the paper explores the application of the ANT process of Translation comprised of: problematization, interessement, enrolment and mobilization, to distinguish how heterogeneous becoming (Chia & King, 1998) of entities is reified within the context of terrorism and radicalization.

Global systemic factors, state structural, and socio-cultural factors (Davis & Cragin, 2009, p. xxiii) provide a macro-perspective on the problem space of terrorism. By opening these 'blackbox' (macro-perspectives) and 'following the actors' (Latour, 2005) we trace a complex, interconnected network that is best characterized as a rhizome (Deleuze & Guattari, 1987). The network schema that is revealed through ANT has far reaching implications beyond the visual representations to include how we understand the temporal and spatial heterogeneity that is resident within the actor network. As such, this challenges our notions of far/close, small scale/ large scale and inside/outside (Latour, 1996, p. 370) and to think in terms of associations and relations. This inherent relationality is central to our understanding of ANT and presents a unique systems lens on terrorism. The spatial and temporal implications are profound for our understanding of radicalization and terrorism. The actor network recognizes that 'what is acting at the same moment in any place is coming from many other places, many distant materials, and many faraway actors' (Latour, 2005, p. 200).

This conceptual model that leverages ANT, draws upon the literature on terrorism and radicalization as well as contextually on recent terrorist events: the terrorist attack on an oil processing plant in Algeria (January 2013); the Boston Marathon terrorist bombing (April 2013), attacks in Paris (2016), Belgium (2016), USA (2016), Manchester (2017), London (2017). Leveraging the ANT lens, this chapter applies Functional Resonance Accident Model (FRAM) methodology (Hollnagel, 2012). FRAM is a method for modelling complex socio-technical systems (the hallmark of the radicalization process). The FRAM method was used to analyse how radicalization activities (as described through ANT) take place and where and how intervention strategies can be designed to interfere with the radicalization process.

Terrorism

The complexities associated with the definition of terrorism transcend the single-discipline approaches to understanding the motivations for terrorism. Within the context of terrorism, Bjelopera and Randol, (2011, p. 3) define radicalization as a '... process of acquiring and holding extremist beliefs. ... violent extremism describes violent action taken on the basis of radical or extremist beliefs'. Bjelopera and Randol, (2011, p. 10) argue that radicalization stems from the exposure of individuals to ideological messages and the movement of those individuals from mainstream beliefs to extremist viewpoints. Rabasa, Pettyjohn, Ghez and Boucek (2010, p. 1) define radicalization as ' ...the process of adopting an extremist belief system, including the willingness to use, support or facilitate violence, as a method to effect societal change', thereby purporting the connection between extremist beliefs and violence.

Regarding the complex aetiology associated with radicalization and terrorism, studies by the DHS Office of Intelligence and Analysis indicate that radicalization varies across ideological and ethno-religious spectrums, different geographic regions and socio-economic conditions (Bjelopera & Randol, 2011, p. 12). It is important to note as argued by TNO (2008, p. 5) that although not every radical becomes a terrorist, every terrorist has gone through a radicalization process. What becomes apparent is that the process of radicalization emerges from the complex interrelationships and interdependencies of factors (actors).

Of particular interest given the recent terrorist events around the world is the strategic importance of home-grown recruits and facilitators of radicalization and recruitment. As described in Davis and Cragin (2009) and TNO (2008, p. 3), although such external factors as political, economic and cultural conditions may influence an individual's environment, they do not have a direct cause and effect on individual behaviour. This implies that independent factors are insufficient to result in radicalization and a systems perspective is required to better understand the inherent complexity associated with radicalization and recruitment.

With regards to homegrown radicalization and recruitment, Jenkins (2011, p. 15) argues that the internet has influenced and supported radicalization by providing a vehicle whereby actors have found resonance and reinforcement for their frustration and anger. In this sense it is this actor (the socio-technical) that is shaping the nature of the struggle and creating opportunities (socialization, facilitating justification, support (fueling)) for the terrorist agenda. The report highlights that al Qaeda today does 99 percent of its work on the Internet. Radicalization thereby requires other actors (technology for example) as an enabler. A Senate Report argues that 'the Internet serves as a source of inspiration, but it may also become a substitute for action, allowing would-be terrorists to safely engage in vicarious terrorism while avoiding the risks of real action. Some have suggested that al Qaeda is employing a strategy of "gamification," providing forums where followers share violent fantasies as a means of encouraging them toward real violence' (Senate report: HSGAC, 2012).

Radicalization is characterized by complexity emerging from the factors that cut across social, political, economic, ideological domains. The complexity of the problem space stems from the inherent interconnectivity and interdependency associated with the actor as well as the various scales of resolution that it can be examined and the spatial and temporal heterogeneity inherent in terrorism. A systems lens is thereby required to open the black box of terrorism.

Systems Thinking

The application of systems thinking crosses many domains resulting from various contributions of key thought leaders such as those cited in Mingers (2006, p. 1): General Systems Theory (Von Bertalanffy,1971), Cybernetics (Weiner, 1948), dialectical systems (Churchman, 1968,1971), purposeful systems (Ackoff & Emery, 1972), autopoiesis and cognition (Checkland, 1981), social system theories (Buckley, 1967; Luhmann, 1995), critical systems thinking (Flood & Jackson, 1991; Midgley, 2000) chaos and complexity (Kaufmann, 1995).

Anderson and Johnson (1997, p. 2) define a system as: '...a group of interacting, interrelated, or interdependent components that form a complex and unified whole. A system's components can be physical objects...and can also be intangible, such as processes; relationships, policies, information flows; interpersonal interactions; and internal states of mind such as feelings, values, and beliefs'. This challenges the reductionist paradigm that fails to capture more non-linear and holistic properties. As described in Masys (2010), central to systems thinking is the concept of interrelationships of objects that form a whole and in which show a property unique to the whole that is not a property of the components. This is significant because it suggests that although events and objects may appear distinct and separate in space and time, they are all interconnected. Senge (1990) remarks that, because the world exhibits qualities of wholeness, our investigation of it should stem from a paradigm of the whole.

The strength of Systems thinking is derived from mapping of inter-relationships (Masys, 2010; Sterman, 2000) as a means to inform decision making and understanding regarding 'causality', intervention strategies and unintended consequences. Within the application domain of terrorism studies, systems thinking recognizes that the 'whole is greater than the sum of the parts' and that understanding terrorism requires a holistic perspective stemming from the interrelationships, interconnectivity and complex feedback and feed forward systems that so characterizes the problem space.

Actor Network Theory

As described in Masys (2010, 2012), the foundational development of ANT stems from the contributions of Latour (1987, 1988, 1999, 2005), Callon (1986ab, 1999) and Law (1987, 1992, 1999). ANT treats both human and non-human elements in a symmetrical manner, thereby facilitating the examination of a situation where Callon (1999) argues, '... it is difficult to separate humans and non-humans', and where 'the actors have variable forms and competencies'.

Yeung (2002) notes that much of the work that draws on actor network theory places its analytical focus on unearthing the complex web of relations between humans and non-humans. It is recognized within ANT that the interaction of non-human actors with the human actors gives shape and definition to identity and action.

ANT revisits the etymology of the word 'social' through which Latour (2005, p. 5) realigns the definition of the word social with its origins associated with a 'trail of associations'. In this sense he describes the 'social' not as a designated thing among other things, but rather as a '...type of connection between things that are not themselves social'. This has a significant ontological, epistemological and methodological impact on how one is to conduct 'social' analysis and presents ANT as a systems thinking framework. Latour (2005, p. 29) argues for an approach that 'follows the actors' rather than beginning with any apriori concept of the social. 'Following the actors' (Callon, 1986a; Callon, 1991; Latour, 2005) allows the researcher to investigate those actors that have been 'silenced or deleted'.

Masys (2010) argues that the network schema that emerges from the ANT lens is characterized by relations, fluidity and dynamics, which has far reaching implications with regards to the spatial and temporal nonlinearity. The traditional conceptualization of networks views them as a collection of nodes and connections, which form a web-like structure (Barab, Hay, & Yamagata-Lynch, 2001). ANT focuses on the inherent relationality within the actor network thereby capturing the fluid topology that characterizes the complexity. Within this topological construct, Urry (2003, p. 122) describes how the micro/macro distinction loses its meaning since '….both micro and macro are local effects of hooking up to circulating entities (Latour, 1999b, p. 19)'. One may say that the relationality is brought about '… through a wide array of networked or circulating relationships that are implicated within different overlapping and increasingly convergent material worlds' (Urry, 2005, p. 245). This inherent relationality is central to our understanding of ANT and is foundational to our understanding of terrorism and radicalization.

Fundamental to our understanding of ANT are the conceptualization of the Actor and the Network. An actor-network, as cited in Aanestad (2003, pp. 6-7), '… is a heterogeneous network of human and non-human actors… where the relations between them are important, rather than their essential or inherent features (Latour, 1987; Callon, 1986, 1991)'. Latour (1987, p. 180) defines the word network as that which '…indicates that resources are concentrated in a few places- the knots and the nodes-which are connected with one another- the links and the mesh: these connections transform the scattered resources into a net that may seem to extend everywhere'. The Actor Network becomes a network of aligned interests formed by the heterogeneous actors, characterized as full of hybrid entities (Latour, 1993) comprised of both human and non-human elements. Action thereby takes place in 'hybrid collectifs' (Callon & Law, 1995) that entangle human actors as well as non-human actants in multiple ways. This has significant implications for how we frame terrorism and radicalization.

As described in Masys (2010), fundamental processes within ANT are inscription and translation. Inscription refers to the way technical artifacts embody patterns of use: Technical objects thus simultaneously embody and measure a set of relations between heterogeneous elements (Akrich, 1992, p. 205). Monterio (2000, p. 77) argues that although 'inscription' might sound deterministic, '…the artifact is always interpreted and appropriated flexible, the notion of inscription may be used to describe how concrete anticipations and restrictions of future patterns of use are involved in the development and use of a technology'. Inscriptions enable action at a distance by creating 'technical artefacts' that ensure the establishment of an actor's interests such that it can travel across space and time and thereby influence other work (Latour, 1987).

The process of translation has been described as pivotal in any analysis of how different elements in an actor network interact (Somerville, 1997). As a transformative process, translation emphasizes '...the continuous displacements, alignments and transformations occurring in the actor network' (Visue, 2005, p. 115). Translation rests on the idea that actors within a network will try to enroll (manipulate or force) the other actors into positions that suit their purposes. When an actor's strategy is successful and it has organized other actors for its own benefit, it can be said to have translated them. Translation as argued by Callon (1991, p. 143) '...are embodied in texts, machines, bodily skills [which] become their support, their more or less faithful executive'. This process of translation comprises undertones of power mechanisms as described in Callon (1986a). Through ANT one recognizes that power resides within the network of heterogeneous elements and is characterized as relational, emergent and distributed. Power is perceived as the capacity to influence that is realized only through the process of exercising this influence. In this sense, considering the relationality inherent within ANT, power can be conceived as a practice rather than a position. Willcocks (2004, p. 255) argues that '...power must be analyzed as something that circulates...that functions only when it is part of a chain. It is never localized here or there, it is never in the hands of the some, and it is never appropriated in the way that wealth or a commodity can be appropriated'. Stemming from the temporal and spatial heterogeneity of power, the actor network is analyzed within a 'relational space' that transcends far/close, large/small characterization.

Within the context of the terrorism, an examination of actors such as those characterized by ideology and politics, by globalization and localization, by matters pertaining to human insecurity, by kinships to culture facilitates an exploration of how these "actors" mediate action and how they are entangled in local socio-cultural/economic/political/technical and ideological configurations. ANT 'following the actors' permits the various scaling and spatial/temporal heterogeneity as an ontological quality and an epistemological necessity. Multi-vocality thereby emerges from the analysis methodology supporting the employment of the systems thinking.

DISCUSSION

The study of terrorism and radicalization requires a cross disciplinary approach leveraging psychology, social psychology, sociology, religious studies, organizational theory, political science, game theory, complexity science, network science and economics. What is unique about ANT is that it does not apriori designate what is social nor by that argument apriori designate the disciplinary lens in which to view the problem space. Rather the actors themselves reveal the necessity for multi-vocality in tracing the relationality and the transformation of heterogeneous networks (Law 1992) that are made up of people, organizations, agents, machines and many other 'objects'. ANT explores the ways that the networks of relations are composed, emerge constructed and maintained, how they compete with other networks. It is this very approach that makes ANT a powerful systems lens for examining radicalization and the terrorist agenda. Through the lens of ANT, the concepts of radicalization and extremism themselves emerge from multiple perspective. Bartolucci and Gallo (2013) argue that '…terrorism cannot be understood outside a context, language and culture'. That is '...implicit in the act of labeling is the silencing of alternative framings' (Bartolucci & Gallo, 2013). The implication this has on the analysis is significant. It reveals alternative framings regarding the definition of radicalization and extremism.

The complexity associated with understanding terrorism is entangled within '…a value framework that may be political, religious, social or economic' (Ackoff & Strumpfer, 2003, p. 287) Analysis conducted by Davis and Cragin (2009), Jenkins (2011) and TNO (2008) highlight key causal factors with regards to terrorism. When we deploy an ANT approach to this body of work leveraging insights and data from recent terrorist events described, we recognize the extent to which the actors emerge from their associative (relational) interdependencies that characterize the process of radicalization. In particular, events in the past become embedded in the relational actor network and thereby become an actor in the process of translation. As described in Hernes (2010), Callon (1986) offers a conceptual framework for explaining processes of becoming. This relational 'mapping' is described by Callon (1986) in terms of the translation process (problematization; interessement; enrolment; and mobilization). Applying the definition and characterization of radicalization in the introduction as a starting point in this analysis, ANT facilitates an examination of how 'actors enlist other actors into their world and how they bestow qualities, desires, visions and motivations on these actors (Latour 1996)', as cited in Powell and Owen, (2011, p. 141). In this sense, the focus is on the process of enlisting actors for a common purpose. This is how ANT acts as the systems lens. Whereas traditional analysis examined individual pieces through a reductionist lens, systems thinking (enacted through ANT) focuses on how the actors interact with the system and the resulting emergent behavior. In essence, systems thinking takes into account the complex interconnectivity and interdependencies as the issue being examined.

Leveraging Callon (1986), problematization serves to help the emergence and seeding of a terrorist agenda (it shapes the raison d'etre thereby making the actors the indispensible architects of the process). Interessement takes the form of using different actors to secure a functioning network. It is about what and how are the actors connected to ensure viability of the network. Through the processes of *enrolment*, actors try to interrelate the roles of respective actors in the network. Drawing upon Callon, enrolment takes place through multilateral work of fitting actors' roles to one another. The processes of *mobilization* ensures the adhesion of the collectivities to the network. Mobilization takes place, for example, through working with actors of the terrorist network space. It is important to recognize the actor network is defined not by human but by the network of heterogeneous elements that reside within it and how they shape behaviour, action and inaction. *The four processes—problematization, interessement, enrolment, and mobilization*—serve to distinguish activities by which the heterogeneous becoming associated with radicalization and terrorism is put into motion and upheld.

Emerging from the thematic analysis (Boyatzis, 1998; Braun & Clarke, 2006) of the contextual case studies and the literature such as Davis and Cragin, (2009); Coolsaet (2008); Forest (2007); Vidino and Brandon (2012), are 4 high-level attributes (actors) that characterize the radicalization and recruitment process: socialization; justification; support and *'perception'* of rewards. Figure 1 provides a conceptual mapping using rich pictures (Checkland,1981; Checkland & Scholes, 1990) of the relational heterogeneity. This conceptual mapping illustrates how power (influence) permeates the actor network from dynamic distributed heterogeneous entities that transcend the linear paradigm of space and time. The translation process applied is very telling. The four attributes and artifacts are interconnected creating a complex network of influence with inherent positive and negative feedback that affects and shape the central actor.

With the cases of homegrown terrorism described in Bjelopera (2013), Senate Committee Report (2012), Vidino and Brandon (2012), the process of translation beginning with problematization focuses on power and influence and how the actor submits to being integrated into the radicalization process, or conversely, refuse the transaction by defining its identity, its goals, projects, orientations, motivations, or interests in another manner. Identity and goals emerge from interdependent influences. They are formed and are adjusted through the complex process of translation.

Interessement is the group of actions by which an entity attempts to impose and stabilize the identity of the other actors it defines through its problematization. A diverse mix of actors is used to implement these actions as noted in the literature regarding gamification and the internet. They influence all elements of socialization, rewards, support and justification. The process of interessement focuses on ways to 'interest' other actors by placing themselves between other actors that would challenge the radicalization and who would therefore want to define their identities otherwise. Figure 2 illustrates this process. As described in Callon (1986b) 'A' interests 'B' by cutting or weakening all the links between 'B' and the invisible (or at times quite visible) group of other entities 'C, D, E' who may want to link themselves to 'B'.

Figure 1. Conceptual mapping (rich picture) of translation process

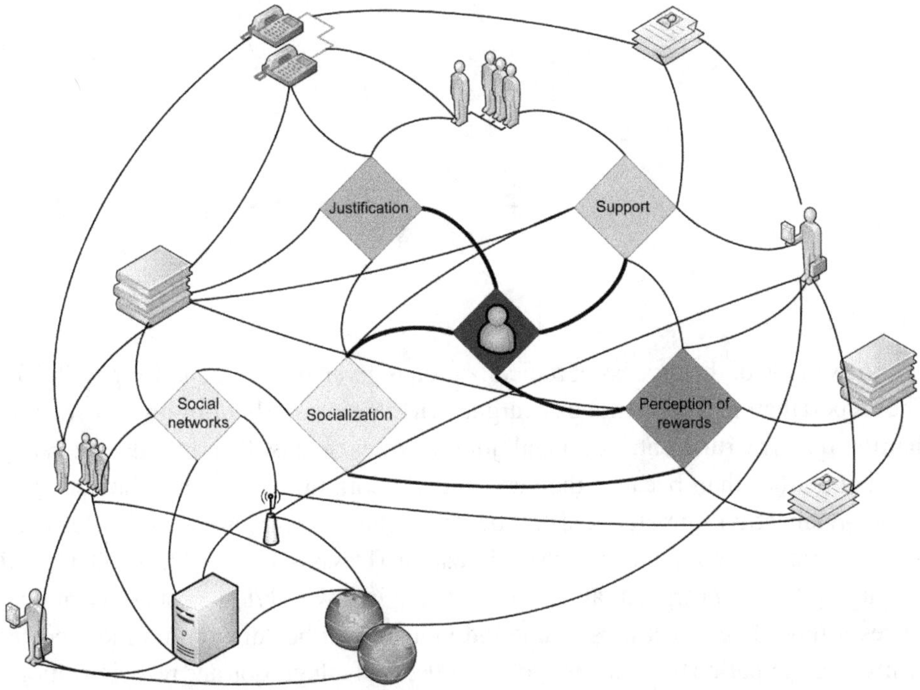

The properties and identity of 'B' in terms of radicalization and violent extremisim are consolidated and/or redefined during the process of interessement. 'B' is a 'result' of the association which links it to 'A'. This link disassociates 'B' from all the 'C, D, and E's (if they exist) that attempt to give it another definition. 'A' itself is an actor network that has been 'punctualized' and black boxed to represent the terrorist agenda. 'B' is redefined in essence through the effect of power and influence of 'A'. The identity and 'geometry' of the interested entities is dynamic and is modified all along the process of interessement. Through interressment, competing associations with other actors are interrupted to support the intention (influence) of the terrorist agenda. It entails a process of becoming and therefore success in this complex non-linear process is never guaranteed. Hence the complexity of power and influence emerge within the dynamics of the translation process in the actor network space. Interrelated and interdependent roles are defined and attributed to actors through power and influence. Interessement achieves enrolment if it is successful through interdependent negotiations (influence). One can say therefore that translation is a process before it is a result. It is about dynamic and nonlinear negotiation, influence and power.

Figure 2. a) Pre-translation of 'B' by 'A'; b) Translation of 'B' by 'A'.

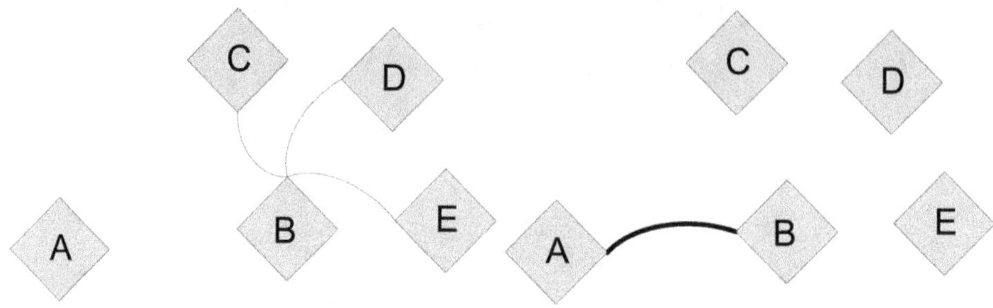

An example of the process of radicalization is given by Davis and Cragin (2009). They report that '...many scholars ...argue that there is more bottom-up "enlistment" than top-down recruitment; individuals join activities because their friends are already members rather than because they are initially committed to either the activities or the organization's greater goals'. It becomes characterized as "low-cost/low-risk activism" that is facilitated through socialization (Davis & Cragin, 2009). Davis and Cragin (2009, p. 37) argue that '...Participation in low-risk/low-cost activism then makes it more likely that those same individuals will be "drawn into more costly forms of participation through the cyclical process of integration and resocialization'. From the ANT perspective, the path to radicalization through low-risk/low-cost actors is fostered through the process of problematization, interessement, enrolment and mobilization on the actor. The high level factors of socialization, perception of rewards, support and justification figure prominently in this process.

As a conceptual process model of these factors, the agent based percolation model illustrates how varying the parameters of the actor network can affect the distribution and 'percolation' of radicalization and material, social, political support for a terrorist act. Galam (2002, 2003) and Galam and Mauger (2003) describe how percolation informs our understanding of terrorism. Working in a 'relational space', percolation theory and modeling provide an insightful approach to examining the translation process as it applies to radicalization.

Figure 3 represent the translation process for radicalization mapped on an actor network space. When the clusters cross the entire ***actor network space***, it is said to be translated (successful problematization, interessement, enrolment and mobilization). Because of the shifting alliances within the power structure of the actor network, radicalization emerges as a complex property. Incomplete translation is depicted in figure 3a. Radicalization resulting from successful translation is depicted in figure 3b. This conceptual modeling highlights how interventions can be developed and applied to the 'system'. Terrorism and radicalization thereby are not static but through this systems lens of ANT '...works from an ontology of becoming rather

Figure 3. a) Incomplete translation; b) Successful translation

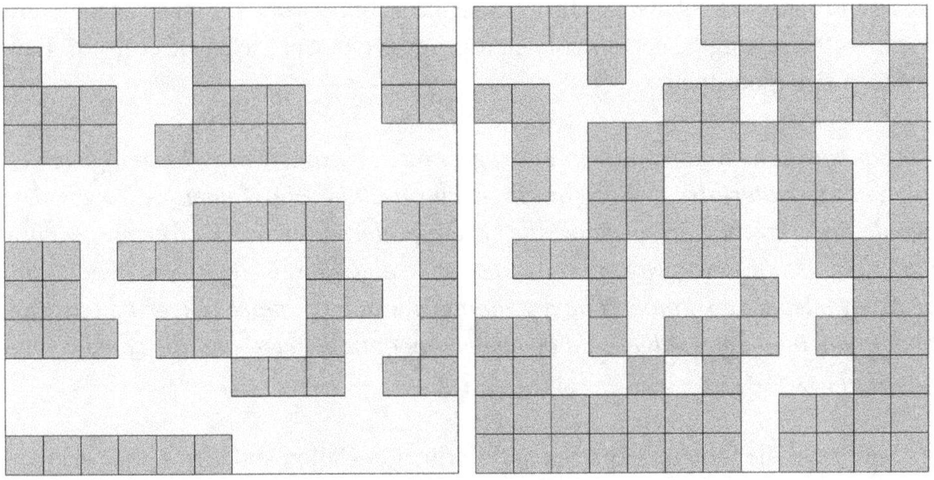

than assuming that entities can be defined in terms of pre-given competencies and capabilities' (Hernes, 2010). In accordance with Galam (2003), the percolation critical value depends upon two quantities: the connections and the dimensionality of the space. Galam (2003) described dimensionality in terms of the dimensions of the 'social space'. Within the context of the actor network space it is the complex 'social' parameters that contribute to radicalization.

Because radicalization and terrorism is context specific, the ANT 'following the actor', brings in a multi-vocality perspective that recognizes context and how the actor is implicated through relational analysis. The paradigm of systems thinking permits a view of the world as a complex system in which as noted by Sterman (2000, p. 10) we come to the understanding that 'you can't do just one thing' and the 'everything is connected to everything else'. This is supported by Senge (1990, p. 73) who is of the opinion that the discipline of the systems approach lies in a shift of mind: in seeing interrelationships rather than linear cause-effect chains and seeing processes of change rather than snapshots. Systems thinking thereby is an appropriate approach for communicating such complexities and interdependencies.

The conceptual model suggests that there exists a disproportionality of 'causes and effects', in which as Urry (2002, p. 59) remarks, past events are never 'forgotten', they are seeded in the actor network. Open the black box of terrorism and you may find centuries old cultural alienation and oppression seeded within the actor network creating a power structure within the translation process. These are inherent within the actor network space and shape the translation process. Social identity approaches described in TNO (2008, p. 25) argue 'that people define themselves not so much in terms of self, but in terms of group membership'. Further TNO (2008,

p. 26) argues that '...radicals...are embedded in complex interaction systems that shape and constrain behaviour'. In defining the actor network, the terrorist therefore becomes this entangled network of heterogeneous elements, a hybrid collectif. Law (1992, p. 3) argues that:

...what counts as a person is an effect generated by a network of heterogeneous, interacting, materials. ... But converted into a claim about humans it says that people are who they are because they are a patterned network of heterogeneous materials. If you took away my computer, my colleagues, my office, my books, my desk, my telephone I wouldn't be a sociologist writing papers, delivering lectures, and producing "knowledge". I'd be something quite other -- and the same is true for all of us.

Similar to the thought experiment described in Callon and Law (1997, p. 171) describing 'Andrew the strategist', the terrorist ceases to be enabled as such by removing key actors that shape the radicalization process through translation, thereby affecting the circulating power (influence) in the actor network that reifies as illustrated in the percolation model.

It is this very relational network of actors that creates the possibility for action (from recruitment to radicalization, to the act of terrorism). What this reveals is that action, as seen in the recent terrorist events described, takes place in a 'hybrid collectif' that is comprised of entangled human actors as well as non-human actors in multiple ways. Viewed from this perspective, tools (such as the hardware, financing, passive and active support) that are embedded in the actor network are as Callon and Caliskan, (2005, p. 18) remark not just things that are used to achieve certain ends: 'They contribute to the making of the universe of possibilities that make action itself'. For example, the internet as an actor in the actor network plays a significant role in the radicalization process as described by Davis and Cragin (2009), Jenkins (2011) and TNO (2008). TNO (2008, p. 27) argue that '...that the internet is a prominent facilitator of network formation and interpersonal interaction...radical movement can exist as a virtual group...connected through shared attitudes and ideology'.

As described in Masys (2010) and contextualized with regards to radicalization and recruitment for terrorism, the analysis reveals that through the process of translation:

1. Non-humans have significance and are not simply resources or constraints
    a. Non humans intervene actively to push action in unexpected directions, integrated into the translation process.
2. Entities are interactive effects
    a. They are networks of associations of human and non-human.

3. Action results from the complex interactions resident within the actor network that is dynamically shaped by inscription and translation processes.
4. The actor network lens reveals that action cannot be explained in a reductionist manner, as a firm consequence of any particular previous action (Callon and Law, 1997, p. 172).

This radical relationality characterizes the hybrid collectif such that as noted in Lockie (2004, p. 50) '…Action, intentionality…derive from relations between entities rather than from either individuals or totalities'. The interaction between causes and catalysts in the radicalization (translation) process is embedded within the actor network. Understanding the power and influence structures within the actor network reveals as Senge (2006, p. 64) remarks:

…that small, well-focused actions can sometimes produce significant, enduring improvements…this is the principle of leverage. …The only problem is that high-leverage changes are usually highly non-obvious to most participants in the system. They are not 'close in time and space' to obvious symptoms.

This necessitates that we think in terms of '…processes of change rather than snapshots'. This systems perspective that is realized by ANT reveals that the actor (human or non-human) is part of the feedback process thereby shifting our awareness from linear causality to complex causality whereby the actor is influenced and influencing the problem space. Senge (2006, p. 78) remarks that 'a linear view always suggests a simple locus of responsibility. When things go wrong, there is either blame…or guilt'. In defining the actor network as described, what emerges from the analysis are complex webs and networks composed of non-linear heterogeneous associations that give rise to insights into how power permeates the networks and how time and space are bound into the networks thereby necessitating a systems lens. From the ANT analysis, radicalization emerges as a network process of translation shaped by the inherent power and influence that circulates within the actor network. Counter-radicalization strategies that leverage insights from the systems lens of ANT may facilitate disruption of the translation.

FUNCTIONAL RESONANCE ACCIDENT MODEL (FRAM)

The lens of ANT provides insights into the radicalization process through the ANT process of 'translation'. The resilience of terrorist networks and the radicalization process necessitates the requirement to examine the functional elements characterizing how the work of radicalization is done. The FRAM is a method to analyse how work activities take place. This is done by *analysing* work activities in order to produce a *model* or representation of how 'work is done'. Within the context of radicalization, this model can then be used to determine how the radicalization went right (worked), as well as to look for possible opportunities for interventions strategies to create a 'failure' of the radicalization process. The FRAM method 'has previously been used to conduct qualitative system analyses: initially as a system accident investigation technique ... and more recently as a self- contained qualitative risk assessment method to inform design activities for large distributed systems .... To date, FRAM analyses have been undertaken for air traffic management, rail transport, financial market, and nuclear waste transport systems' (Frost & Mo, 2014).

The basis of the FRAM is the description of the functions that make up an activity or a process. In this application of FRAM, the translation processes characterized by: problematization; interessment; enrolment; mobilization, represent the functions associated with radicalization. As discussed, the foundation of ANT is a relational analysis. As described in Hollnagel et al (2014, p. 30) in FRAM, the relationship is not described directly by a graphical rendering, such as an event tree, but indirectly as relations defined by aspects of functions. The common technical term for such relations is coupling. If, for example, the Output of a function A also is defined as a Precondition for another function B, then the two functions are potentially coupled. One can also say that the function B uses the Output of function A as a Precondition. The couplings in a FRAM model are generally n-to-n (or many-to-many) rather than 1-to-1. For example, the Output from function A may also be Input to a function E. Function B can also have another Precondition that is Output from a function H'. This is critical to our understanding of radicalization. It presents opportunities to initiate a 'resonance' within the radicalization process such that it disrupts 'work as done'.

FRAM in Context

As a qualitative analysis technique, FRAM supports modeling of the inherent complexity of an actor network, including the relations between '...adaptable human agents and technology, coupling and dependence effects, non- linear dependencies between sub-systems, and functional performance variability' (Frost & Mo, 2014). Through this, FRAM supports a systemic analysis that describes the functionality associated with an actor network objectives and the inherent functional variation that supports the achievement of these objectives. Key to understanding the radicalization process and development of intervention strategies is the characterizing of the variability associated with 'translation process' and how interactions and relationships within the actor network reconfigure, leading to undesirable functional variability and resonance within the system (Hollnagel, 2012).

Radicalization does not follow a 'linear like' approach. Rather it is dynamic nonlinear and adaptive approach of events and interventions that is responsive to the changing landscape. Work as done may with regards to the radicalization process differ from work as imagined. Therefore success or failure in radicalization and any intervention strategies must consider the inherent complexity (figure 4).

The principle of resonance applied to radicalization and translation process

As described by Hollnagel et al. (2014, p. 20), 'the variability within the functions associated with *'radicalization'* may sometimes coincide, i.e., they may mutually influence each other. This can cause the amplitude of one or more functions to become unusually large (leading to either positive or negative outcomes). The consequences of such increased variability may spread to other functions in analogy with the phenomenon of resonance. It is thus no longer a question of single or multiple cause-effect chains, because that would imply that one could speak of one or more specific and recognisable causes (something that went wrong). The impacts instead emerge, i.e., they appear in a manner that cannot be explained by or reduced to linear causality'.

Hollnagel et al. (2014, pp. 20-21) describes three types of resonance:

Figure 4. The FRAM principles of approximate adjustment (Hollnagel et al., 2014, p. 17) alongside the ANT process of Translation

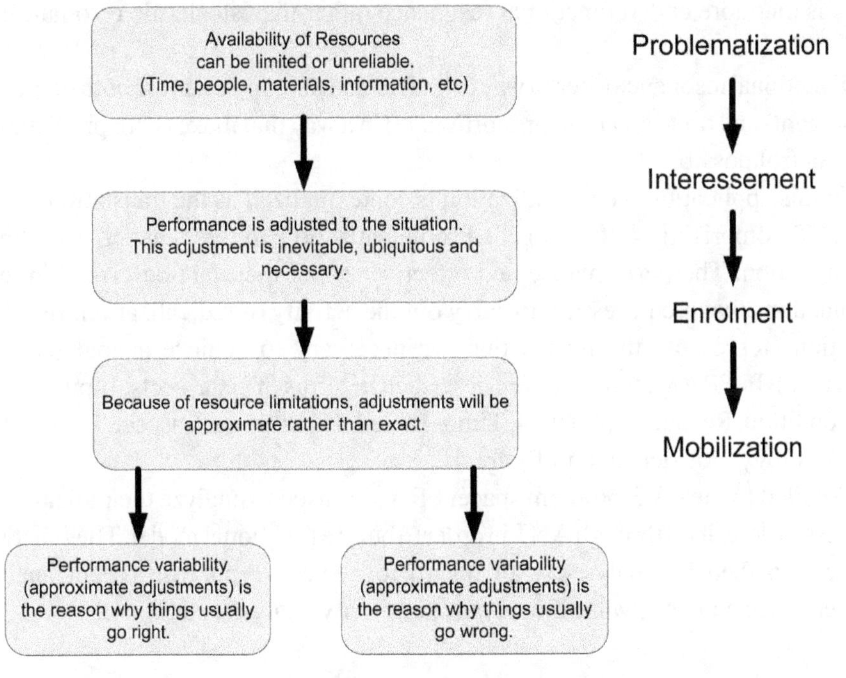

1. Physical systems, classical resonance is the phenomenon that a system can oscillate with larger amplitude at some frequencies than at others. These are known as the system's resonant (or resonance) frequencies. At these frequencies even small external forces that are applied repeatedly can produce large amplitude oscillations, which may seriously damage or even destroy the system.
2. In stochastic resonance there is no forcing function, but rather random noise, which every now and then pushes a subliminal signal over the detection threshold. Stochastic resonance can be defined as the enhanced sensitivity of a device to a weak signal that occurs when random noise is added to the mix. The outcome of stochastic resonance is non-linear, which simply means that the output is not directly proportional to the input. The outcome can also occur instantaneously, unlike classical resonance which must be built-up over time.
3. A third type of resonance, which is central to the FRAM, is called functional resonance. As a phenomenon, functional resonance describes the noticeable performance variability (in this application variability in the radicalization process) that is the outcome of the multiple approximate adjustments that are the basis for everyday work activities. Functional resonance is the detectable outcome (or signal) that emerges from the unintended interaction of the everyday variability of multiple signals. There is a regularity in how people behave and in how they respond to unexpected situations – including those that arise from how other people behave. The resonance effects that occur can be seen as a consequence of the ways in which the system functions, and the phenomenon is therefore called functional resonance rather than stochastic resonance.

Functional resonance offers a way to understand outcomes that are both non-causal (emergent) and non-linear (disproportionate) in a way that makes both predictability and control possible.

In this application the radicalization is contextualized as the translation process of ANT comprising 4 'functions': Problematization; Interessement; Enrolment; Mobilization. The performance and effectiveness of these are described in terms of functions that are necessary to carry out the activity of radicalization. In FRAM functions represents the means that are necessary to achieve a goal (acts and activities). In FRAM, functions are described in terms of six aspects: Input; Output; Precondition; Resources; Control; Time. Table 1 describes each aspect. In the FRAM methodology, it is depicted in figure 5.

Applied to the CVE problem space, FRAM is used to analyze the radicalization process as described through ANT in order to build a functional model. The functional model can then be analysed with regard to a specific purpose (event analysis, consequence analysis, what-if analysis, sensitivity analysis, etc.).

Table 1. Description of the six defining FRAM aspects

Aspect	Description
Input	Input as that which activates or starts a function (matter, energy, information)
Output	The result of what the function does. The Output describes a change of state – of the system or of one or more output parameters.
Precondition	In many cases a function cannot begin before one or more Preconditions have been established. These Preconditions can be understood as system states that must be [True], or as conditions that ought to be verified before a function is carried out
Resources	A Resource is something that is needed or consumed while a function is carried out. A Resource can represent matter, energy, information, competence, software, tools, manpower, etc.
Control	Control, or control input, is that which supervises or regulates a function so that it produces the desired Output. Control can be a plan, a schedule, a procedure, a set of guidelines or instructions, a program (an algorithm), a "measure and correct " functionality, etc. Another, less formal type of control is social control or expectations to how the work should be done. Social control can be external, such as the expectations of others (management, organisation, co-workers) or a person's own expectations. Social control can also be internal, for example, when we plan a job and mentally go through when and how to do it, or when we imagine what others expect of us.
Time	Time can affect how a function is carried out.

(Hollnagel et al., 2014)

This model depicted in figure 6 is based upon analysis of transcripts from various case studies regarding radicalized individuals. Through the lens of Resonance, we have insights into how actual works get down. From figure 3a and figure 3b we can see how those self radicalizing or being radicalized must be agile in terms of their actions in order to support the radicalization process. There are many pathways and variability. Matters pertaining to security of identity and intentions, with regards to terrorism and radicalization, are fragile and therefore require adjustments to behaviours. With that being said, a successful radicalization, through this lens, is subject to a 'resonance' that can as easily accelerate the radicalization as it can disintegrate it. As discussed earlier, through ANT one recognizes that power resides within the network of heterogeneous elements and is characterized as relational, emergent and distributed. Power is perceived as the capacity to influence that is realized only through the process of exercising this influence. In this sense, considering the relationality inherent within ANT, power can be conceived as a practice rather than a position. This is key to understanding functional resonance. For Counter Violent Extremism (CVE) interventions, understanding this idea of resonance as it pertains to 'work as done', is an opportunity to formulate strategies and feedback awareness. Figure 6 shows the multiple couplings that exist across the radicalization process

Figure 5. The six aspects of a function or activity (Hollnagel et al., 2014)

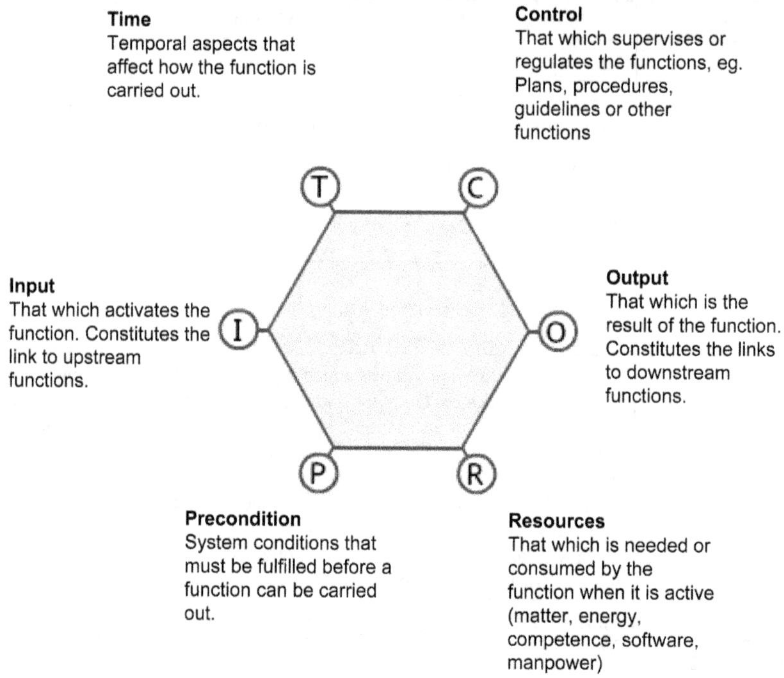

Figure 6. FRAM model of radicalization

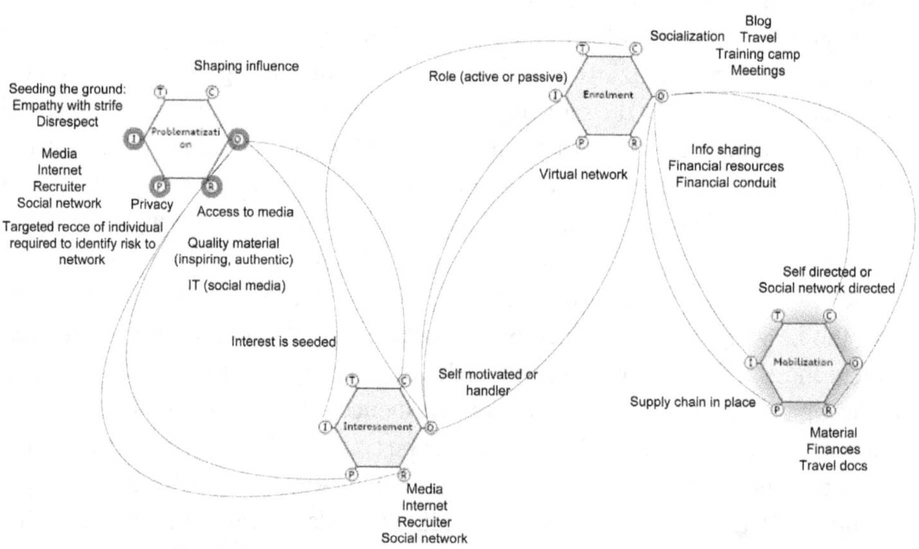

and how sensitivity to the environment (whether conducting radicalization or CVE interventions) is an essential quality to prevent adverse outcomes. The volatility of the actors, such as access to social media or recruiters for example, can have cascading effects across the process (affecting time, control, resources, preconditions, inputs and outputs). As such with further resolution on the process and its characteristics, either party (radicalizing actor or CVE intervention actor) can disrupt or enhance the effectiveness of the translation process. As discussed earlier, FRAM then can support event analysis, consequence analysis, what-if analysis, and sensitivity analysis.

Emerging from this analysis with the context of CVE, FRAM is based on four principles: (Hollnagel et al., 2014)

- **'Equivalence of successes and failures':** Radicalization agents must adjust to the current conditions in everything they do. The adjustments are the reason why everyday work is effective, but also the reason why things sometimes go wrong.
- **'Approximate adjustments':** Because time, information and resources always are finite, radicalization agents must balance between efficiency and thoroughness. Due to this balance, the adjustments will always be approximate.
- **'Emergence':** Outcomes of an actual work situation (radicalization) is emergent because the conditions that could explain them are short-lived, e.g. a pattern that existed at one point in time. Therefore, outcomes cannot be traced back to the original conditions.
- **'Functional Resonance':** The variability of a work supporting radicalization can merge or interact with the variability of what others do in the system, and increase the variability of the person's work, known as functional resonance.

As FRAM defines a system in terms of how it operates rather than in terms of its architecture and components, the aim of FRAM method is to represent the system's dynamics by modeling the non-linear interactions that are part of the system. FRAM thereby provides a way of understanding how performance variations may propagate within these tasks. In order to assess the effectiveness of a system, the main query is whether its functions (or tasks) are producing an acceptable outcome under existing conditions and how all the functions may interact to produce an unintended or unwanted outcome (Pereira, 2013).

CONCLUSION

When actors influence others through the translation process, goals (intended and unintended) emerge from the 'hybrid collectif'. Power and influence becomes the prime mover that shapes the outcome in this heterogeneous actor network space. If radicalization and violent extremism are recast as a collective force then the implications of associations are rooted far deeper than mere focus of the terrorist event.

A key attribute of ANT is concerned with how power permeates and influences the network dynamics. This requires tracing the transformation of these heterogeneous networks (Law, 1992) that are made up of the actors. Through the systems lens of ANT we explore the ways that the networks of relations emerge, how they are constructed and maintained. It examines how actors enlist other actors through the process of translation and thereby influence the qualities, desires, visions, and motivations on these actors (Latour, 1996). Bjelopera and Randol, (2011, p. 11) argue that as the terrorist threat becomes increasingly homegrown, a key way to fight it is to develop an understanding of how radicalization works and formulate ways to prevent the radicalization from morphing into violent extremism. The application of FRAM in this instance provides greater insights into 'work (meaning radicalization) as done' in terms of variability from 'work as imagined'. This FRAM approach provides greater resolution into the radicalization process and can support event analysis, consequence analysis, what-if analysis, and sensitivity analysis.

ANT's potential as a lens on the theory of radicalization and the application of FRAM lies primarily in its focus on how material and human actors connect into durable wholes and how relations across the actor network are nonlinear, volatile and impactful.

ACKNOWLEDGMENT

This work leverages theoretical thoughts and insights derived from Masys (2010) and expands on the original paper (Masys, 2014) with the application of FRAM.

REFERENCES

Aanestad, M. (2003). The Camera as an Actor Design-in-Use of Telemedicine Infrastructure in Surgery. *Computer Supported Cooperative Work, 12*(1), 1–20. doi:10.1023/A:1022492210898

Ackoff, R., & Emery, F. (1972). *On Purposeful Systems*. London: Tavistock.

Ackoff, R., & Strümpfer, J. P. (2003). Terrorism: A Systemic View. *Systems Research and Behavioral Science, 20*(3), 287–294. doi:10.1002res.544

Akrich, M. (1992). The De-Scription of Technical Objects. In W. Bijker & J. Law (Eds.), *Shaping Technology, Building Society: Studies in Sociotechnical Change* (pp. 205–224). Cambridge, MA: MIT Press.

Anderson, V., & Johnson, L. (1997). *Systems Thinking Basics: from concepts to causal loops*. Waltham, MA: Pegasus Communications, Inc.

Barab, S., Hay, K. E., & Yamagata-Lynch, L. C. (2001). Constructing Networks of Action-Relevant Episodes: An In situ Research Methodology. *Journal of the Learning Sciences, 10*(1&2), 63–112. doi:10.1207/S15327809JLS10-1-2_5

Bartolucci, V., & Gallo, G. (2013). Terrorism, System Thinking and Critical Discourse Analysis. *Systems Research and Behavioral Science*.

Bjelopera, J.P. (2013). *American Jihadist Terrorism: Combating a Complex Threat*. CRS Report for Congress.

Bjelopera, J. P., & Randol, M. A. (2011). American Jihadist terrorism: combating a complex threat. In *S.J. Perkins Homegrown terror and American jihadists: addressing the threat* (pp. 1–118). New York: Nova Science Publishers Inc.

Boyatzis, R. E. (1998). *Transforming qualitative information: thematic analysis and code development*. Thousand Oaks, CA: Sage Publications.

Braun, V., & Clarke, V. (2006). Using thematic analysis in psychology. *Qualitative Research in Psychology, 3*(2), 77–101. doi:10.1191/1478088706qp063oa

Buckley, W. (1967). *Sociology of modern systems theory*. Englewood Cliffs, NJ: Prentice Hall.

Callon, M. (1986b). Some Elements of a Sociology of Translation: Domestication of the Scallops and the Fishermen of St Brieuc bay. In J. Law (Ed.), *Power, Action and Belief. A New Sociology of Knowledge?* (pp. 196–233). London: Routledge & Kegan Paul.

Callon, M. (1991). Techno-economic Networks and Irreversibility. In J. Law (Ed.), *A Sociology of Monsters? Essays on Power, Technology and Domination, Sociological Review Monograph* (pp. 132–161). London: Routledge.

Callon, M. (1999). Actor-Network Theory: the Market Test. In J. Law & J. Hassard (Eds.), *Actor Network and After* (pp. 181–195). Oxford, UK: Blackwell and the Sociological Review.

Callon, M., & Caliskan, K. (2005). *New and old directions in the anthropology of markets*. Paper presented at the Conference New Directions in the Anthropology of Markets, New York, NY.

Callon, M., & Law, J. (1995). Agency and the Hybrid *Collectif. The South Atlantic Quarterly*, *94*(2), 481–507.

Callon, M., & Law, J. (1997). After the individual in society: Lessons on collectivity from science, technology and society. *Canadian Journal of Sociology*, *22*(2), 165–182. doi:10.2307/3341747

Checkland, P. (1981). *Systems Thinking, Systems Practice*. Chichester, UK: Wiley.

Checkland, P., & Scholes, J. (1990). *Soft Systems Methodology in Action*. Chichester, UK: John Wiley & Sons.

Chia, R., & King, I. W. (1998). The Organizational Structuring of Novelty. *Organization*, *5*(4), 461–478. doi:10.1177/135050849854002

Churchman, C. W. (1968). *The Systems Approach*. New York: Dell Publishing.

Churchman, C. W. (1971). *The design of enquiring systems*. New York: Basic Books.

Coolsaet, R. (2008). *Jihadi Terrorism and the Radicalization Challenge in Europe*. Aldershot: Ashgate Publishing.

Davis, P.K., & Cragin, K. (2009). *Social Science for Counter-terrorism: putting the pieces together*. RAND MG849.

Deleuze, G., & Guattari, F. (1987). *A Thousand Plateaus: Capitalism & Schizophrenia*. Minneapolis, MN: University of Minnesota Press.

Flood, R., & Jackson, M. (1991). *Creative problem solving: Total systems intervention*. London: Wiley.

Forest, J. J. (2007). *Countering Terrorism and Insurgency in the 21st Century: International Perspectives*. Westport, CT: Praeger Security International.

Frost, B., & Mo, J. P. T. (2014). *System Hazard Analysis of a Complex Socio-Technical System: The Functional Resonance Analysis Method in Hazard Identification. Australian System Safety Conference*, Melbourne, Australia.

Galam, S. (2002). The September 11 Attack: A Percolation of Individual Passive Support. *The European Physical Journal B*, *26*(3), 269–272. doi:10.1140/epjb/e20020089

Galam, S. (2003). Global physics: From percolation to terrorism, guerilla warfare and clandestine activities. *Physica A*, *330*(1-2), 139–149. doi:10.1016/j.physa.2003.08.035

Galam, S., & Mauger, A. (2003). On reducing terrorism power: A hint from physics. *Physica A*, *323*, 695–704. doi:10.1016/S0378-4371(03)00006-2

GTI. (2017). *Global Terrorism Index*. Retrieved from http://economicsandpeace.org/wp-content/uploads/2016/11/Global-Terrorism-Index-2016.2.pdf

Hall, A. D. (1962). *A methodology for systems engineering*. New York: Van Nostrand Reinhold.

Hernes, T. (2010). Actor-network theory, Callon's Scallops, and process-based organization studies. In T. Hernes & S. Maitlis (Eds.), *Process, sensemaking and organizing*. Oxford, UK: Oxford University press. doi:10.1093/acprof:oso/9780199594566.003.0009

Hollnagel, E. (2012). *FRAM: The functional resonance analysis method for modelling complex socio-technical systems*. Farnham, UK: Ashgate.

Hollnagel, E., Hounsgaard, J., & Colligan, L. (2014). *FRAM – the Functional Resonance Analysis Method: a handbook for the practical use of the method*. Centre for Quality, in the Southern Region of Denmark. Retrieved from www.centerforkvalitet.dk/framhandbook

Jackson, M. (2000). *Systems approaches to management*. Dordrecht: Kluwer Academic.

Jenkins, B. M. (2011). *Stray Dogs and Virtual Armies Radicalization and Recruitment to Jihadist Terrorism in the United States Since 9/11*. RAND Occasional Paper 343.

Kaufmann, S. (1995). *At home in the universe: the search for the laws of complexity*. London: Penguin.

Latour, B. (1987). *Science in Action: How to Follow Scientists and Engineers Through Society*. Milton Keynes, UK: Open University Press.

Latour, B. (1988). *The Pasteurization of France*. Cambridge, MA: Harvard University Press.

Latour, B. (1993). *We have never been modern*. Hemel Hempstead, UK: Harvester Wheatsheaf.

Latour, B. (1996). On actor-network theory. A few clarifications. *Soziale Welt, 47*, 369–381.

Latour, B. (1999a). *Pandoras Hope: Essays on the reality of science studies*. Cambridge, MA: Harvard University Press.

Latour, B. (1999b). On Recalling ANT. In J. Law & J. Hassard (Eds.), *Actor Network and After* (pp. 15–25). Oxford, UK: Blackwell and the Sociological Review.

Latour, B. (2005). *Reassembling the Social: An Introduction to Actor network theory*. Oxford, UK: Oxford University Press.

Law, J. (1987). Technology and heterogeneous Engineering: The Case of Portuguese Expansion. In W. E. Bijker, T. P. Hughes, & T. J. Pinch (Eds.), *The Social Construction of technological Systems: New Directions in the Sociology and History of Technology* (pp. 111–134). Cambridge, MA: MIT Press.

Law, J. (1992). Notes on the theory of the actor network: Ordering, strategy and heterogeneity. *Systems Practice, 5*(4), 379–393. doi:10.1007/BF01059830

Law, J. (1999). After ANT: complexity, naming and topology. In J. Law & J. Hassard (Eds.), *Actor Network Theory and After* (pp. 1–14). Oxford, UK: Blackwell.

Lockie, S. (2004). Collective Agency, Non-human Causality and Environmental Social Movements: A case study of the Australian Landcare movement. *Journal of Sociology (Melbourne, Vic.), 40*(1), 41–58. doi:10.1177/1440783304040452

Luhmann, N. (1995). *Social systems*. Stanford, CA: Stanford University Press.

Masys, A. J. (2010). *Fratricide in Air Operations: Opening the black box- revealing the social* (PhD Dissertation). University of Leicester, UK.

Masys, A. J. (2012). 'The emergent nature of risk as a product of 'heterogeneous engineering- A relational analysis of the Oil and Gas Industry safety culture. In S. Bennett (Ed.), *Innovative thinking in risk, crisis and disaster management*. Gower Publishing.

Masys, A. J. (2014). Radicalization and Recruitment: A systems approach to understanding violent extremism. *International Journal of Systems and Society, 1*(2), 51–65. doi:10.4018/ijss.2014070104

Midgely, G. (2000). *Systemic intervention: philosophy, methodology and practice*. New York: Kluwer Plenum. doi:10.1007/978-1-4615-4201-8

Mingers, J. (2006). *Realising Systems Thinking: Knowledge and Action in Management Science*. Canterbury, UK: Springer Publishing.

Monteiro, E. (2000). Actor Network Theory and Information Infrastructure. In C. Ciborra (Ed.), *From control to drift. The dynamics of corporate information infrastructures*. Oxford, UK: Oxford University Press. Retrieved on July 11, 2005 from www.idi.ntnu.no/~ericm/ant.FINAL.htm

Pereira, A. G. A. A. (2013). Introduction to the use of FRAM on the effectiveness assessment of a radiopharmaceutical dispatches process. *2013 International Nuclear Atlantic Conference – INAC*.

Powell, J. L., & Owen, T. (2011). Actor network theory and social science: possibilities and implications. *Journal of Public Administration and Governance, 1*(2), 140-157. Retrieved from www.macrothink.org/jpag

Rabasa, A., Pettyjohn, S.L., Ghez, J.J., & Boucek, C. (2010). *Deradicalizing Islamist Extremists*. RAND_ MG1053.

Senate Report. (2012). *Zachary Chesser: A Case Study in Online Islamist Radicalization and Its Meaning for the Threat of Homegrown Terrorism*. A Report by Majority and Minority Staff Senate Committee on Homeland Security and Governmental Affairs. Retrieved from http://www.hsgac.senate.gov/imo/media/doc/CHESSER%20FINAL%20REPORT(1).pdf

Senge, P. (1990). *The Fifth Discipline: The Art and Practice of the Learning Organization*. New York: Doubleday Currency.

Senge, P. (2006). *The Fifth Discipline: The Art and Practice of the Learning Organization*. New York: Doubleday Currency.

Somerville, I. (1997). *Actor network theory: a useful paradigm for the analysis of the UK cable/on-line socio-technical ensemble?* Retrieved from http://hsb/baylor.edu/eamsower/ais.ac.97/papers/somervil.html,

Sterman, J. D. (2000). *Business Dynamics: Systems Thinking and Modeling for a Complex World*. Boston: McGraw-Hill Publishing.

TNO. (2008). *Radicalisation, Recruitment and the EU Counter-radicalisation Strategy*. Retrieved from http://www.transnationalterrorism.eu/tekst/publications/WP4%20Del%207.pdf

Urry, J. (2002). The Global Complexities of September 11th. *Theory, Culture & Society, 19*(4), 57–69. doi:10.1177/026327640201900400

Urry, J. (2003). *Global Complexity*. Cambridge, UK: Polity Press.

Urry, J. (2005). The Complexities of the Global. *Theory, Culture & Society, 22*(5), 235–254. doi:10.1177/0263276405057201

US Department of State Report. (2011). *National Counterterrorism Center: Annex of Statistical Information Office of the Coordinator for Counterterrorism.* Office of the Coordinator for Counterterrorism Country Reports on Terrorism 2011 Report July 31, 2012. Retrieved from http://www.state.gov/j/ct/rls/crt/2011/195555.htm

Vidino, L., & Brandon, J. (2012). *Countering Radicalization in Europe. Policy report published by the Centre for the Study of Radicalization and Political Violence (ICSR).* London: Kings College.

Viseu, A. A. B. (2005). *Augmented Bodies: The Visions and realities of wearable computers* (PhD Dissertation). University of Toronto.

Von Bertalanffy, L. (1971). *General Systems Theory*. London: Penguin.

Weiner, N. (1948). *Cybernetics: or Communication and Control in the Animal and the Machine.* Cambridge, MA: MIT Press.

Willcocks, L. P. (2004). Foucault, Power/Knowledge and Information Systems: Reconstructing the Present. In J. Mingers & L. P. Willcocks (Eds.), *Social Theory and Philosophy for Information Systems* (pp. 238–296). Chichester, UK: John Wiley & Sons, Ltd.

Yeung, H. W. C. (2002). *Towards a relational economic geography: old wine in new bottles?* Paper presented at the 98th Annual meeting of the Association of American Geographers, Los Angeles, CA. Retrieved from http://courses.nus.edu.sg/course/geoywc/publication/Yeung_AAG.pdf

Compilation of References

(2005). Theory of Constraints: practice and research. InRonen, B. (Ed.), *The Theory of Constraints: practice and research* (pp. 1–2). Amsterdam: IOS Press, Editorial.

A megaton bomb, the circulation of the second list. (1997). *Joongang Ilbo.* (in Korean)

Aanestad, M. (2003). The Camera as an Actor Design-in-Use of Telemedicine Infrastructure in Surgery. *Computer Supported Cooperative Work, 12*(1), 1–20. doi:10.1023/A:1022492210898

Abramo, C. (2005). How Far. *Transparencia Brasil.* Available online at www.tranparencia.org.br/ docs/Howfar.pdf

Ackoff, R. (1974). *Systems, Messes, and Interactive Planning, Redesigning the Future.* New York, London: Wiley.

Ackoff, R., & Emery, F. (1972). *On Purposeful Systems.* London: Tavistock.

Ackoff, R., & Strümpfer, J. P. (2003). Terrorism: A Systemic View. *Systems Research and Behavioral Science, 20*(3), 287–294. doi:10.1002res.544

Adams, W. M., & Jeanrenaud, S. J. (2008). *Transition to Sustainability: Towards a Humane and Diverse World.* International Union for Conservation and Nature, Publications Services.

Ahmed, N. (2010). *A User's Guide To The Crisis of Civilization.* London: Pluto Press.

Ahn, J.H. (2001, January 5). CCEJ Says they won't take Corporate Funds. *Hankuk Ilbo.* (in Korean)

Akrich, M. (1992). The De-Scription of Technical Objects. In W. Bijker & J. Law (Eds.), *Shaping Technology, Building Society: Studies in Sociotechnical Change* (pp. 205–224). Cambridge, MA: MIT Press.

Alatas, S. H. (1990). *Corruption: Its Nature, Causes and Functions.* Brookfield, VT: Ashgate.

Allvin, M., Aronsson, G., Hagström, T., Johansson, G., & Lundberg, U. (2006). *Gränslöst arbete – socialpsykologiska perspektiv på det nya arbetslivet.* Malmö: Liber.

Almond, G. A., & Powell, G. B. Jr. (1966). *Comparative Politics: A Developmental Approach.* Boston: Little, Brown and Company.

Alvesson, M., & Spicer, A. (2016). Conditional Surrender? Why do Professionals Comply with Managerialism? *Journal of Organizational Change Management*, *29*(1), 29–45. doi:10.1108/JOCM-11-2015-0221

Anderson, V., & Johnson, L. (1997). *Systems Thinking Basics: from concepts to causal loops*. Waltham, MA: Pegasus Communications, Inc.

Anti-Corruption and Civil Rights Commission. (2010). Retrieved from http://www.acrc.go.kr/eng/index.do

Arenalilla. (2007). *Doma Vaquera*. Retrieved on November 23, 2017 from: http://www.youtube.com/watch?v=nk_n1yQlmT0

Argyris, C. (1988). Crafting a theory of practice: The case of organizational paradoxes. Academic Press.

Argyris, C. (1986). Reinforcing organizational defensive routines: An unintended human resources activity. *Human Resource Management*, *25*(4), 541–555. doi:10.1002/hrm.3930250405

Argyris, C. (2005). Double-loop learning in organizations: a theory of action perspective. In *Great minds in management: the process of theory development* (pp. 261–279). Oxford, UK: Oxford University Press.

Aristotle. (1981). *The Politics*. Penguin Classics.

Ashby, R. W. (1964). *An Introduction to Cybernetics*. London: Methuen.

Ashford, S. J., Lee, C., & Bobko, P. (1989). Content, cause, and consequences of job insecurity: A theory-based measure and substantive test. *Academy of Management Journal*, *32*(4), 803–829. doi:10.2307/256569

Ashlag, Y. (2014). *Removing constraints for business growth*. Great Barrington, MA: North River Press.

Aubert, N., & Pagès, M. (1989). *Occupational stress*. Paris: Klincksieck edition.

Aubert, N. (2003). *Le Culte de l'urgence : La Société Malade du Temps*. Flammarion.

Aubert, N., & Gaulejac, V. (1991). *Le coût de l'excellence*. Paris: Éditions Du Seuil.

Avery, D., & Thomson, M. (2017). *The resilient farmer: Weathering the challenges of life and the land*. Penguin Random House.

Avgerou, C., & Mcgrath, K. (2007). Power, Rationality, and the Art of Living through Socio-Technical Change. *Management Information Systems Quarterly*, *31*(2), 295–315. doi:10.2307/25148792

Avril, E., & Zumello, C. (Eds.). (2013). *New technology, organizational change and governance*. Macmillan. doi:10.1057/9781137264237

Axelrod, R. M. (1997). *The Complexity of Cooperation: Agent-Based Models of Competition and Cooperation*. Princeton, NJ: Princeton University Press.

Bae, H. S. (1997). Laws Related to Public Agency Freedom of Information Laws. [in Korean]. *Goshi Yungusa*, *275*, 124–129.

Baker, V. E., & Sinkula, J. M. (1999). The Synergistic Effect of Market Orientation and Learning Orientation on Organizational Performance. *Journal of the Academy of Marketing Science*, *27*(4), 411–427. doi:10.1177/0092070399274002

Ballard, D. I., & Ramgolam, D. I. (2011). *Connecting through Smartphones and open doors: A spatiotemporal perspective on communication technologies at work*. In The annual meeting of the NCA 95th Annual Convention, Chicago, IL.

Bammer, G. (2003). *Integration and Implementation Sciences: Building a New Specialisation*. Retrieved from http://www.anu.edu.au/iisn

Barab, S., Hay, K. E., & Yamagata-Lynch, L. C. (2001). Constructing Networks of Action-Relevant Episodes: An In situ Research Methodology. *Journal of the Learning Sciences*, *10*(1&2), 63–112. doi:10.1207/S15327809JLS10-1-2_5

Barley, S. R. (1990). Images of imaging: Notes on Doing Longitudinal Fieldwork. *Organization Science*, *1*(3), 220–247. doi:10.1287/orsc.1.3.220

Bartolucci, V., & Gallo, G. (2013). Terrorism, System Thinking and Critical Discourse Analysis. *Systems Research and Behavioral Science*.

Bassett, K. (1995). On reflexivity: Further comments on Barnes and the sociology of science. *Environment & Planning*, *27*, 1527–1538. doi:10.1068/a271527

Bateson, G., Jackson, D. D., Haley, J., & Weakland, J. (1956). Toward a theory of schizophrenia. *Behavioral Science*, *1*(4), 251–254. doi:10.1002/bs.3830010402

Baucher, F. (1919). *The Principles of Horsemanship*. London: Vinton & Co.

Bayareabert. (2009). *Spanish Riding School Vienna Philharmonic Beliebte Annen Polka Johann Strauss Elisen Polka Op 151*. Retrieved on November 23, 2017 from http://www.youtube.com/watch?v=Pn6Jmqs9IHQ

Beau, P. (2015). Evaluer La Performance Individuelle Sur Les Resultats: De La Recherche D'Efficience Au Stress Professionnel. In Comptabilité, Contrôle et Audit des invisibles, de l'informel et de l'imprévisible. Academic Press.

Beck, D., & Cowan, C. (2006). *Spiral Dynamics. Mastering values, leadership and change*. Moscow: Exmo.

Beck, U. (1986). *Risk Society: Towards a New Modernity*. Sage Publications.

Beer, S. (1981) Brain of the Firm (2nd ed.). John Wiley.

Beer, S. (1972). *Brain of the Firm; Allen Lane*. London: The Penguin Press.

Beer, S. (1979). *The Heart of Enterprise*. Chichester, UK: John Wiley.

Beer, S. (1979). *The Heart of the Enterprise*. Chichester, UK: John Wiley and Sons.

Beer, S. (1985). *Diagnosing the System for Organizations*. Chichester, UK: John Wiley.

Belasik, P. (1993). *An Overview of the Classical Principles behind the Training of Horse and Rider: An interview with Paul Belasik, Part One: Breaking the Young Horse* [Audio cassette]. Telford, PA: Base 10 Studios.

Belasik, P. (1990). *Riding towards the light: An apprenticeship in the art of dressage riding*. London: J.A. Allen.

Belasik, P. (2001). *Dressage for the 21st Century*. London: J.A. Allen.

Belasik, P. (2009). *A Search for Collection: Science and Art in Riding*. London: J.A. Allen.

Belasik, P. (2012). *Nature, Nurture and Horses: A Journal of Four Dressage Horses – From Birth through the First Year of Training*. Vermont: Trafalga Square Books.

Bellinger, G. (2012). *Enabling a Better World*. Retrieved from http://insightmaker.com/insight/3395

Benkler, Y. (2006). *The Wealth of Networks: how social production transforms markets and freedom*. New Haven, CT: Yale University Press.

Bertalanffy von, L. (1968). General System theory: Foundations, Development, Applications. New York: George Braziller.

Bieling, P. J., Israeli, A. L., & Antony, M. M. (2004). Is perfectionism good, bad, or both? Examining models of the perfectionism construct. *Personality and Individual Differences*, *36*(6), 1373–1385. doi:10.1016/S0191-8869(03)00235-6

Billet, B. L. (1990). South Korea at the Crossroads. *Asian Survey*, *30*(3), 300–311. doi:10.2307/2644567

Bjelopera, J.P. (2013). *American Jihadist Terrorism: Combating a Complex Threat*. CRS Report for Congress.

Bjelopera, J. P., & Randol, M. A. (2011). American Jihadist terrorism: combating a complex threat. In *S.J. Perkins Homegrown terror and American jihadists: addressing the threat* (pp. 1–118). New York: Nova Science Publishers Inc.

BjerkeP. (2004) Retrieved from http://www.de-facto.no/page/de-facto-kunnskapssenter-for-fagorganiserte/publikasjoner/bestille-publikasjon/det-nye-arbeidslivet/index.html

Bjørndall, T., Herrero, I., Newman, A., Romero, C., & Weintraub, A. (2011). Operations research in the natural resource industry. In *International Transactions in Operational Research*. International Federation of Operational Research Societies. Retrieved October 2012 from http://onlinelibrary.wiley.com/ doi/10.1111/j.1475-3995.2010.00800.x/

Bloomberg Businessweek. (2010). *Dow Jones Industrial Average*. Retrieved from: https://www.bloomberg.com/quote/INDU:IND/members

Bock, G., Zmud, R., & Lee, J. (2005). Behavioral Intention Formation in Knowledge Sharing: Examining the Roles of Extrinsic Motivators, Social-Psychological Forces, and organisational Climate. *Management Information Systems Quarterly*, *29*(1), 87–111. doi:10.2307/25148669

Bosselmann, K. (2002). The concept of sustainable development. In K. Bosselmann & D. Grinlinton (Eds.), *Environmental law for a sustainable society* (Vol. 1). Auckland: The New Zealand Centre for Environmental Law.

Bourdin, S. (2008). Approche communicationnelle du changement organisationnel: Organisations pédagogiques, TICE et situations de communication. *Science and Society*, (74): 123–140.

Bourgelat, C. (1754). A New System of Horsemanship: from the French of Monsieur Bourgelat by Richard Berenger. LaVergne, TN: Ecco Print Editions. doi:10.5962/bhl.title.33714

Boyatzis, R. E. (1998). *Transforming qualitative information: thematic analysis and code development*. Thousand Oaks, CA: Sage Publications.

Braun, V., & Clarke, V. (2006). Using thematic analysis in psychology. *Qualitative Research in Psychology*, *3*(2), 77–101. doi:10.1191/1478088706qp063oa

Bresnahan, T., & Trajtenberg, M. (1995). General Purpose Technologies "Engines of Growth?". Stanford University.

Bresnahan, T., Trajtenberg, M., & Helpman, E. (Eds.). (1998). General Purpose Technologies and Economic Growth. The MIT Press.

Brödner, P., & Forslin, J. (2002). *O tempora, O mores! Work intensity-why again an issue. In Creating Sustainable Work Systems. Emerging Perspectives and Practice* (pp. 15–26). London: Routledge.

Bryman, A. (2004). *Social Research Methods*. Oxford, UK: Oxford University Press.

Buckley, W. (1967). *Sociology of modern systems theory*. Englewood Cliffs, NJ: Prentice Hall.

Buckminster Fuller, R. (1975). *SYNERGETICS. Explorations in the Geometry of Thinking*. Macmillan Publishing.

Bueno de Mesquita, B., Morrow, J. D., Siverson, R. M., & Smith, A. (2000). "Political Institutions, Political Survival and Policy Success," chap. In B. De Mesquita & H. Root (Eds.), *Governing For Prosperity*. New Haven, CT: Yale University Press.

Bueno de Mesquita, B., Smith, A., Siverson, R. M., & Morrow, J. D. (2003). *The Logic of Political Survival*. Cambridge, MA: MIT Press.

Bunce, V. (2003). Rethinking Recent Democratization: Lessons from the Post-communist Experience. *World Politics*, *55*(2), 167–192. doi:10.1353/wp.2003.0010

Burgess, R. G. (1984). *In the Field: An introduction to Field Research*. London: George Allen and Unwin. doi:10.4324/9780203418161

Burke, R. J., & Cooper, C. L. (2000). *The organization in crisis: Downsizing, restructuring, and privatization.* Blackwell Publishing.

Burton-Houle, T. (2001). What if my organization's goal is not to make money? *TOC Times Quarterly Newsletter.* Retrieved 5 April 2011 from http://www.goldratt.com/for-cause/partnerperspsep2001toct.htm

Business International Corporation. (1984). *Introduction to the country assessment service.* New York: Business International Corporation.

Cabrera, A., & Cabrera, E. (2002). Knowledge-sharing Dilemmas [Electronic version]. *organisation. Studies, 23*(5), 687–710.

Cabrera, D., & Cabrera, L. (2015). *Systems Thinking Made Simple: New Hope for Solving Wicked Problems.* Ithaca, NY: Odyssean.

Call, C. T. (2010). Beyond the 'failed state': Toward conceptual alternatives. In *European Journal of International Relations.* Washington, DC: US Institute of Peace. doi:10.1177/1354066109353137

Callon, M., & Caliskan, K. (2005). *New and old directions in the anthropology of markets.* Paper presented at the Conference New Directions in the Anthropology of Markets, New York, NY.

Callon, M. (1986). Some elements of a sociology of translation: domestication of the scallops and the fishermen of St Brieuc Bay. In J. Law (Ed.), *Power, Action and Belief* (pp. 196–233). London: Routledge and Keegan Paul.

Callon, M. (1986b). Some Elements of a Sociology of Translation: Domestication of the Scallops and the Fishermen of St Brieuc bay. In J. Law (Ed.), *Power, Action and Belief. A New Sociology of Knowledge?* (pp. 196–233). London: Routledge & Kegan Paul.

Callon, M. (1991). Techno-economic Networks and Irreversibility. In J. Law (Ed.), *A Sociology of Monsters? Essays on Power, Technology and Domination, Sociological Review Monograph* (pp. 132–161). London: Routledge.

Callon, M. (1999). Actor-Network Theory: the Market Test. In J. Law & J. Hassard (Eds.), *Actor Network and After* (pp. 181–195). Oxford, UK: Blackwell and the Sociological Review.

Callon, M., & Law, J. (1995). Agency and the Hybrid *Collectif. The South Atlantic Quarterly, 94*(2), 481–507.

Callon, M., & Law, J. (1997). After the individual in society: Lessons on collectivity from science, technology and society. *Canadian Journal of Sociology, 22*(2), 165–182. doi:10.2307/3341747

Cameron, K., Quinn, R., & DeGaff, J. (2005). *The competing values framework: creating value through purpose, practices, and people.* Ann Arbor, MI: Addison-Wesley.

Cappelli, P. (1999). *The New Deal at Work: Managing the Market-Driven Workforce.* Boston, MA: Harvard Business School Press.

Capra, F., & Luisi, P. L. (2014). The Systems View of Life: A Unifying Vision. Cambridge University Press.

Capra, F., & Luisi, P. (2014). *The Systems View of Life, A Unifying Vision*. Cambridge University Press. doi:10.1017/CBO9780511895555

Carayol, V. (2005). Principes de contrôle, communication et temporalités organisationnelles. *Études de Communication.*

Carayol, V. (2004). *Communication organisationnelle: une perspective allagmatique*. Paris: L'harmattan.

Carse, J. (1986). *Finite and Infinite Games*. New York: Random House.

Casti, J. (1997). *Would-be Worlds*. John Wiley & Sons.

Cavendish, W. (2000). *Duke of Newcastle, (1743). A General System of Horsemanship*. London: J.A. Allen.

Cederman, L. E. (1997). *Emergent Actors in World Politics: How States and Nations Develop and Dis- solve*. Princeton, NJ: Princeton University Press.

Cerbone. (2008). *Heidegger, A Guide for the Perplexed*. Continuum.

Chaebol and Chaebol Families. (1992). Seoul: Knowledge Industries Publications.

Champion, D., & Stowell, F. A. (2001). PEArl; A Systems Approach to Demonstrating Authenticity in Information System Design. *Journal of Information Technology*, *16*(1), 3–12. doi:10.1080/02683960010028438

Champion, D., & Stowell, F. A. (2003, February). Validating Action Research Field Studies: PEArL. *Systemic Practice and Action Research*, *16*(1), 21–36. doi:10.1023/A:1021928511690

Chang, H. J. (2006). *The East Asian Development Experience: the Miracle, the Crisis and the Future*. St. Martin's Press.

Charging Regime Advisory Group (CRAG). (2016). *Tariff review March 2016.* Retrieved from http://www.kapiticoast.govt.nz/contentassets/3f9411b3fb4a4092a115c89c4207ff79/crag_report_2016_review.pdf

Charpentier, P. (2004). La gestion du changement dans les organisations. *CAHIERS FRANCAIS-PARIS*, 29–36.

Checkland, P. (1991). From framework through experience to learning: the essential nature of action research. *Information systems research: Contemporary approaches and emergent traditions*, 397-403.

Checkland, P. (1981). *Systems Thinking, Systems Practice*. Chichester, UK: John Wiley.

Checkland, P. (1985). From optimizing to learning: A development of systems thinking for the 1990s. *The Journal of the Operational Research Society*, *36*(9), 757–767. doi:10.1057/jors.1985.141

Checkland, P. (1999). *Systems Thinking, Systems Practice, 30 year retrospective*. Chichester, UK: Wiley.

Checkland, P. (1999). *Systems Thinking, Systems Practice. Includes a 30 year Retrospective*. Wiley.

Checkland, P. B. (1981). *Systems theory, systems practice*. New York: Wiley.

Checkland, P. B. (1999). *Systems Thinking Systems Practice. Includes a 30 year retrospective*. Chichester, UK: John Wiley and Sons.

Checkland, P. B., & Holwell, S. E. (1998). *Information, Systems and Information Systems*. Chichester, UK: Wiley and Sons.

Checkland, P. B., & Poulter, J. (2006). *Learning for Action*. Chichester, UK: John Wiley and Sons.

Checkland, P. B., & Poulter, J. (2006). *Learning for Action: A Short Definitive Account of Soft Systems Methodology and its use for Practitioners, Teachers and Students*. Chichester, UK: John Wiley and Sons, Ltd.

Checkland, P. B., & Scholes, J. (1990). *Soft Systems Methodology in Action*. Chichester, UK: Wiley and Sons.

Checkland, P., & Holwell, S. (1998). *Information, systems and information systems: making sense of the field*. Chichester, UK: John Wiley & Sons.

Chen, S., Chuang, Y., & Chen, P. (2012). Behavioral intention formation in knowledge sharing: Examining the roles of KMS quality, KMS self-efficacy, and organisational climate. *Knowledge-Based Systems*, *31*, 106–118. doi:10.1016/j.knosys.2012.02.001

Chia, R., & King, I. W. (1998). The Organizational Structuring of Novelty. *Organization*, *5*(4), 461–478. doi:10.1177/135050849854002

Chiu, C.-M., Hsu, M.-H., & Wang, E. (2006). Understanding knowledge sharing in virtual communities: An integration of social capital and social cognitive theories. *Decision Support Systems*, *42*(3), 1872–1888. doi:10.1016/j.dss.2006.04.001

Cho, Y. H., & Choi, B. (2004). E-government to combat corruption: The case of Seoul metropolitan government. *International Journal of Public Administration*, 27.

Christensen, C. M., Grossman, J. H., & Hwang, J. (2009). *'The Innovator's Prescription: A disruptive Solution for Health Care*. McGraw Hill.

Chun Gathered 950 billion won. (1996a). *Joongang Ilbo*. (in Korean)

Churchman, C. W. (1968). *The Systems Approach*. New York: Dell Publishing.

Churchman, C. W. (1971). *The design of enquiring systems*. New York: Basic Books.

Churchman, C. W. (1971). *The Design of Inquiring Systems, Basic Concepts of Systems and Organizations*. New York: Basic Books.

Ciborra, C. (1996). Improvisation and Information Technology in Organizations. *Proceedings of the International Conference on Information Systems*, 369-380.

Ciborra, C. (2002). *The Labyrinths of Information: Challenging the Wisdom of Systems*. Oxford, UK: Oxford University Press.

Cillers, P. (1998). *Complexity and Postmodernism: Understanding Complex Systems*. New York: Routledge.

Cisco Systems. (2007). *Benchmarking a Global Business: Semiconductor Manufacturer Infineon Leverages Cisco's Experience*. Retrieved from http://www.cisco.com/web/about/ac79/docs/success/Infineon_Final.pdf

Clarke, J., & Newman, J. (1997). *The managerial State: Power, Politics and Ideology in the Remaking of Social Welfare*. London: Sage.

Classical versus classique: with Christoph Hess and Philippe Karl. (2007). [DVD]. Thomas Vogel, pferdia tv.

Clayton, E. (2005). *International Encyclopaedia of Philosophy*. Academic Press.

Clayton, H. M., Singleton, W. H., Lanovaz, J. L., & Cloud, G. L. (2003). Measurement of rein tension during horseback riding using strain gage transducers. *Experimental Techniques*, *27*(3), 34–36. doi:10.1111/j.1747-1567.2003.tb00112.x

Clegg, S. (1990). Modern organizations: Organization studies in the postmodern world. *Sage (Atlanta, Ga.)*.

Clegg, S. R. (1989). *Frameworks of Power*. London: Sage.

Clegg, S., & Dunkerley, D. (1980). *Organization, class and control*. Taylor & Francis.

Cohen, M., March, J., & Olsen, J. (1972, March). A Garbage Can Model of Organizational Choice. *Administrative Science Quarterly*, *17*(1), 1–25. doi:10.2307/2392088

Collins, C., & Smith, K. (2006). Knowledge exchange and combination: The role of human resource practices in the performance of high-technology firms. *Academy of Management Journal*, *49*(3), 544–560. doi:10.5465/AMJ.2006.21794671

Connelly, C., Zweig, D., Webster, J., & Trougakos, J. (2012). Knowledge Hiding in Organisations. *Journal of Organizational Behavior*, *33*(1), 64–88. doi:10.1002/job.737

Constitutional Court Ruling 89 Hunma 31, Third Public Report (cited as Yang vs. MOF Ruling).

Constitutional lawsuit concerning an infringement of property rights through execution of government authority: Yang Jung Moe vs. Ministry of Finance. (1993). (in Korean)

Cook, S., & Brown, J. (1999). Bridging Epistemologies: The Generative Dance Between organisational Knowledge and organisational Knowing. *Organisation Science*, *10*(4), 381–400.

Coolsaet, R. (2008). *Jihadi Terrorism and the Radicalization Challenge in Europe*. Aldershot: Ashgate Publishing.

Coombs, R., Knights, D., & Willmott, H. C. (1992). Culture, Control and Competition; Towards a Conceptual Framework for the study of Information Technology in Organizations. *Organization Studies*, *13*(1), 51–72. doi:10.1177/017084069201300106

Cooray. (2010). *End User Driven Development of Information Systems –Revisiting Vickers Notion of Appreciation* (PhD Thesis). University of Portsmouth, Portsmouth, UK.

Cox, James, Boyd, Sullivan, Reid, & Cartier. (2012). *The Theory of Constraints International Certification Organization Dictionary* (2nd ed.). Retrieved from http://www.tocico.org/?page=dictionary 2

Cox, J. F., Blackstone, J. H., & Schleier, J. G. (2003). *Managing operations: A focus on excellence* (Vol. 1). Great Barrington, MA: North River Press.

Cox, J. F., & Schleier, J. G. (Eds.). (2010). *Theory of Constraints handbook*. New York: McGraw-Hill.

Crabtree, B. F., & Miller, W. L. (1999). Doing Qualitative Research. *Sage (Atlanta, Ga.)*, 1999.

Crane, L. (2013). A New Taxonomy Of Knowledge Management Theory: The Turn To Knowledge As Constituted In Social Action. *Journal of Knowledge Management Practice*, *14*(1). Retrieved from http://www.tlainc.com/articl332.htm

Cunha, M. P., Clegg, S. R., & Kamoche, K. (2012). Improvisation as "real time foresight". *Futures*, *44*(3), 265–272. doi:10.1016/j.futures.2011.10.009

Cunha, M. P., Viera Da Cunha, J., & Kamoche, K. (1999). Organizational improvisation: What, when, how and why. *International Journal of Management Reviews*, *1*(3), 299–341. doi:10.1111/1468-2370.00017

D'Endrody, A. L., & Col, Lt. (1959). Give your horse a chance: the classic work on the training of horse and rider for all disciplines. London: J.A. Allen.

Dahl, R. A., & Tufte, E. R. (1973). *Size and Democracy*. Palo Alto, CA: Stanford University Press.

Dalton, R. J. (2008). The Quantity and Quality of Party Systems. *Comparative Political Studies, 41*(7). Retrieved from http://www.socsci.uci.edu/~rdalton/archive/CPS(2008.pdf

Davenport, T. H. (1993). Reengineering the corporation. *MIT Sloan Management Review*, *35*(1), 103.

Davis, P.K., & Cragin, K. (2009). *Social Science for Counter-terrorism: putting the pieces together*. RAND MG849.

Dawes, S., Cresswell, A., & Pardo, T. (2009). From 'Need to Know' to 'Need to Share': Tangled Problems, Information Boundaries, and the Building of Public Sector Knowledge Networks. *Public Administration Review*, *69*, 392–402. doi:10.1111/j.1540-6210.2009.01987_2.x

de Bragança, D. (1975). *L'Equitation de Tradition Francaise*. Paris: Belin.

De Kunffy, C. (1993). *The Ethics and Passions of Dressage*. Middletown, MD: Half-Halt Press.

de la Guérinière, F. R. (1733). School of Horsemanship (T. Boucher, Trans.). London: J.A. Allen.

De Long, D., & Fahey, L. (2000). Diagnosing cultural barriers to knowledge management. *The Academy of Management Executive*, *14*(4), 113–127.

de Pluvinel, A. (1626). The Maneige Royal. London: J.A. Allen.

De Witte, H. (1999). Job insecurity and psychological well-being: Review of the literature and exploration of some unresolved issues. *European Journal of Work and Organizational Psychology*, *8*(2), 155–177. doi:10.1080/135943299398302

Deem, R. (2004). The knowledge worker, the manager- academic and the contemporary UK university:new and old forms of public management. *Financial Accountability & Management*, *20*(2), 107–128. doi:10.1111/j.1468-0408.2004.00189.x

Deem, R., & Brehony, K. J. (2005). Management as ideology: The case of 'new managerialism' in higher eduction. *Oxford Review of Education*, *31*(2), 217–235. doi:10.1080/03054980500117827

Deetz, S. A. (1982). Critical interpretive research in organizational communication. *Western Journal of Communication*, *46*(2), 131–149. doi:10.1080/10570318209374073

Dejours, C. (1980). *Work: Mental wear test psychopathology of work.* Centurion.

Deleuze, G., & Guattari, F. (1987). *A Thousand Plateaus: Capitalism & Schizophrenia*. Minneapolis, MN: University of Minnesota Press.

Derks, D., Duin, D., Tims, M., & Bakker, A. B. (2015). Smartphone use and work–home interference: The moderating role of social norms and employee work engagement. *Journal of Occupational and Organizational Psychology*, *88*(1), 155–177. doi:10.1111/joop.12083

Desanctis, G., & And Poole, M. S. (1994). Capturing the Complexity in Advanced Technology Use: Adaptive Structuration Theory. *Organization Science*, *5*(2), 121–147. doi:10.1287/orsc.5.2.121

Dettmer, H. W. (2011). *Our goal is... What is our goal?* Retrieved 21 September 2014 from http://www. goalsys.com/books/documents/WhatisOurGoal- v5_000.pdf

Dettmer, H. W. (2007). *The logical thinking process: a systems approach to complex problem solving*. Milwaukee, WI: Quality Press.

Dewing, C., & Jones, T. (2016). *Future Agenda – Six Challenges for the Next Decade*. London: Third Millennium Publishing.

Dixon, N. (2010). *Common Knowledge: How Companies Thrive by Sharing What They Know*. Cambridge, MA: Harvard Business School Press.

Dobbs, R., Manyika, J., & Woeetzel, J. (2015). *No Ordinary Disruption: The Four Global Forces Breaking All the Trends*. PublicAffairs.

Dong, L., Neufeld, D., & Higgins, C. (2009). Top management support of enterprise systems implementations. *Journal of Information Technology*, *24*(1), 55–80. doi:10.1057/jit.2008.21

Doolin, B. (1998). Information technology as disciplinary technology: Being critical in interpretive research on Information Systems. *Journal of Information Technology*, *13*(4), 301–311. doi:10.1057/jit.1998.8

Downs, A. (1957). *An Economic Theory of Democracy*. New York: Harper & Brothers.

Du Gay, P. (1996). Consumption and identity at work. *Sage (Atlanta, Ga.)*.

Duarte, King of Portugal (1438) The Royal Book of Horsemanship, Jousting and Knightly Combat (Livro Da Ensinanca De Bem Cavalgar Toda sela). (2005). (A. F. Preto & L. Preto, Trans.). The Chivalry Bookshelf.

Dupuy, F. (2004). *Sociologie du changement: pourquoi et comment changer les organisations*. Dunod.

Durkheim, E. (1893). The division of Labour in Society. New York: Free Press.

Durkheim, E. (1900). *Sociology and its Scientific Domain*. Moscow: INFRA-M.

Dyer, J., & Nobeoka, K. (2002). Creating and managing a high-performance knowledge-sharing network: The Toyota case. *Strategic Management Journal*, *21*(3), 345–367. doi:

Easterby-Smith, M., & Lyles, M. (Eds.). (2011). *Handbook of organisational Learning and Knowledge Management* (2nd ed.). Chichester, UK: John Wiley & Sons.

Easton, D. (1953). *The Political System*. Knopf.

Easton, D. (1965a). *A Systems Analysis of Political Life*. New York: John Wiley and Sons.

Easton, D. (1965b). *A Framework for Political Analysis*. Englewood Cliffs, NJ: Prentice-Hall, Inc.

Egenvall, A., Rhodin, M., Roepstorff, L., Eisersiö, M., & Clayton, H. M. (2016). Rein tension peaks within canter (poster 39). *Proceedings 12th International Equitation Science Conference: Understanding horses to improve training and performance*. Retrieved on November 23, 2017 from: http://equitationscience.com/previous-conferences/2016-12th-international-conference

Eisenhardt, K. M., & Martin, J. A. (2000). Dynamic capabilities: What are they? *Strategic Management Journal*, *21*(10-11), 1105–1121. doi:

Elbanna, A. R. (2006). The validity of the improvisation argument in the implementation of rigid technology. *Journal of Information Technology*, *21*(3), 165–175. doi:10.1057/palgrave.jit.2000069

Elbanna, A. R. (2007). Implementing an integrated system in a socially dis-integrated enterprise: A critical view of ERP enabled integration. *Information Technology & People*, *20*(2), 121–139. doi:10.1108/09593840710758040

Elias, A. A., Cavana, R. Y., & Jackson, L. S. (2002). Stakeholder analysis for R&D project management. *R & D Management*, *32*(4), 301–310. doi:10.1111/1467-9310.00262

Elliott, R. (1938). The Wave Principle. Gainesville, GA: New Classics Library.

Epona, T. V. (2017). Retrieved on November 23, 2017 from: https://epona.tv/farewell-falsterbo

Epstein, D. L., Bates, R., Goldstone, J., Kristensen, I., & O'Halloran, S. (2006). Democratic Transitions. *American Journal of Political Science*, *50*(3), 551–569. doi:10.1111/j.1540-5907.2006.00201.x

Équitation - Le hibou vert. (2012). *Nuno Oliveira resumen.mpg.mp4*. Retrieved on November 23, 2017 from: https://www.youtube.com/watch?v=o53YxuOCHuY

Espejo, R. (1989a). PM Manufacturers: The VSM as a diagnostic tool. In R. Espejo & R. Harnden (Eds.), The Viable System Model: Interpretations and Applications of Stafford Beer's VSM. John Wiley.

Espejo, R. (2003). *The Viable System Model; A Briefing About Organisational Structure*. Syncho Ltd. Retrieved from www.syncho.com

Espejo, R., Bowling, D., & Hoverstadt, P. (1999). The Viable System Model and the Viplan Software. *Kybernetes, 28*(6-7), 661-678.

Espejo, R. (1989b). Cybernetic Method to Study Organisations. In R. Espejo & R. Harnden (Eds.), *The Viable System Model: Interpretations and Applications of Stafford Beer's VSM*. Chichester, UK: John Wiley.

Espejo, R. (1993). Managing Complexity in Problem Solving. In R. Espejo & M. Schwaninger (Eds.), *Organizational Fitness, Corporate Effectiveness Through Management Cybernetics*. Frankfurt: Campus Verlag.

Espejo, R. (2008a). Observing organisations: The use of identity and structural archetypes, *Int. J. Applied Systemic Studies*, *2*(1/2), 6–24. doi:10.1504/IJASS.2008.022791

Espejo, R. (2015a, January-June). An Enterprise Complexity Model: Variety Engineering and Dynamic Capabilities. *International Journal of Systems and Society*, *2*(1), 1–22. doi:10.4018/ijss.2015010101

Espejo, R. (2015b). Performance for viability: Complexity and variety management. *Kybernetes*, *44*(6/7), 1020–1029. doi:10.1108/K-02-2015-0044

Espejo, R., & Reyes, A. (2011). *Organizational Systems: Managing Complexity with the Viable System Model*. Heidelberg, Germany: Springer. doi:10.1007/978-3-642-19109-1

Evenstad, S. B. N. (2016, December 12). *Pour une approche communicationnelle du burnout les apports de l'Analyse Phénoménologique Interprétative. Le rôle des communications stressantes dans l'épuisement professionnel; une analyse comparée France -Norvège de groupes d'informaticiens*. Retrieved January 20, 2017, from http://www.theses.fr/s170697

Evenstad, S. B. N. (2015). "Be Excellent and Do More with Less": A Paradox behind Job Burnout Threatening Organizational Sustainability. *International Journal of Systems and Society*, *2*(2), 52–67. doi:10.4018/IJSS.2015070104

Feldman, R. A., & Watson, C. M. (n.d.). *Central Europe: From Transition to EU Membership.* Retrieved from http://www.imf.org/external/pubs/ft/fandd/2000/09/feldman.htm

Ferneley, E., & And Bell, F. (2006). Using bricolage to integrate business and information technology innovation in SMEs. *Technovation*, *26*(2), 232–241. doi:10.1016/j.technovation.2005.03.005

Fernie, S., Green, S. D., & Weller, S. J. (2003). Dilettantes, discipline and discourse: Requirements management for construction. *Engineering, Construction, and Architectural Management*, *10*(5), 354–367. doi:10.1108/09699980310502964

Fillis, J. (1911). Breaking and Riding: With Military Commentaries. Lincoln, NE: University of Nebraska Press.

Finchelstein, G. (2011). *La dictature de l'urgence*. Fayard.

Flood, R., & Jackson, M. (1991). *Creative problem solving: Total systems intervention*. London: Wiley.

Foote, J. L., Gregor, J. E., Hepi, M. C., Baker, V. E., Houston, D. J., & Midgley, G. (2007). Systemic problem structuring applied to community involvement in water conservation. *The Journal of the Operational Research Society*, *58*(5), 645–654. doi:10.1057/palgrave.jors.2602248

Forest, J. J. (2007). *Countering Terrorism and Insurgency in the 21st Century: International Perspectives*. Westport, CT: Praeger Security International.

Foucault, M. (1975). *Discipline and Punish: The Birth of the Prison* (A. Sheridan, Trans.). Harmondsworth, UK: Penguin Books.

Foucault, M. (1977). *Discipline and Punishment: The birth of the prison*. London: Penguin.

Foucault, M. (1988). Technologies of the Self. In *Hutton* (pp. 16–49). Amherst, MA: The University of Massachusetts Press.

Fowler, A., & And Gilfillan, M. (2003). A framework for stakeholder integration in higher education information systems projects. *Technology Analysis and Strategic Management*, *15*(4), 467–489. doi:10.1080/095373203000136051

Freeman, R. E. (1984). *Strategic management: a stakeholder approach*. Boston, MA: Pitman.

Frost, B., & Mo, J. P. T. (2014). *System Hazard Analysis of a Complex Socio-Technical System: The Functional Resonance Analysis Method in Hazard Identification. Australian System Safety Conference*, Melbourne, Australia.

Frykt Kultur Google Search. (2018). Retrieved from https://www.google.no/search?q=fryktkultur&ie=utf-8&oe=utf-8&client=firefox-b-ab&gfe_rd=cr&ei=4qknWKTSGMmq8wfi9bXwAQ

Futures Studies. (2017). Retrieved from https://en.wikipedia.org/wiki/Futures_studies

Gadamer, H. G. (2004). *Truth and Method* (2nd ed.). New York: Continuum.

Galam, S. (2002). The September 11 Attack: A Percolation of Individual Passive Support. *The European Physical Journal B*, *26*(3), 269–272. doi:10.1140/epjb/e20020089

Galam, S. (2003). Global physics: From percolation to terrorism, guerilla warfare and clandestine activities. *Physica A*, *330*(1-2), 139–149. doi:10.1016/j.physa.2003.08.035

Galam, S., & Mauger, A. (2003). On reducing terrorism power: A hint from physics. *Physica A*, *323*, 695–704. doi:10.1016/S0378-4371(03)00006-2

Garvin, D. A., & Roberto, M. A. (2001). What you don't know about making decisions. *Harvard Business Review*, (Sept): 108–116. PMID:11550627

German National Equestrian Federation. (1997). *The Principles of Riding: the official instruction handbook of the German National Equestrian Federation – Book I* (C. Belton, Trans.). Buckingham, UK: Kenilworth Press.

Giddens, A. (2001). *Sociology* (4th ed.). Cambridge, UK: Polity.

Gigch, J. P. (1978). *Applied General Systems Theory*. New York: Harper&Row, Publishers.

Gilding, P. (2011). *The Great Disruption, How the Climate Crisis Will Transform the Global Economy*. London: Bloomsbury Publishing.

Ginsberg, B. (2011). *The Fall of the Faculty: The Rise of the all administrative university and why it matters*. Oxford, UK: Oxford University Press.

Gliek, J. (1987). *Chaos: Making a New Science*. National Book Foundation.

Glißmann, C., & König von Borstel, U. (2012). Alternatives to conventional evaluation of rideability: rein tension and behavior from performance test dressage training. 8th International Equitation Science Conference, Edinburgh, UK.

Goldratt, E. M. (1997). *Critical chain*. Great Barrington, MA: The North River Press.

Goldratt, E. M. (2010). Introduction to TOC – my perspective. In J. F. Cox & J. G. Schleier (Eds.), *Theory of Constraints handbook* (pp. 3–9). New York: McGraw-Hill.

Goldratt, E. M., & Cox, J. (1984). *The Goal: A process of ongoing improvement*. Great Barrington, MA: The North River Press.

Gore, A. (2013). *The Future*. New York: Random House.

Gorton, G., Kahl, M., & Rosen, R. (2005). *Eat or Be Eaten: A Theory of Mergers and Waves*. Cambridge, MA: National Bureau of Economic Research. doi:10.3386/w11364

Goudie, A. W., & Stasavage, D. (2000). A Framework for the Analysis of Corruption. In Explaining Corruption. Elgar Reference Collection.

Government of Korea. (n.d.). *Measurement Results of the Integrity Index for Public Institutions*. Retrieved from http://kostat.go.kr/portal/korea/kor_ip/6/1/index.board?bmode=read&aSeq=309650&pageNo=9&rowNum=10&amSeq=&sTarget=&sTxt= (in Korean)

Government of the Republic of Korea, Ministry of Finance (MOF). (1988). *Reorganization of infeasible firms (a compilation of documents from the MOF, Bank of Korea (BOK), Bank Supervisory Board, and main banks)*. Government of the Republic of Korea, Ministry of Finance (MOF).

Green, F. (2006). *Demanding work: The paradox of job quality in the affluent economy*. Princeton University Press.

Greenwood, J. (1997). *The third industrial revolution: technology, productivity, and income inequality*. American Enterprise Institute.

Grisone, F. (1558). *Ordini di Cavalcare*. Retrieved on November 23, 2017 from: http://www2.vetagro-sup.fr/bib/fondsancien/ouvonline/1355-2/1355-2.htm

Grove, N. (1974, June). Oil, the Dwindling Treasure. *National Geographic*.

GTI. (2017). *Global Terrorism Index*. Retrieved from http://economicsandpeace.org/wp-content/uploads/2016/11/Global-Terrorism-Index-2016.2.pdf

Gunther, Diamandouros, & Puhle. (1995). The Politics of Democratic Consolidation. Baltimore, MD: Johns Hopkins University Press.

Habermas, J. (1984). *The theory of communicative action; Volume one: Reason and the rationalization of society*. Boston, MA: Beacon Press.

Haggard, S., & Kaufman, R. (1995). *The Political Economy of Democratic Transitions*. Princeton, NJ: Princeton University Press.

Hajkowicz, S. (2015). *Global Megatrends, Seven Patterns of Change Shaping Our Future*. Clayton South, Australia: CSIRO Publishing.

Haken, H. (2000). Information and Self-Organization. Macroscopic Approach to complex System. Springer-Verlag.

Hall, A. D. (1962). *A methodology for systems engineering*. New York: Van Nostrand Reinhold.

Hallsten. (2005). Performance-based self-esteem: A driving force in burnout processes and its assessment. *Arbete och hälsa vetenskaplig skriftserie*.

Hammer, M., & Champy, J. (1993). *Reengineering the corporation. 1993*. New York: HaperCollins.

Hansen, M., More, M., & Lovas, B. (2005). Knowledge sharing in organisations: Multiple networks, multiple phases. *Academy of Management Journal*, *48*(5), 776–793. doi:10.5465/AMJ.2005.18803922

Hardy, C. (1996). *Understanding Power: Bringing about Strategic Change*. Wiley.

Harris, J. (2009). *Tönnies*. Cambridge, UK: Cambridge University Press.

Harrison, N. E. (2006). *Complexity in World Politics*. State University of New York Press Albany.

Hart, P. J. (2014). *Investigating Issues Influencing Knowledge Sharing in a Research Organisation Using AIM* (Thesis). University of Portsmouth, Portsmouth, UK.

Hector, C. (2017). *Guarding the Classical Principles*. Retrieved on November 23, 2017 from: http://www.horsemagazine.com/thm/2017/06/guarding-the-classical-principles/

Heeks, R. (2006). Health information systems: Failure, success and improvisation. *International Journal of Medical Informatics*, *75*(2), 125–137. doi:10.1016/j.ijmedinf.2005.07.024 PMID:16112893

Heidegger, M. (1962). *Being and Time*. Malden, MA: Blackwell Publishing.

Heidenheimer, A. J. (1989). Terms, Concepts, and Definitions: An Introduction. In A. J. Heidenheimer, M. Johnston, & V. T. LeVine (Eds.), *Political Corruption: a Handbook*. New Brunswick, NJ: Transaction.

Heller, Huët, & Vidaillet. (2013). Introduction. Du sens et du rôle de la critique. In *Communication et organisation: Perspectives critiques*. Presses du Septentrion.

Henriquet, M. (1991). Henriquet on Dressage (H. Nelson, Trans.). London: J.A. Allen.

Heo, U., & Stockton, H. (2005). The Impact of Democratic Transition on Elections and Parties in South Korea. *Party Politics*, *11*(6), 675–688. doi:10.1177/1354068805057604

Herbert, L. (Ed.). (1939). *Henry, Elizabeth and George: Letters and Diaries of Henry, Tenth Earl of Pembroke and his Circle* (Vol. 1). London: Jonathan Cape.

Hernes, T. (2010). Actor-network theory, Callon's Scallops, and process-based organization studies. In T. Hernes & S. Maitlis (Eds.), *Process, sensemaking and organizing*. Oxford, UK: Oxford University press. doi:10.1093/acprof:oso/9780199594566.003.0009

Hetland, H., Sandal, G. M., & Johnsen, T. B. (2007). Burnout in the information technology sector: Does leadership matter? *European Journal of Work and Organizational Psychology*, *16*(1), 58–75. doi:10.1080/13594320601084558

Heuschmann, G. (2006). *Tug of War: Classical versus ""Modern"" Dressage - Why Classical Training Works and How Incorrect "Modern" Riding Negatively Affects Horses' Health*. Trafalga Square, VT: Reina Abelshauser.

Heuschmann, G. (2012). *Balancing Act: The Horse in Sport – an Irreconcilable Conflict?* London: J.A. Allan.

High Court of the Government of Korea. (1999). *Chun Doo Hwan trial documents*. High Court of the Government of Korea.

Hirschheim, R., & And Newman, M. (1991). Symbolism and information systems development: Myth, metaphor and magic. *Information Systems Research*, *2*(1), 29–62. doi:10.1287/isre.2.1.29

Hodgson, A. (2011). *Ready for Anything: Designing Resilience for a Transforming World.* Triarchy Press.

Hodgson, A. (2016). *International Futures Forum.* Retrieved from http://www.internationalfuturesforum.com/

Hohmann, C. (2017). *What is a goal tree?* Retrieved 30 August 2017 from https://hohmannchris.wordpress.com/goal-tree/

Holland, J. H. (1995). *Hidden Order: How Adaptation Builds Complexity.* Reading, MA: Perseus Books.

Hollnagel, E., Hounsgaard, J., & Colligan, L. (2014). *FRAM – the Functional Resonance Analysis Method: a handbook for the practical use of the method.* Centre for Quality, in the Southern Region of Denmark. Retrieved from www.centerforkvalitet.dk/framhandbook

Hollnagel, E. (2012). *FRAM: The functional resonance analysis method for modelling complex socio-technical systems.* Farnham, UK: Ashgate.

Holste, S., & Fields, D. (2010). Trust and tacit knowledge sharing and use. *Journal of Knowledge Management, 14*(1), 128–140. doi:10.1108/13673271011015615

Holzer, M., & Kim, B. J. (2002). *Building Good Governance: Reforms in Seoul.* Seoul: Seoul Development Institute.

Homer-Dixon, T. (2001). *The Ingenuity Gap.* London: Vintage.

Honneth, A. (1996). *The Struggle for Recognition: The Moral Grammar of Social Conflicts.* Cambridge, MA: MIT Press.

Hornborg, A. (2014). Does the Anthropocene Really Imply the End of Culture/Nature and Subject/Object Distinctions? Conference The Thousand Names of Gaia, From the Anthropocene to the Age of the Earth, Rio de Janerio, Brazil.

Horn, R., & Weber, R. (2007). *New Tools For Resolving Wicked Problems: Mess Mapping and Resolution Mapping Processes.* Strategy Kinetics.

Horseherovideo. (2010). *Advanced work in hand with Francisco de Bragança.* Retrieved on November 23, 2017 from: https://www.youtube.com/watch?v=2DnQ1s-Fj6E

HorseOsteopath. (2010). *Pedro Torres and Oxidado Style Section Working Equitation European Championships 2009 - Medium.m4v.* Retrieved on November 23, 2017 from: http://www.youtube.com/watch?v=DzvopwrzjIo

Hoverstad, P. (2008). *The Fractal Organization: Creating sustainable organizations with the Viable System Model.* Chichester, UK: Wiley.

Howard, A. (1995). A framework for work change. *The Changing Nature of Work, 3,* 44.

Hsu, C., Hsu, C., Ju, T., Yen, C., & Chang, C. (2007). Knowledge sharing behavior in virtual communities: The relationship between trust, self-efficacy, and outcome expectations. *International Journal of Human-Computer Studies*, *65*(2), 153–169. doi:10.1016/j.ijhcs.2006.09.003

Huarng, A. (2001). Burnout Syndrome among Information System Professionals. *Information Systems Management*, *18*(2), 1–6. doi:10.1201/1078/43195.18.2.20010301/31272.3

Huberman, A. M., & Miles, M. B. (1998). Data management and analysis methods. In N. K. Denzin & Y. S. Lincoln (Eds.), *Collecting and interpreting qualitative materials*. Thousand Oaks, CA: Sage.

Huntington, S. (1968). *Political Order in Changing Societies*. New Haven, CT: Yale University Press.

Husky005reloaded. (2010). *Cadre Noir de Saumur @ Paris – Bercy*. Retrieved on November 23, 2017 from: http://www.youtube.com/watch?v=1WlcXhTjqi0

Husserl, E. (2002). Pure Phenomenology, its Method and its Field of Investigation. In D. Moran & T. Mooney (Eds.), *The Phenomenology Reader* (pp. 124–133). London: Routledge.

Huysman, M., & Wulf, V. (2006). IT to support knowledge sharing in communities, towards a social capital analysis. *Journal of Information Technology*, *21*(1), 40–51. doi:10.1057/palgrave.jit.2000053

Hyland, A. (1998). *The Warhorse: 1250-1600*. Stroud: Sutton Publishing Ltd.

Igbaria, M., & Greenhaus, J. H. (1992). Determinants of MIS employees' turnover intentions: A structural equation model. *Communications of the ACM*, *35*(2), 34–49. doi:10.1145/129630.129631

Igbaria, M., & Siegel, S. R. (1992). The reasons for turnover of information systems personnel. *Information & Management*, *23*(6), 321–330. doi:10.1016/0378-7206(92)90014-7

IKT Norge. (n.d.). *About ICT Norway*. Retrieved from https://www.ikt-norge.no/english/

Im, H. B. (2004). Faltering Democratic Consolidation in South Korea. *Democratization*, *11*(5), 179–198. doi:10.1080/1351034041233l304642

Inglehart, R. F. (2003). How Solid is Mass Support for Democracy — And How Can We Measure It? *PS, Political Science & Politics*, *36*(1), 51–57. doi:10.1017/S1049096503001689

Insights, H. (2018). *Compare Countries*. Retrieved from https://geert-hofstede.com/countries.html

International Society for Equitation Science. (2018). Retrieved on November 23, 2017 from: (http://equitationscience.com/

Investigation of Chun's use of slush funds. (1996b). *Joongang Ilbo*. (in Korean)

Ipe, M. (2003). Knowledge Sharing in organisations: A Conceptual Framework. *Human Resource Development Review*, *2*(4), 337–359. doi:10.1177/1534484303257985

Jackson, M. (1982). The Nature of Soft Systems Thinking: The Work of Churchman, Ackoff and Checkland. *Journal of Applied Systems Analysis*, *9*, 17–29.

Jackson, M. (2000). *Systems approaches to management*. Dordrecht: Kluwer Academic.

Jang, J. C. (2001, February 16). Seoul and the Local Governments Establish the Roots of an Integrity Contract System. *Sehgeh Ilbo*. (in Korean)

Jary, D., & Jary, J. (1995). *Sociology, Collins Dictionary*. Glasgow, UK: Harper Collins.

Jenkins, B. M. (2011). *Stray Dogs and Virtual Armies Radicalization and Recruitment to Jihadist Terrorism in the United States Since 9/11*. RAND Occasional Paper 343.

Jervis, R. (1997). *System Effects: Complexity in Political and Social Life*. Princeton, NJ: Princeton University Press.

Jian, G. (2008). Identity and technology: Organizational control of knowledge-intensive work. *Qualitative Research Reports in Communication*, *9*(1), 62–71. doi:10.1080/17459430802400365

Johnson, S. (2012). *Future perfect*. Allen Lane.

Johnston, M. (2005). *Syndromes of Corruption: Wealth, Power, and Democracy*. Cambridge, UK: Cambridge University Press. doi:10.1017/CBO9780511490965

Jump, P. (2015). Academics in the Minority at more than two thirds of UK Universities. *Times Higher Education*. Retrieved on July 11, 2017 from https://www.timeshighereducation.com/news/academics-minority-more-two-thirds-uk-universities#survey-answer

Kakabadse, N., Kakabadse, A., & Kouzmin, A. (2003). Reviewing the knowledge management literature: Towards a taxonomy. *Journal of Knowledge Management*, *7*(4), 75–91. doi:10.1108/13673270310492967

Kamoche, K., & And Cunha, M. P. (2001). Minimal Structures: From Jazz Improvisation to Product Innovation. *Organization Studies*, *22*(5), 733–764. doi:10.1177/0170840601225001

Kamp, A., Lambrecht Lund, H., & Søndergaard Hvid, H. (2011). Negotiating time, meaning and identity in boundaryless work. *Journal of Workplace Learning*, *23*(4), 229–242. doi:10.1108/13665621111128655

Kang, D. (2002). *Crony Capitalism*. Cambridge, UK: Cambridge University Press. doi:10.1017/CBO9780511606175

Kāpiti Coast District Council (KCDC). (2009). *Kapiti water supply LTCCP review: water supply options for the Waikanae Paraparaumu Raumati area (2 April 2009)*. Retrieved 15 May 2009 from http:// www.kapiticoast.govt.nz/NR/rdonlyres/

Kāpiti Coast District Council (KCDC). (2010). *Residents opinion survey August 2010*. Retrieved from http://www.kapiticoast.govt.nz/globalassets/internal-page-images-and-documents/resident-opinion-surveys/residentopinionsurvey-august-2010.pdf

Kāpiti Coast District Council (KCDC). (2017a). *Residents opinion survey May 2017*. Retrieved from http://www.kapiticoast.govt.nz/globalassets/your-council/planning/documents/residents-opinion-survey---annual-report-13-07-17-v5.pdf

Kāpiti Coast District Council (KCDC). (2017b). *Kāpiti Coast water conservation report 2016 / 17*. Retrieved from http://www.kapiticoast.govt.nz/contentassets/d76c873551904e66bb160442 4127b027/2016_17_waterconservationreport_v1bt.docx_.pdf

Kāpiti Coast District Council (KCDC). (2017c). *Annual Reports (on water)*. Retrieved from http://www.kapiticoast.govt.nz/services/A---Z-Council-Services-and-Facilities/Water/Water-Supply-Annual-Reports/

Kāpiti Coast District Council. (2012). *Annual Plan 2011/12 – Part 1*. Retrieved 15 May 2013 from http:// www.kapiticoast.govt.nz/Documents/Downloads/ annual-plan/annual-plan-1112-part1.pdf

Kāpiti Coast District Council. (2012a). *Water – what will it cost?* Retrieved from http://www. kapiticoast.govt.nz/Docu- ments/NewslettersAndPeriodicals/water-project/ water-update/2012/ water-update-9May2012.pdf

Kāpiti Coast District Council. (2012b). *Water meter survey*. Retrieved 27 May 2013 from http:// www.ka- piticoast.govt.nz/Projects/Water-Meter-Installation/ water-meter-survey/

Kāpiti Coast District Council. (2013). *Our district: Statistics and demographics*. Retrieved January 2013 from http://www.Kāpiticoast.govt.nz/Our-District/The-Kāpiti-Coast/Statistics-and-Demographics/

Kāpiti Coast District Council. (2013a) *Water supply project*. Retrieved 11 May 2013 from Error! Hyperlink reference not valid.

Kāpiti Coast District Council. (2013b). *Water matters – Kāpiti Coast District Sustainable Water Management Strategy, 2003*. Retrieved from http://www.kapiticoast.govt.nz/ Documents/ Downloads/Strategies/Water-Matters- Sustainable-Water-Management-Strategy.pdf

Kāpiti Observer. (2013). The 2013/14 programme: Financial strategy and rates impact. *Kāpiti Coast District Council – Summary – Draft Annual Plan 2013/14*. Retrieved 11 May 2013 from http://kapitiobserver.realviewdigital. com/?iid=75522&startpage=page0000027

Kāpiti Observer. (2013, May 27). *Council backing water meters*. Author.

Karl, P. (2008). *Twisted truths of modern dressage: A search for a classical alternative*. Brunsbek: Cadmos Verlag.

Kaufmann, S. (1995). *At home in the universe: the search for the laws of complexity*. London: Penguin.

Kerbrech, General Faverot de. (1891). *Dressage Méthodique du cheval de selle: d'après les dernieres enseignements de F. Baucher*. LaVergne: Nabu Public Domain Reprints.

Kim, S. M. (1999, August 12). OPEN is a Top 10 Korean Website. *Munhwa Ilbo*. (in Korean)

Kimani, S. (2015). *Exploring Quality of Learning and Teaching Experiences in Higher Education using the Theory of Constraints.* Victoria University of Wellington.

Kim, B. J. (1995). Bringing an end to Faceless Administration through the RNAS [in Korean]. *Jibang Jachee*, *86*, 15–19.

Kimble, C., Grenier, C., & Goglio-Primard, K. (2010). Innovation and knowledge sharing across professional boundaries: Political interplay between boundary objects and brokers. *International Journal of Information Management*, *30*(5), 437–444. doi:10.1016/j.ijinfomgt.2010.02.002

Kim, S. H. (2000). Civic Mobilization for Democratic Reform. In L. Diamond & D. C. Shin (Eds.), *Institutional Reform and Democratic Consolidation in Korea*. Stanford, CA: Hoover Press.

Kim, S., Mabin, V. J., & Davies, J. (2008). The Theory of Constraints Thinking Processes: Retrospect and Prospect. *International Journal of Operations & Production Management*, *28*(2), 155–184. doi:10.1108/01443570810846883

Kindleberger, C.P., & Aliber, R.Z. (2005). *Manias, panics and crashes. A history of financial crisis.* Academic Press.

King, J. (1995). Stress rattles "help!" desks. *Computerworld*, *29*, 1.

King, N. (2004). Using templates in the thematic analysis of text. In C. Cassell & G. Symon (Eds.), *Essential Guide to Qualitative Methods in Organisational Research* (pp. 256–270). London: Sage. doi:10.4135/9781446280119.n21

Kitchin, J. (1923). Cycles and Trends in Economic Factors. *The Review of Economics and Statistics*, *5*(1), 10–16. doi:10.2307/1927031

Kolstad, I., & Wiig, A. (2011). *Does democracy reduce corruption?* CMI Working Paper 4. Available from: http://www.cmi.no/publications/file/4315-does-democracy-reduce-corruption.pdf

Kondratiev, N. (1935). The Major Economic Cycles. *The Review of Economics and Statistics*, *18*, 103–115.

Korea Stock Exchange Data Base. (1996). Seoul: Korean Stock Exchange. (in Korean)

Kottas-Heldenberg, A. (2010). *Kottas on Dressage*. Shrewsbury, UK: Kenilworth Press.

Kucharska, W., & Kowalczyk, R. (2016). Trust, Collaborative Culture and Tacit Knowledge Sharing in Project Management – a Relationship Model. In *Proceedings of the 13th International Conference on Intellectual Capital, Knowledge Management & Organisational Learning: ICICKM 2016* (p. 159-166). Humanistic Management Network, Research Paper Series No. 49/16.

Kunda, G. (1992). *Engineering culture: Control and commitment in a high-tech organization.* Philadelphia: Temple University Press.

Kuznets, S. (1973). *Population, Capital and Growth*. Norton.

Kuznets, S. (1971). *Economic Growth of Nations: Total Output and Production Structure*. Cambridge.

Kwahk, K. Y., & Ahn, H. (2010). Moderating Effects of Localization Differences on ERP use: A Socio-Technical Systems Perspective. *Computers in Human Behavior*, *26*(2), 186–198. doi:10.1016/j.chb.2009.10.006

Kwon, T. H. (1995, June 21). New Breeze Blowing in Elections with Citizen Group Candidates. *Hankyulleh*. (in Korean)

Lamsdorff, J. G. (2001). *Background paper to the 2001 corruptions perceptions index*. Transparency International and Gottingen University.

lanchonman. (2006). *Garrocha*. Retrieved on November 23, 2013 from: http://www.youtube.com/watch?v=H9Y1F_f87F0

Land and Water Forum. (2010). *Report of the land and water forum: a fresh start for freshwater.* Retrieved 8 October, 2010 from http://www.landandwater.org.nz

Land and Water Forum. (2011). *Land and Water forum recommendations – Implementation.* Retrieved 25 January 2013 from http://www.landandwater.org.nz/includes/ download.aspx?ID=118917

Land and Water Forum. (2012). *Third report of the land and Water forum*. Retrieved December 2012 from http://www.landandwater.org.nz/includes/download.aspx?ID=124767

Latour, B., (1986). Visualization and Cognition: Thinking with Eyes and Hands. *Knowledge and Society: Studies in the Sociology of Culture Past and Present, 6,* 1-40.

Latour, B. (1987). *Science in Action*. Cambridge, MA: Harvard University Press.

Latour, B. (1987). *Science in Action: How to Follow Scientists and Engineers Through Society*. Milton Keynes, UK: Open University Press.

Latour, B. (1988). *The Pasteurization of France*. Cambridge, MA: Harvard University Press.

Latour, B. (1993). *We have never been modern*. Hemel Hempstead, UK: Harvester Wheatsheaf.

Latour, B. (1996). On actor-network theory. A few clarifications. *Soziale Welt*, *47*, 369–381.

Latour, B. (1999a). *Pandoras Hope: Essays on the reality of science studies*. Cambridge, MA: Harvard University Press.

Latour, B. (1999b). On Recalling ANT. In J. Law & J. Hassard (Eds.), *Actor Network and After* (pp. 15–25). Oxford, UK: Blackwell and the Sociological Review.

Latour, B. (2005). *Reassembling the Social: An Introduction to Actor network theory*. Oxford, UK: Oxford University Press.

Laurioux, A., & Henry, G. (2009). *The Great European Schools of Classical Dressage: Vienna, Saumur, Jerez, Lisbon.* Brunsbeck: Cadmos.

Law, J. (1991). Power, discretion and strategy. In A Sociology of monsters: Essays on power, technology and domination (pp. 165-191). Routledge.

Law, J. (1986). *Power, Action and Belief.* London: Routledge and Kegan Paul.

Law, J. (1987). Technology and heterogeneous Engineering: The Case of Portuguese Expansion. In W. E. Bijker, T. P. Hughes, & T. J. Pinch (Eds.), *The Social Construction of technological Systems: New Directions in the Sociology and History of Technology* (pp. 111–134). Cambridge, MA: MIT Press.

Law, J. (1992). Notes on the theory of the actor network: Ordering, strategy and heterogeneity. *Systems Practice*, *5*(4), 379–393. doi:10.1007/BF01059830

Law, J. (1999). After ANT: complexity, naming and topology. In J. Law & J. Hassard (Eds.), *Actor Network Theory and After* (pp. 1–14). Oxford, UK: Blackwell.

Lee, Y. S. (1996, July 1). One Year Survey of Citizens and Officials Regarding Local Government. *Hankuk Ilbo*. (in Korean)

Leemans, I., Willemsen, M. E., Douwes, B., Steenbergen, M. Y., van Iwaarden, S., & Rettig, A. (2016). Influence of rider's actiontype profile on rein tension, Poster 41. *Proceedings 12th International Equitation Science Conference: Understanding horses to improve training and performance*. Retrieved on November 23, 2017 from: http://equitationscience.com/previous-conferences/2016-12th-international-conference

Levett, S. (2009). *Tradition and Change at the Spanish Riding School.* Retrieved on November 23, 2017 from: http://www.viennareview.net/news/ideas-and-trends/tradition-and-change-at-the-spanish-riding-school

Levin, K., Cashore, B., Bernstein, S., & Auld, G. (2009). Playing it forward: Path dependency, progressive incrementalism and the "Super Wicked" problem of global climate change. *IOP Conference Series: Earth and Environmental Science, 50*(6).

Lewis, J., & Weigert, A. (1985). Trust as a Social Reality. *Social Forces*, *63*(4), 967–985. doi:10.1093f/63.4.967

Liao, L.-F. (2008). Knowledge-sharing in R&D departments: A social power and social exchange theory perspective. *International Journal of Human Resource Management*, *19*(10), 1881–1895. doi:10.1080/09585190802324072

Li, E. Y., & Shani, A. B. R. (1991). *Stress Dynamics of Information Systems Managers: A Contingency Model*. Polytechnic State University.

Lijphart, A. (1977). *Democracy in Plural Societies: A Comparative Exploration*. New Haven, CT: Yale University Press.

Li, M., & Gao, F. (2003). Why Nonaka highlights tacit knowledge: A critical review. *Journal of Knowledge Management*, *7*(4), 6–14. doi:10.1108/13673270310492903

Lin, H. (2007). Effects of extrinsic and intrinsic motivation on employee knowledge sharing intentions. *Journal of Information Science*, *33*(2), 135–149. doi:10.1177/0165551506068174

Lipset, S. M. (1960). *Political Man: The Social Bases of Politics*. Garden City, NY: Doubleday & Company, Inc.

Lipsey, R. G., Carlaw, K. I., & Bekar, C. T. (2005). *Economic Transformations*. General Purpose Technologies and Long-Term Economic Growth.

Loch, S. (1986). *The Royal Horse of Europe: The Story of the Andalusian and Lusitano*. London: J.A. Allen.

Loch, S. (1988). *The Classical Seat*. London: Unwin Hyman.

Loch, S. (1990). *Dressage: The Art of Classical Riding*. London: The Sportsman's Press.

Loch, S. (1997). *The Classical Rider: Being at one with your horse*. London: J.A. Allen.

Loch, S. (2003). *Invisible riding: The secret of balance for you and your horse*. Surry, UK: Horse & Rider Magazine.

Lockie, S. (2004). Collective Agency, Non-human Causality and Environmental Social Movements: A case study of the Australian Landcare movement. *Journal of Sociology (Melbourne, Vic.)*, *40*(1), 41–58. doi:10.1177/1440783304040452

Lockwood, D. (1964). Social and System Integration. In G. K. Zollschan & W. Hirsch (Eds.), *Exploration in Social Change* (pp. 244–257). London: Routledge and Kegan Paul.

Logout before you burnout. (2018). Retrieved from https://www.reddit.com/r/programming/comments/ua27q/logout_before_you_burnout/

Luhmann, N. (1995). *Social systems*. Stanford, CA: Stanford University Press.

Lyytinen, K., & Newman, M. (2008). Explaining Information Systems Change: A Punctuated Socio-technical Change Model. *European Journal of Information Systems*, *17*(6), 589–613. doi:10.1057/ejis.2008.50

Lyytinen, K., Newman, M., & Al-Muharfi, A. R. A. (2009). Institutionalising Enterprise Resource Planning in the Saudi Steel Industry: A Punctuated Socio-Technical Analysis. *Journal of Information Technology*, *24*(4), 286–304. doi:10.1057/jit.2009.14

Mabin, V. J., & Davies, J. (2010). The TOC thinking processes: Their nature and use – reflections and consolidation. In J. F. Cox & J. G. Schleier (Eds.), *Theory of Constraints handbook* (pp. 631–669). New York: McGraw-Hill.

Mabin, V., Davies, J., & Cox, J. (2006). Using the Theory of Constraints Thinking Processes to Complement Systems Dynamics' Causal Loop Diagrams in Developing Fundamental Solutions. *International Transactions in Operational Research*, *13*(1), 33–57. doi:10.1111/j.1475-3995.2006.00532.x

Mainwaring, S., & Torcal, M. (2005). Party System Institutionalization and Party System Theory After the Third Wave of Democratization. *Working Paper #319.*

Mainwaring, S., & Scully, T. (Eds.). (1995). *Building Democratic Institutions: Party Systems in Latin America*. Stanford, CA: Stanford University Press.

Mairinger, F. (1983). *Horses are made to be horses: A personal philosophy of horsemanship.* Adelaide: Rigby Publishers.

Makinson, P., Hundley, S., Feldhaus, C., & Fernandez, E. (2012). Mobile communications anytime, anywhere: The impact on work-life balance and stress. In *2012 Frontiers in Education Conference Proceedings* (pp. 1–6). IEEE.

Mangham, I. L., & Pye, A. (1991). *The Doing of Managing*. Oxford, UK: Basil Blackwell.

Marabelli, M., & Galliers, R. D. (2016). A Reflection on Information Systems Strategizing: The Role of Power and Everyday Practices. *Information Systems Journal.*

Marabelli, M., & Newell, S. M. (2014). Knowing, Power and Materiality: A Critical Review and Reconceptualization of Absorptive Capacity. *International Journal of Management Reviews*, *16*(4), 479–499. doi:10.1111/ijmr.12031

Marshall, J. (2004, Summer). Living systemic thinking: Exploring quality in first-person action research. *Action Research*, *2*(3), 309–329. doi:10.1177/1476750304045945

Marsick, V. (2009). Toward a unifying framework to support informal learning theory, research and practice. *Journal of Workplace Learning*, *21*(4), 265–275. doi:10.1108/13665620910954184

Martin, B. R. (2016). What's Happening with our Universities? *Prometheus*, *34*(1), 7–24. doi:1 0.1080/08109028.2016.1222123

Maslach, C., & Leiter, M. P. (1997). *The Truth About Burnout: How Organizations Cause Personal Stress and What to Do About It.* San Francisco, CA: Jossey-Bass.

Masys, A. J. (2010). *Fratricide in Air Operations: Opening the black box- revealing the social* (PhD Dissertation). University of Leicester, UK.

Masys, A. J. (2012). 'The emergent nature of risk as a product of 'heterogeneous engineering- A relational analysis of the Oil and Gas Industry safety culture. In S. Bennett (Ed.), *Innovative thinking in risk, crisis and disaster management*. Gower Publishing.

Masys, A. J. (2014). Radicalization and Recruitment: A systems approach to understanding violent extremism. *International Journal of Systems and Society*, *1*(2), 51–65. doi:10.4018/ijss.2014070104

McGee, M. K. (2006). Time for a career change. *Information.*

McGee, M. K., Khirallah, D. R., & Lodge, M. (2000). Backlash. *Information Week*, *805*, 50–56.

McGinn, M. (1997). *Wittgenstein, and the Philosophical Investigations*. London: Routledge.

Mcgrath, K. (2005). Doing Critical Research in Information Systems: A Case of Theory and Practice Not Informing Each Other. *Information Systems Journal*, *15*(2), 85–101. doi:10.1111/j.1365-2575.2005.00187.x

Meadows, D. (1972). *The Limits to Growth; A Report for the Club of Rome's Project on the Predicament of Mankind*. New York: Universe Books.

Mensch, G. (1979). *Stalemate in Technology: Innovations Overcome the Depression*. Ballinger Pub. Co.

Merton, R. K. (1949). *Social Theory and Social Structure*. New York: Free Press.

Midgely, G. (2000). *Systemic intervention: philosophy, methodology and practice*. New York: Kluwer Plenum. doi:10.1007/978-1-4615-4201-8

Midgley, G., & Ochoa-Arias, A. (Eds.). (2004). *Community Operational Research: OR and Systems Thinking for Community Development*. New York: Kluwer; doi:10.1007/978-1-4419-8911-6

Miller, E. F. (1971). *David Easton's political theory*. The Political Science Reviewer.

Miller, W. L., Godeland, A. B., & Koshechnkina, T. (2002). Values and Norms versus Extortion and Temptation. In D. Della Porta & S. R. Ackerman (Eds.), *Corrupt Exchanges: Empirical Themes in the Politics and Political Economy of Corruption*. Baden-Baden: Nomos Verlagsgesellschaft.

Mingers, J. (2006). *Realising Systems Thinking: Knowledge and Action in Management Science*. Canterbury, UK: Springer Publishing.

Mingers, J., & Brocklesby, J. (1997). Multi-methodology: Towards a framework for mixing methodologies. *Omega-International Journal of Management Science*, *25*(5), 489–509. doi:10.1016/S0305-0483(97)00018-2

Mingers, J., & Rosenhead, J. (2004). Problem structuring methods in action. *European Journal of Operational Research*, *152*(3), 530–554. doi:10.1016/S0377-2217(03)00056-0

Ministry for the Environment (MfE). (2003). *Dairying and Clean Streams Accord between Fonterra Co-operative Group, Regional Councils, Ministry for the Environment, and Ministry of Agriculture and Forestry*. Retrieved 28 January 2013 from http://www.mfe.govt.nz/issues/land/rural/dairying- accord-may03.pdf

Ministry for the Environment (MfE). (2011). *National Policy Statement Freshwater Management 2011*. Retrieved October 2012 from http://www.mfe.govt.nz/ publications/rma/nps-freshwater-management-2011/ docs/nps-freshwater-mgnt-2011.pdf

Mintzberg, H. (1982). A note on that dirty word "Efficiency." *Interfaces*, *12*(5), 101–105. doi:10.1287/inte.12.5.101

Mintzberg, H. (1994). The Fall and Rise of Strategic Planning. *Harvard Business Review*, (Jan-Feb): 107–114.

Mirzaei, M. (2016). *Exploring Applicability of TOC to projects: Critical Chain Project Management and Beyond.* Wellington: Victoria University of Wellington.

Mitchell, R. K., Agle, B. R., & Wood, D. J. (1997). Towards a theory of stakeholder identification and salience: Defining the principle of who and what really counts. *Academy of Management Review*, *22*(4), 853–886.

Mo, J., & Moon, C. I. (2003). Business Government Relations under Kim Dae Jung. In S. Haggard, W. Lim, & E. Kim (Eds.), *Economic Crisis and Corporate Restructuring in Korea.* Cambridge, UK: Cambridge University Press; doi:10.1017/CBO9781139085083.008

Monteiro, E. (2000). Actor Network Theory and Information Infrastructure. In C. Ciborra (Ed.), *From control to drift. The dynamics of corporate information infrastructures.* Oxford, UK: Oxford University Press. Retrieved on July 11, 2005 from www.idi.ntnu.no/~ericm/ant.FINAL.htm

Moore, R. M. (2009). *Shaping more sustainable communities: a case study in urban water management* (Master's thesis). Victoria University of Wellington. Retrieved from http://www.j.co.nz/robyn/Robyn_Moore_%20Shap- ing_More_Sustainable_Communities.pdf

Moore, R. M. (2009a). Shaping more sustainable communities: a case study in urban water management. In *Operations Research Society of New Zealand, Proceedings of the 44th Annual Conference ORSNZ 09*. University of Canterbury. Retrieved from http:// www.j.co.nz/robyn/Robyn_Moore_ORSNZ_2009_ conference_paper.pdf

Moore, J. E. (2000). One Road to Turnover: An Examination of Work Exhaustion in Technology Professionals. *Management Information Systems Quarterly*, *24*(1), 141–168. doi:10.2307/3250982

Moore, R. M. (2009b). *Shaping more sustainable communities - Developing sustainable environmental and sound business agendas. Power-Point Presentation. ORSNZ 09 conference, December 3 – 4th 2009.* New Zealand: University of Canterbury. Retrieved from http:// www.j.co.nz/robyn/Robyn_Moore_ORSNZ_2009_ thesis_presentation.pdf

Moorman, C., & Miner, A. S. (1998). The Convergence of Planning and Execution: Improvisation in New Product Development. *Journal of Marketing*, *62*(July), 1–20. doi:10.2307/1251740

Morch, A. I., & Mehandjiev, N. D. (2000). Tailoring as collaboration: The mediating role of multiple representation and application units. *Computer Supported Cooperative Work*, *9*(1), 75.

Morgan, G. (2006). *Images of Organization* (2nd ed.). Beverly Hills, CA: Sage.

Mumford, E. (2003). *Redesigning Human Systems*. Hershey, PA: Information Science Publishing. doi:10.4018/978-1-59140-118-6

Mumford, E. (2006). The Story of Socio-Technical Design: Reflections on its Successes, Failures and Potential. *Information Systems Journal*, *16*(4), 317–342. doi:10.1111/j.1365-2575.2006.00221.x

Mundate, L., & Bennebroek Gravenhorst, K. M. (2003). Power Dynamics and Organisational Change: *An Introduction, Applied Psychology. International Review (Steubenville, Ohio)*, *52*(1), 1–13.

Müseler, W. (1937). *Riding Logic*. London: Eyre Methuen.

Nachira, F., Nicolai, A., Dini, P., Le Louarn, M., & Rivera Leon, L. (2007). Digital Business Ecosystems. *European Commission, Information Society and Media*. Retrieved from www.dogital-ecosystems

National Assembly. (1988). *Investigation into Fifth Republic corruption*. National Assembly.

National Tax Office. (1989). *Confirmation of the record of the tax treatment of political donations*. National Tax Office.

New Zealand Landcare Trust. (2011). *Review of the Starborough Flaxbourne Soil Conservation Group Project.* Retrieved December 2012 from http://www. landcare.org.nz/files/file/566/Starborough%20Flaxbourne%20Review%20Dec2011.pdf

New Zealand Treasury. (2016). *New Zealand Economic and Financial Overview 2016*. Retrieved from http://www.treasury.govt.nz/economy/overview/2016/nzefo-16.pdf

Nicolini, D. (2007). Stretching out and expanding medical practices: The Case of Telemedicine. *Human Relations*, *60*(6), 889–920. doi:10.1177/0018726707080080

Nisbett, R. E., & Ross, L. (1980). *Human Inference: Strategies and Shortcomings of Social Judgment* (p. 18). Englewood Cliffs, NJ: Prentice Hall.

Nonaka, I. (1994). A Dynamic Theory of organisational Knowledge Creation [Electronic version]. *organisation. Science*, *5*(1), 14–37.

Noonan, W. R. (2007). *Discussing the undiscussable: a guide to overcoming defensive routines in the workplace*. San Francisco, CA: Jossey-Bass.

Nordrik, B., & Bjerke, P. (2011). *De-Facto, Oslo*. Retrieved from http://www.de-facto.no/wp-content/uploads/2015/06/193927fleksibelt-for-hvem.pdf

Nsouli, S. M. (1999). *A Decade of Transition An Overview of the Achievements and Challenges*. Retrieved from http://www.imf.org/external/pubs/ft/fandd/1999/06/nsouli.htm

Nye, J. S. (1967). Corruption and Political Development: A Cost-Benefit Analysis. *The American Political Science Review*, *61*(2), 417–427. doi:10.2307/1953254

Oldroyd, J., & Shad, S. (2012). Catching Falling Stars: A Human Resource Response to Social Capital's Detrimental Effect of Information Overload on Star Employees. *Academy of Management Review*, *37*(3), 396–418. doi:10.5465/amr.2010.0403

Oliveira, N. (1965). Reflections on Equestrian Art. London: J.A. Allen.

Oliver, D., & Romm, C. (2009). ERP adoption –What do they say about it? In C. Brooke (Ed.), *Critical Management Perspectives on Information Systems* (pp. 103–131). Oxford, UK: Butterworth-Heinmann.

Olson, M. (1965). *The Logic of Collective Action*. Cambridge, MA: Harvard University Press.

Organization for Economic Co-operation and Development (OECD). (n.d.). OECD Factbook 2015-2017 Economic, Environmental and Social Statistics [Data file]. Retrieved from: http://www.oecd-ilibrary.org/economics/oecd-factbook_18147364

Orlikowski, W. (2002). Knowing in Practice: Enacting a Collective Capability in Distributed Organizing. *Organisation Science*, *13*(3), 249–273.

Orlikowski, W. J. (1996). Improvising Organizational Transformation over Time: A Situated Change Perspective. *Information Systems Research*, *7*(1), 63–92. doi:10.1287/isre.7.1.63

Orlikowski, W. J., & Hofman, J. B. (1997, Winter). An Improvisational Model for Change Management: The Case of Groupware Technology. *Sloan Management Review*, 11–21.

Osborne, J. W. (1994). Some similarities and differences among phenomenological and other methods of psychological qualitative research. *Canadian Psychology*, *35*(2), 167–189. doi:10.1037/0708-5591.35.2.167

Parker, M. (2000). *Organizational Culture and Identity*. London: Sage.

Parliamentary Commissioner for the Environment. (2000). *Aging pipes and murky waters. Urban water system issues for the 21st century*. Wellington: Parliamentary Commissioner for the Environment. June.

Parliamentary Commissioner for the Environment. (2001). *Whose water is it? The sustainability of urban water systems on the Kāpiti Coast*. Wellington: Parliamentary Commissioner for the Environment. May.

Parliamentary Commissioner for the Environment. (2015). *Preparing New Zealand for rising seas: Certainty and Uncertainty*. Retrieved from http://www.pce.parliament.nz/publications/preparing-new-zealand-for-rising-seas-certainty-and-uncertainty

Parliamentary Commissioner for the Environment. (2015a). *Water quality in New Zealand: Land use and nutrient pollution*. Retrieved from http://www.pce.parliament.nz/media/1008/update-report-water-quality-in-new-zealand-web.pdf

Parliamentary Commissioner for the Environment. (2016). *The state of New Zealand's environment: Commentary by the Parliamentary Commissioner for the Environment on Environment Aotearoa 2015*. Retrieved from http://www.pce.parliament.nz/publications/the-state-of-new-zealands-environment-commentary-by-the-parliamentary-commissioner-for-the-environment-on-environment-aotearoa-2015

Parsons, T. (1937). *The Structure of Social Action*. Moscow: Ecsmo.

Patočka, J. (2016). The Natural World as a Philosophical Problem. Northwestern University Press.

Patomaki, H., & Wright, C. (2000). After Post- positivism? The Promises of Critical Realism. *International Studies Quarterly*, *44*(2), 213–237. doi:10.1111/0020-8833.00156

Paugam, S. (2005). *Les formes élémentaires de la pauvreté*. Presses universitaires de France Paris.

Pauleen, D., & Wang, W. (2017). Does big data mean big knowledge? KM perspectives on big data and analytics. *Journal of Knowledge Management, 21*(1), 1–6. doi:10.1108/JKM-08-2016-0339

Pawlowski, S. D., Kaganer, E. A., & Cater, J. J. III. (2007). Focusing the research agenda on burnout in IT: Social representations of burnout in the profession. *European Journal of Information Systems, 16*(5), 612–627. doi:10.1057/palgrave.ejis.3000699

Peet, M. (2012). Leadership transitions, tacit knowledge sharing and organisational generativity. *Journal of Knowledge Management, 16*(1), 45–60. doi:10.1108/13673271211198936

Pembroke, H. (1978). *Earl of (1773). Military equitation or a method of breaking horses and teaching soldiers to ride.* London: J.A. Allen.

Pereira, A. G. A. A. (2013). Introduction to the use of FRAM on the effectiveness assessment of a radiopharmaceutical dispatches process. *2013 International Nuclear Atlantic Conference – INAC.*

Pereira, C. H. (2001). *Étude du premier traité d'équitation portugais: Livro da ensinça de bem cavalgar toda sela du roi Dom Duarte*. Paris: L'Harmattan.

Perret & Girod-Séville. (1999). *Fondements épistémologique de la recherche.* Thiétart.

Pesqueux, Y. (2008). Un modèle organisationnel du changement ? *Communication et organisation, 33.*

Peters, T. J., & Waterman, R. H. (1982). *In Search of Excellence: Lessons from America's Best-Run Companies.* New York: Harper & Row.

Philippe, B. (2004). *Sociologie du changement dans les entreprises et les organisations.* Le Seuil.

Phillips, A. (1996). Why does local democracy matter? In Local Democracy and Local Government. London: Macmillan. doi:10.1007/978-1-349-25022-6_2

Pirsig, R. (2013). *Lila: An Inquiry Into Morals.* New York: Bantam Books.

Podhajsky, A. (1967). The Complete Training of Horse and Rider. London: The Sportsman's Press.

Polanyi, M., & Prosch, H. (1975). *Meaning*. Chicago: University of Chicago Press.

Pollock, N., & And Cornford, J. (2004). ERP systems and the university as a "unique" organisation. *Information Technology & People, 17*(1), 31–52. doi:10.1108/09593840410522161

Pongsart, G. (2015). *Theory of Constraints (TOC) and Appreciative Inquiry (AI): a comparative study of their effectiveness in improving master's thesis students' performance (Doctoral thesis).* Wellington: Victoria University of Wellington. Retrieved from http://researcharchive.vuw.ac.nz/xmlui/bitstream/handle/10063/4743/thesis.pdf?sequence=1

Powell, J. L., & Owen, T. (2011). Actor network theory and social science: possibilities and implications. *Journal of Public Administration and Governance, 1*(2), 140-157. Retrieved from www.macrothink.org/jpag

Prediction. (2017). Retrieved from https://ipfs.io/ipfs/QmXoypizjW3WknFiJnKLwHCnL72vedxjQkDDP1mXWo6uco/wiki/Prediction.html

Prigogine, I., & Nicolis, G. (1989). *Exploring complexity.* Saint-Petersburg: Piter.

Prigogine, I., & Stengers, I. (1994). *Time, Chaos, Quantum. Saint-Petersburg*: Piter.

Przeworski, A., Cheibub, J. A., Alvarez, M. E., & Limongi, F. (2000). *Democracy and Development: Political Institutions and Material Well-being in the World, 1950-1990.* Cambridge, UK: Cambridge University Press. doi:10.1017/CBO9780511804946

Putnam, R. D. (1988). Diplomacy and Domestic Politics: The Logic of Two Level Games. *International Organization, 42*(3), 427–460. doi:10.1017/S0020818300027697

Rabasa, A., Pettyjohn, S.L., Ghez, J.J., & Boucek, C. (2010). *Deradicalizing Islamist Extremists.* RAND_ MG1053.

Racinet, J-C. (2009). *Falling for Fallacies: Misleading Commonplace Notions of Dressage Riding*. Cadmos: Brunsbeck.

Raven, B. H. (1992). A Power/Interaction Model of Interpersonal influence: French and Raven Thirty years Later. *Journal of Social Behavior and Personality, 7*, 217–244.

Rawlins, B. L. (2006). *Prioritizing Stakeholders for Public Relations.* Retrieved from http://www.instituteforpr.org/research_single/prioritizing_stake-holders

Resource Management Act (RMA). (1991). *Resource Management Act 1991.* Public Act 1991, no. 69. Date of assent 22 July 1991.

Richards, D. (2000). *Political Complexity: Nonlinear Models of Complexity.* Ann Arbor, MI: University of Michigan Press. doi:10.3998/mpub.16428

Riege, A. (2005). Three-dozen knowledge-sharing barriers managers must consider. *Journal of Knowledge Management, 9*(3), 18–35. doi:10.1108/13673270510602746

Rifkin, J. (2011). *The third industrial revolution: how lateral power is transforming energy, the economy, and the world.* Macmillan.

Riker, W. H. (1962). *The Theory of Political Coalitions.* New Haven, CT: Yale University Press.

Rittel, H., & Webber, M. (1973). *Dilemmas in a General Theory of Planning. In Policy Sciences* (Vol. 4). Amsterdam: Elsevier Scientific Publishing Company, Inc.

Rock, M. T. (2009). Corruption and democracy. *The Journal of Development Studies, 45*(1), 55–75. doi:10.1080/00220380802468579

Rockström, J., Steffen, W. L., Noone, K., Persson, Å., Chapin, F. S. III, Lambin, E. F., ... Foley, J. (2009). Planetary Boundaries: Exploring the Safe Operating Space for Humanity. *Ecology and Society, 14*(2), 32. doi:10.5751/ES-03180-140232

Roeder, P. G. (1993). *Red Sunset: The Failure of Soviet Politics*. Princeton, NJ: Princeton University Press.

Root, H. L. (1994). *The Fountain of Privilege: Political Foundations of Markets in Old Regime France and England*. Berkeley, CA: University of California Press.

Rosa, H. (2012). *Aliénation et accélération. Vers une théorie critique de la modernité tardive*. La: Découverte.

Ross, M. L. (2001). Does Oil Hinder Democracy? *World Politics*, *53*(April), 325–361. doi:10.1353/wp.2001.0011

Royal, C. (2012). Kaitiakitanga – guardianship and conservation. *Te Ara - the Encyclopedia of New Zealand*. Retrieved 18 January 2013 from http://www.TeAra.govt.nz/en/kaitiakitanga-guardianship-and-conservation/page-4

Rpgdressage. (2012). *Paulo Caetano Bullfight 2012*. Retrieved on November 23, 2017 from: http://www.youtube.com/watch?v=Vz14t8Wuek&playnext=1&list=PL246A51242E32977A&feature=results_main0

Rustow, D. A. (1970). Transitions to democracy: Toward a dynamic model. *Comparative Politics*, *2*(3), 337–363. doi:10.2307/421307

Ryan, R. M., & Deci, E. L. (2000). Self-determination theory and the facilitation of intrinsic motivation, social development, and well-being. *The American Psychologist*, *55*(1), 68–78. doi:10.1037/0003-066X.55.1.68 PMID:11392867

Ryle, G. (1990). *The Concept of Mind*. London: Penguin Classics.

Saarinen, E., & Hämäläinen, R. (2007). *Systems Intelligence: Connecting Engineering Thinking with Human Sensitivity*. Academic Press.

Salamon, L. M., Sokolowski, S. W., & ... (2004). *Global Civil Society: Dimensions of the Nonprofit Sector* (Vol. 2). Bloomfield, CT: Kumarian Press.

Samuelson, P. A. (1954). Diagrammatic Exposition of a Theory of Public Expenditure. *The Review of Economics and Statistics*, *37*(4), 350–356. doi:10.2307/1925849

Sarker, S., & Valacich, J. (2010). An Alternative To Methodological Individualism: A Non-Reductionist Approach To Studying technology Adoption By Group. *Management Information Systems Quarterly*, *34*(4), 779–808. doi:10.2307/25750705

Sayah, S. (2013). Managing work–life boundaries with information and communication technologies: The case of independent contractors. *New Technology, Work and Employment*, *28*(3), 179–196. doi:10.1111/ntwe.12016

Scheinkopf, L. (1999). *Thinking for a change, putting the TOC thinking processes to use*. St Lucie Press/ CRC Press; doi:10.1201/9781420049046

Schmitter, P. C. (2001). Parties Are Not What They Once Were. In L. Diamond & R. Gunther (Eds.), *Political Parties and Democracy*. Baltimore, MD: Johns Hopkins University Press.

Schopf, J. C. (2017). The Governance Effects of Korea's Leading E-Government Websites. *Journal of Cases on Information Technology, 19*(2).

Schopf, J. C. (2001). An Explanation for the End of Political Bank Robbery in the Republic of Korea. *Asian Survey, 41*(5), 693–715. doi:10.1525/as.2001.41.5.693

Schragenheim, E. (1998). *Management dilemmas, the Theory of Constraints approach to problem identification and solutions*. St Lucie Press.

Schragenheim, E. (1999). *Management Dilemmas: The Theory of Constraints approach to problem identification and solutions*. Boca Raton, FL: CRC Press LLC.

Schumpeter, J. (1942). Capitalism. In *Socialism and Democracy*. Moscow: Ecsmo.

Schumpeter, J. A. (1949). *Theory of Economic Development*. Cambridge, MA: Harvard University Press.

Schuurman, E. (1997). Philosophical and ethical problems of technicism and genetic engineering. *Techné: Research in Philosophy and Technology, 3*(1), 27–44. doi:10.5840/techne19973111

Schwab, B (2017). *The Fourth Industrial Revolution*. Penguin.

Schwaber, K., & Beedle, M. (2002). *Agile software development with Scrum* (Vol. 1). Prentice Hall Upper Saddle River.

Seligman, M. E. P. (1972). Learned helplessness. *Annual Review of Medicine, 23*(1), 407–412. doi:10.1146/annurev.me.23.020172.002203 PMID:4566487

Senate Report. (2012). *Zachary Chesser: A Case Study in Online Islamist Radicalization and Its Meaning for the Threat of Homegrown Terrorism*. A Report by Majority and Minority Staff Senate Committee on Homeland Security and Governmental Affairs. Retrieved from http://www.hsgac.senate.gov/imo/media/doc/CHESSER%20FINAL%20REPORT(1).pdf

Senge, P. (1990). *The Fifth Discipline*. Random House.

Senge, P. (1990). *The Fifth Discipline: The Art and Practice of the Learning Organization*. New York: Doubleday Currency.

Senge, P. M. (1994). *The fifth discipline fieldbook: strategies and tools for building a learning organization*. London: Nicholas Brealey Publishing Ltd.

Senge, P. M., Smith, B., Kruschwitz, N., Laur, J., & Schley, S. (2008). *The necessary revolution: how individuals and organizations are working together to create a sustainable world*. London: Nicholas Brealey Publishing.

Seong, K. R. (1999). Civil Society and Democratic Consolidation in South Korea: Great Achievement and Remaining Problems. In *Consolidating Democracy in South Korea*. Retrieved from Seoul City Integrity Index (in Korean): http://data.seoul.go.kr/openinf/linkview.jsp?infId=OA-727

Seunig, W. (1961). The Essence of Horsemanship. London: J.A. Allen.

Shand, M. (2017). *The Concordia Connection – putting the horse first*. Retrieved October 27, 2017 from: https://www.facebook.com/groups/Concordiaequestrians

Sharp, B. (2013). *Three Horizons: The Patterning of Hope*. Triarchy Press.

Sheppard, D. F. (2010). *Reaching sustainable management of freshwater*. NZ RMLA Conference 2010: Sustainable freshwater management: Are we there yet? Christchurch, New Zealand.

Shim, J. H. (1981). An Opposing Voice in South Korea. *Far Eastern Economic Review*, *111*(12).

Shirk, S. L. (1993). *The Political Logic of Economic Reform in China*. Berkeley, CA: University of California Press.

Shleifer, A., & Vishney, R. (1993). Corruption. *The Quarterly Journal of Economics*, *108*(3), 599–617. doi:10.2307/2118402

Shoemaker, T. E., & Reid, A. (2005). Applying the TOC Thinking Process: A case study in the government sector. *Human Systems Management*, *24*, 21–37.

Shropshire, J., & Kadlec, C. (2012). I'm Leaving the IT Field: The Impact of Stress, Job Insecurity, and Burnout on IT Professionals. *International Journal of Information and Communication Technology Research*, *1*.

Silva, L. (2002). Outsourcing as an Improvisation: A Case Study in Latin America. *The Information Society*, *18*(2), 129–138. doi:10.1080/01972240290075057

Silva, L. (2007). Epistemological and theoretical challenges for studying power and politics in information systems. *Information Systems Journal*, *17*(2), 165–183. doi:10.1111/j.1365-2575.2007.00232.x

Silva, L., & Backhouse, J. (2003). The circuits of power framework for the study of power in the institutionalisation of information systems. *Journal of the Association for Information Systems*, *4*(6), 294–336. doi:10.17705/1jais.00041

Simard, A., & Jourdeuil, P. (2012). Knowledge Manageability: A New Paradigm. In Y. Al-Bastaki & A. Shajera (Eds.), *Building a Competitive Public Sector with Knowledge Management Strategy*. Idea Group.

Simon, H. (1957). A Behavioral Model of Rational Choice. In *Models of Man, Social and Rational: Mathematical Essays on Rational Human Behavior in a Social Setting*. New York: Wiley.

Simon, H. A. (1981). *The Sciences of the Artificial* (2nd ed.). Cambridge, MA: The MIT Press.

Singer, J. D. (1971). *A General Systems Taxonomy for Political Science*. New York: General Learning.

Smith, J. A., Flowers, P., & Larkin, M. (2009). *Interpretative Phenomenological Analysis: Theory, Method and Research*. London: Sage Publications Ltd.

Somerville, I. (1997). *Actor network theory: a useful paradigm for the analysis of the UK cable/on-line socio-technical ensemble?* Retrieved from http://hsb/baylor.edu/eamsower/ais.ac.97/papers/somervil.html,

Sorokin, P. (2006). *Social & Cultural dynamics: A Study of Change in Major Systems of Art, Truth, Ethics, law and Social Relationships*. Academic Press.

Spencer, H. (1852). *The Development Hypothesis*. Moscow: Ecsmo.

Spender, J.- C. (2009). Organisational Capital: Concept, Measure or Heuristic? *Organisational Capital: Modelling, Measuring and Contextualising*, 5-23.

Spender, J.-C. (1996). Organisational Knowledge, Learning and Memory: Three Concepts in Search of a Theory. *Journal of Organizational Change Management*, *9*(1), 63–78. doi:10.1108/09534819610156813

Spengler, O. (1918, 1922). The Decline of the West. Academic Press.

Stahl, B. C. (2007). Ethics, Morality and Critique: An Essay on Enid Mumford's Socio-Technical Approach. *Journal of the Association for Information Systems*, *8*(9), 479–490. doi:10.17705/1jais.00138

Stainton, R. J. (2010). Contextualism in epistemology and the context sensitivity of 'knows'. In Knowledge and Skepticism (pp. 113–139). MIT Press. doi:10.7551/mitpress/9780262014083.003.0007

Steffen, W. (2015). Planetary Boundaries: Guiding human development on a changing planet. *Science, 347*(6223). DOI: 10.1126/science.1259855

Steinberg, D. I., & Shin, M. (2006). Tensions in South Korean Political Parties in Transition: From Entourage to Ideology? *Asian Survey*, *46*(4), 517–537. doi:10.1525/as.2006.46.4.517

Steinbrecht, G. (1885). The Gymnasium of the Horse. Franktown, VA: Xenophon Press.

Sterman, J. D. (2000). *Business Dynamics: Systems Thinking and Modeling for a Complex World*. Boston: McGraw-Hill Publishing.

Stevens, M. J. (1994). *A Classical Riding Notebook*. The Kenilworth Press.

Stowell, A. F. (2011). *Organising 'e' Waste* (PhD thesis). University of Lancaster, Lancaster, UK.

Stowell, F. A. (1989). *Change, Organisational Power and the Metaphor Commodity* (PhD thesis). University of Lancaster, Lancaster, UK.

Stowell, F. A. (2012). *Peter Checkland, interview*. Retrieved from https:// www.youtube.com/watch?v=Pq90qS5FvBgSchutz

Stowell, F. A. (2013). Peter Checkland Interview. International Journal of Information Technologies and Systems Approach, 6(2), 53 – 59. doi:10.4018/jitsa.2013070105

Stowell, F.A., & Cooray, S. (2017). *Policy, participation and the challenges of virtual teams- a "Soft" approach? An account of a research project into the use of Soft Action Research*. (submitted)

Stowell, F. (2000). Organisational Power. *Systemist*, *22*(2), 106–127.

Stowell, F. A. (2009). Soft Systems and Research. *Kybernetes*, *38*(6), 883–900. doi:10.1108/03684920910973144

Stowell, F. A. (2012). The Appreciative Inquiry Method – A Suitable Framework for Action research? *Systems Research and Behavioral Science*, *30*(1), 15–30. doi:10.1002res.2117

Stowell, F. A. (2014). Organisational Power and the Metaphor Commodity. *International Journal of Systems and Society*, *1*(1), 12–20. doi:10.4018/ijss.2014010102

Stowell, F. A., & Allen, G. A. (1988). Co-operation, Power and the Impact of Information Systems. *Systems Practice*, *1*(2), 181–192. doi:10.1007/BF01059857

Stowell, F. A., & Stowell, A. F. (2011). *Appreciative Inquiry Method – Day 2 Facilitator's notes – Social Services. University of Northumbria*. Workshops.

Stowell, F. A., & Welch, C. (2012). *A Managers Guide to Systems Practice*. Wiley. doi:10.1002/9781119208327

Sverke, M., & Hellgren, J. (2002). The nature of job insecurity: Understanding employment uncertainty on the brink of a new millennium. *Applied Psychology*, *51*(1), 23–42. doi:10.1111/1464-0597.0077z

Swales, S., Wright, J., & Oxenham, D. (2011). *Systems Skills Development – The Yellow Brick Road Approach*. UKSS.

Swinnen, L. (2017). *Water bans looming for Wellington*. Retrieved from https://www.stuff.co.nz/environment/99308577/water-bans-looming-for-wellington-as-region-on-the-cusp-of-a-shortage

Sysadmin Burnout Stories. (2018). Retrieved from https://www.reddit.com/r/sysadmin/comments/tmc28/sysadmin_burnout_stories/

Tan, A., Ho, K., Kang, K. T., & Yu, T. C. (2000). What If We Don't Party? Political Partisanship in Taiwan and Korea in the 1990s. *Journal of Asian and African Studies*, *35*(1), 67–84. doi:10.1177/002190960003500105

Tapscott & Tapscott. (2016). *Blockchain Revolution: How the Technology Behind Bitcoin Is Changing Money, Business and the World*. Penguin.

Tapscott, D. (2009). *Grown up digital*. New York: Mc Graw Hill.

Teece, J. D. (2008). *Technological Know-How, Organizational Capabilities, and Strategic Management: Business Strategy and Enterprise Development in Competitive Environments*. London: World Scientific Publishing Co.; doi:10.1142/6147

Thayer, L. (1998). Leadership/ communication: a critical review and a modest proposal. In G. M. Goldhaber & G. A. Barnett (Eds.), *Handbook of Organizational Communication* (pp. 231–263). Norwood, NJ: Alex.

The Spanish Riding School. (n.d.) Retrieved on November 23, 2017 from http://www.srs.at/en_US/start-en/

Tjørnehøj, G., & And Mathiassen, L. (2010). Improvisation during process-technology adoption: A longitudinal study of a software firm. *Journal of Information Technology*, *25*(1), 20–34. doi:10.1057/jit.2009.20

TNO. (2008). *Radicalisation, Recruitment and the EU Counter-radicalisation Strategy*. Retrieved from http://www.transnationalterrorism.eu/tekst/publications/WP4%20Del%207.pdf

Toffler, A. (1971). *Future Shock. Bantam.*

Tollison, R. D. (1982). Rent seeking: A Survey. *Kyklos*, *35*(4), 575–602. doi:10.1111/j.1467-6435.1982.tb00174.x

Townley, B. (1994). *Reframing Human Resource Management: Power, Ethics and the Subject at Work*. London: Sage.

Transparency International Corruption Perceptions Index. (n.d.). Retrieved from http://www.transparency.org/research/cpi/overview

Trutnevyte, E., Stauffacher, M., & Scholz, R. W. (2012). Linking stakeholder visions with resource allocation scenarios and multi-criteria assessment. *European Journal of Operational Research*, *219*(3), 762–772. doi:10.1016/j.ejor.2012.01.009

Truzzi, M. (1971). *Sociology: The Classic Statements*. New York: Oxford University Press.

Tsai, M. T., & Cheng, N. C. (2010). Programmer perceptions of knowledge-sharing behavior under social cognitive theory. *Expert Systems with Applications*, *37*(12), 8479–8485. doi:10.1016/j.eswa.2010.05.029

Tsai, W. (2002). Social structure of 'coopetition' within a multiunit organisation: Coordination, competition, and intraorganisational knowledge sharing. *Organisational Science*, *13*(2), 179–190.

Tsoukas, H. (1996). The Firm as a Distributed Knowledge System: A Constructionist Approach. *Strategic Management Journal*, *17*(S2), 11–25. doi:10.1002mj.4250171104

Tsoukas, H. (2011). *How should we understand tacit knowledge? A phenomenological view. In Handbook of organisational Learning and Knowledge Management* (2nd ed.). Chichester, UK: Wiley.

Udías, A., Insua, D. R., Cano, J., & Fellag, H. (2012). Cost-efficient equitable water distribution in Algeria: A bi-criteria fair division problem with network constraints. *International Transactions in Operational Research, 19*(5), 675–693. doi:10.1111/j.1475-3995.2012.00846.x

Ulrich, W. (1983). *Critical Heuristics of Social Planning*. Bern: P. Haupt.

Ulrich, W. (1988). Systems Thinking, Systems Practice and Practical Philosophy: A programme of research. *Systems Practice, 1*(2), 137–163. doi:10.1007/BF01059855

Ulrich, W. (2002). Chp: *Boundary critique*. In H. G. Daellenbach & R. L. Flood (Eds.), *The Informed Student Guide to Management Science*. London: Thomson.

UN General Assembly. (n.d.). *Press Release United Nations Division for Sustainable Development*. Author.

United Nations Conference of Trade and Development. (n.d.). World Investment Report, 2016 [Data file]. Available from: http://unctadstat.unctad.org/wds/TableViewer/tableView.aspx?ReportId=96740

United Nations General Assembly. (1987). *Report of the World Commission on Environment and Development: Our Common Future*. Transmitted to the General Assembly as an Annex to document – Development and International Co-operation: Environment. Author.

Urry, J. (2002). The Global Complexities of September 11th. *Theory, Culture & Society, 19*(4), 57–69. doi:10.1177/0263276402019004004

Urry, J. (2003). *Global Complexity*. Cambridge, UK: Polity Press.

Urry, J. (2005). The Complexities of the Global. *Theory, Culture & Society, 22*(5), 235–254. doi:10.1177/0263276405057201

US Department of State Report. (2011). *National Counterterrorism Center: Annex of Statistical Information Office of the Coordinator for Counterterrorism*. Office of the Coordinator for Counterterrorism Country Reports on Terrorism 2011 Report July 31, 2012. Retrieved from http://www.state.gov/j/ct/rls/crt/2011/195555.htm

van Dierendonck, M., van Dalum, M., Beekmans, M., & Christensen, J. W. (2012). Acute stress responses of dressage horses ridden in three different head and neck positions. *8th International Equitation Science Conference*. Retrieved on November 23, 2017 from: http://equitationscience.com/previous-conferences/2012-8th-international-conference

Van Maanen, J. (1979). The Fact of Fiction in Organizational Ethnography. *Administrative Science Quarterly, 24*(4), 539–550. doi:10.2307/2392360

Vickers, G. (1965). *The Art of Judgement*. London: Chapman and Hall.

Vickers, G. (1965). *The Art of Judgment: A Study of Policy Making*. London: Chapman and Hall.

Vickers, G. (1972). *Freedom in a rocking boat: changing values in an unstable society*. Harmondsworth, UK: Penguin Books Ltd.

Vickers, G. (1983). *The Art of Judgment, A Study in Policy Making*. London: Harper and Rowe.

Vidino, L., & Brandon, J. (2012). *Countering Radicalization in Europe. Policy report published by the Centre for the Study of Radicalization and Political Violence (ICSR)*. London: Kings College.

Viseu, A. A. B. (2005). *Augmented Bodies: The Visions and realities of wearable computers* (PhD Dissertation). University of Toronto.

Von Bertalanffy, L. (1971). *General Systems Theory*. London: Penguin.

Von Krogh, G., Roos, J., & Slocum, K. (1994). An essay on corporate epistemology. *Strategic Management Journal, 15*(S2), 53–71. doi:10.1002mj.4250151005

von Neindorff, E. (2005). *The Art of Classical Horsemanship*. Brunsbek: Cadmos.

Waddington, D. (2004). Participant Observation. In C. Cassells & G. Symon (Eds.), *Essential guide to qualitative methods in organizational research* (pp. 154–164). London: Sage. doi:10.4135/9781446280119.n13

Wagner, E. L., Newell, S., & Piccoli, G. (2010). Understanding Project Survival in an ESS Environment: A Sociomaterial Practice Perspective. *Journal of the Association for Information Systems, 11*(5), 276–297.

Waltz, K. N. (1959). *Man, the State, and War*. New York: Columbia University Press.

Waltz, K. N. (1979). *Theory of International Politics*. Reading, MA: Addison-Wesley.

Wang, S., & Noe, R. (2010). Knowledge sharing: A review and directions for future research. *Human Resource Management Review, 20*(2), 115–131. doi:10.1016/j.hrmr.2009.10.001

Waring, T., Wainwright, D., & Skoumpopoulou, D. (2016). Enterprise Systems Adoption: A Power and Improvisation Perspective. *International Journal of Systems and Society, 3*(1), 49–66. doi:10.4018/IJSS.2016010104

Warnecke, H.-J. (1993). *Revolution der Unternehmenskultur. Das Fraktale Unternehmen*. Academic Press.

Wastell, D. (2011). *Managers as Designers in the Public Services – Beyond Technomagic*. Devon, UK: Triarchy Press.

Watson, K. J., Blackstone, J. H., & Gardiner, S. C. (2007). The evolution of a management philosophy: The theory of constraints. *Journal of Operations Management, 25*(2), 387–402. doi:10.1016/j.jom.2006.04.004

Watson, T. J. (2011). Ethnography, Reality, and Truth: The Vital Need for Studies of 'How Things Work' in Organization and Management. *Journal of Management Studies, 48*(1), 202–217. doi:10.1111/j.1467-6486.2010.00979.x

Watzlawick, P., Beavin, J. H., Jackson, D. D., & Alto, M. (1967). Pragmatics of Human Communication. A Study of Interactional Patterns, Pathologies, and Paradoxes. WW Norton.

Weber, M. (2001). *The Protestant Ethic and the Spirit of Capitalism*. New York, NY: Routledge.

Weber, M. (2012). *The Protestant Ethic and the Spirit of Capitalism*. Rennaisance Classics.

Weick, K. E. (1993). The collapse of sensemaking in organizations: The Man Gulch disaster. *American Quarterly, 38*, 628–652.

Weick, K. E. (1995). *Sense Making in Organizations*. Sage.

Weick, K. E. (1998). Introductory Essay: Improvisation as a Mindset for Organizational Analysis. *Organization Science, 9*(5), 543–555. doi:10.1287/orsc.9.5.543

Weick, K. E. (1999). Theory Construction as Disciplined Reflexivity: Trade Offs in the 90s. *Academy of Management Review, 24*(4), 797–806.

Weiner, N. (1948). *Cybernetics: or Communication and Control in the Animal and the Machine*. Cambridge, MA: MIT Press.

Welz, Y. (2010). *What has happened to Dressage?* Retrieved on November 23, 2017 from: https://www.thehorseshoof.com/HLC_dressage.html

Werner, R., Blaas-Franken, J., & Harry, M. (2016). The impact of (low) trust on knowledge sharing. *Journal of Knowledge Management, 2*(20), 199–214. doi:10.1108/JKM-10-2015-0391

West, D. (2012). An application of 'systems science' to classical dressage. *8th International Equitation Science Conference*. Retrieved on November 23, 2017 from: http://equitationscience.com/previous-conferences/2012-8th-international-conference

West, D. (2005). Vickers' concept of 'Relationship-maintenance' as an alternative to 'Goal-seeking' models of organization: A difference in the notion of control. *Systemic Practice and Action Research, 18*(3), 261–274. doi:10.100711213-005-4814-7

West, D. (2015). Classical Dressage: A Systemic Analysis. *International Journal of Systems and Society*, *2*(1), 23–43. doi:10.4018/ijss.2015010102

West, D., & de Bragança, F. (2012). A systemic approach to eliciting and gathering the expertise of a 'knowledge guardian': An application of the appreciative inquiry method to the study of classical dressage. *Systemic Practice and Action Research*, *25*(3), 241–260. doi:10.100711213-011-9223-7

West, J., Salter, A., Vanhaverbeke, W., & Chesbrough, H. (2014). Open innovation: The next decade. *Research Policy*, *43*(5), 805–811. doi:10.1016/j.respol.2014.03.001

Whyte, G., & Classen, S. (2012). Using storytelling to elicit tacit knowledge from SMEs. *Journal of Knowledge Management*, *16*(6), 950–962. doi:10.1108/13673271211276218

Widen-Wulff, G., & Ginman, M. (2004). Explaining knowledge sharing in organizations through the dimensions of social capital. *Journal of Information Science*, *30*(5), 448–458. doi:10.1177/0165551504046997

Wilkinson, A. J., & Willmott, H. (1995). *Making quality critical: New perspectives on organizational change*. Routledge.

Willcocks, L. P. (2004). Foucault, Power/Knowledge and Information Systems: Reconstructing the Present. In J. Mingers & L. P. Willcocks (Eds.), *Social Theory and Philosophy for Information Systems* (pp. 238–296). Chichester, UK: John Wiley & Sons, Ltd.

Williams, D. (Ed.). (1997). Environmental and Resource Management Law (2nd ed.). Wellington: Butterworths of New Zealand.

Willig, C. (2001). Interpretative phenomenology. *Introducing Qualitative Research in Psychology: Adventures in Theory and Method*, 50–69.

Wilson, E. O. (2002). *The Future of Life*. New York: Vintage Books.

Wiseman, H. V. (1965). review of David Easton's "A Framework for Political Analysis," and "A Systems Analysis of Political Life" by David Easton. *The Canadian Journal of Economics and Political Science*, *31*, 4.

Wittgenstein, L. (1978). *The Blue and Brown Books*. Oxford, UK: Basil Blackwell.

World Bank Worldwide Governance Indicators. (n.d.). Retrieved from http://info.worldbank.org/governance/wgi/index.aspx

Xenophon. (2006). *The Art of Horsemanship* (M. H. Morgan Reprint, Trans.). New York: Dover Publications.

Yeung, H. W. C. (2002). *Towards a relational economic geography: old wine in new bottles?* Paper presented at the 98th Annual meeting of the Association of American Geographers, Los Angeles, CA. Retrieved from http://courses.nus.edu.sg/course/geoywc/publication/Yeung_AAG.pdf

Young, D. (2001). *Values as Law - the History and Efficacy of the Resource Management Act.* Wellington: Institute of Policy Studies.

Young, D. (2004). *Our islands, our selves – a history of conservation in New Zealand.* Dunedin: Otago University Press.

Yourdon, E. (2003). *Death march.* Pearson Education.

Zack, M. (2002, April). A strategic pretext for knowledge management. *Proceedings of the Third European Conference on organizational Knowledge, Learning and Capabilities.*

Zettle, W. (1998). *Dressage in Harmony: From Basic to Grand Prix.* Boonsboro, MD: Half Halt Press.

Zettle, W. (2007). *The Circle of Trust: Reflections of the Essence of Horses and Horsemanship.* Boonsboro, MD: Half Halt Press.

Zhang, X., Voegl, D., & Zhou, Z. (2012). Effects of information technologies, department characteristics and individual roles on improving knowledge sharing visibility: A qualitative case study. *Behaviour & Information Technology, 31*(11), 1117–1131. doi:10.1080/0144929X.2012.687770

About the Contributors

Frank Stowell is Emeritus Professor of Systems and Information Systems at the University of Portsmouth. He has a PhD in Organisational Change and his research centers around methods of participative design. He has supervised a number of research projects from modeling complex decision-making in mental health care, knowledge management, through to methods for client-led information systems development. He has been co-chair of a number of research council funded projects notably the Systems Practice for Managing Complexity project, designed to help managers address complex issues, which has developed into a self-sustaining network. His latest publication "The Managers Guide to Systems Practice" (2012, Wiley Chichester) is a text written expressly with the kind of managers in mind who have attended the workshops over the past decade. The text deals with Systems ideas and models that are discussed as potential methods of addressing the complex issues encountered in the workplace. He is past President of the UK Academy of Information Systems and the UK Systems Society (http://www.ukss.org.uk/) he presently occupies the chair of the Council of Information Systems Professors and has recently joined the Board of the World Organisation of Systems and Cybernetics; and an elected Distinguished Fellow of the Schumacher Institute. He has published papers and texts in the field and presented papers at a number of international conferences in Europe and the United States. Prior to his academic career he was employed by central government as a consultant within the Management Systems Development Group and has experience of defining and developing IT supported management information systems.

* * *

Raúl Espejo is President of the World Organization of Systems and Cybernetics (WOSC) and Director of Syncho Research, UK. Appointed full professor of Systems and Cybernetics at the University of Lincoln UK in 1995. from 1977 to 1994, he was senior lecturer at the University of Aston UK. Visiting Professor at several universities, including the universities of Amsterdam and Nijmegen, Holland, the

University of Witten-Herdecke, Germany, the University of Santiago, Chile and the University of Los Andes, Colombia. During the mid and late70s he was research scholar at the Manchester Business School, UK and at the International Institute for Applied Systems Analysis, Laxenburg, Austria. The focus of his research has been organisational cybernetics and systems. His most recent book, with Alfonso Reyes, is Organizational Systems: Managing Complexity with the Viable System Model (Springer, 2011). Professor Espejo has worked with a wide range of organisations worldwide, including national and local governments and small and large enterprises, mainly in aspects of organisational transformation and learning, information management, communications, transparency, and democratic processes for decision making. From 1971 to 1973 he was operations director of the CYBERSYN project - the project of the Chilean Government for the management of the industrial economy, under the scientific direction of Professor Stafford Beer.

Solveig Beyza Evenstad, Doctor of Information and Communication Sciences, obtained her PhD in 2016 at the University of Nice Sophia Antipolis, the research laboratory i3M, the department of Information and Communication Sciences. She works as an Assistant Professor at the Inland Norway University of Applied Sciences, Lillehammer, Norway and she is a Research associate at the SIC.Lab Méditerranée, Université Côte D'Azur. Her PhD thesis was on the role of communication in work-related stress and burnout. She obtained her master degree in Organization and Leadership in 2011 with a specialization in Relationship Management: Coaching, Counseling and Motivation. She holds also a BSc in Informatics (1988, Honors, University of Oslo) and BA in Management (1985, High Honors, Bogazici University, Istanbul). She worked with information and communication systems in global companies as a line manager, consultant, project manager, and system developer in 20 years before embarking upon her academic career.

Penny Hart is a senior lecturer in the School of Computing at the University of Portsmouth, United Kingdom. Her research interests are in knowledge management and sharing, soft systems methodologies and information security.

Victoria J. Mabin is Professor of Management at Victoria Business School, specialising in problem solving/decision making using Theory of Constraints, together with operations research/management science methods. She works with organisations and community groups to help them manage their resources to improve performance. Publications include "The World of the Theory of Constraints", the lead chapter on the Thinking Processes in the "Theory of Constraints Handbook", and many journal articles on TOC, MCDA and related areas. Vicky is a Fellow of the Operational Research Society, past president of the Operational Research Society of New Zealand, and currently serves on the board of TOCICO.

Anthony Masys is a Senior Defence Scientist for the Department of National Defence, Centre for Security Science, Defence Research and Development Canada. A former senior Air Force Officer, Dr Masys has a BSc in Physics and MSc in Underwater Acoustics and Oceanography from the Royal Military College of Canada and a PhD from the University of Leicester. He is Editor in Chief for Springer Publishing book series: Advanced Sciences and Technologies for Security Applications and holds various advisory board positions with academic journals and books series. Dr Masys is an internationally recognized author, speaker and facilitator and has held workshops on security, visual thinking, design thinking and systems thinking in Europe, Canada, South America, West Africa and Asia. He has published extensively in the domains of physics and the social sciences.

Robyn M. Moore is an independent researcher and writer/editor at Julian Meadow Associates in New Zealand, where she is managing partner. She is also managing director of Wellington-based children's retailer, Bambini Ltd, a family-owned enterprise. Robyn undertakes research, transforms research into articles/web content and acts as a facilitator for community-based planning and development. Her work typically involves interdisciplinary action research, with a focus on community concerns. Clients in 2017 have included Porirua City Council, Te Mana o Kupe Trust and Victoria University of Wellington.

Ian Roderick is the director of The Schumacher Institute for Sustainable Systems, an independent think tank promoting the ideas of E. F. Schumacher, author of Small is Beautiful. He has an MSc in Operational Research and an MSc in Responsibility & Business Practice. He started his career at the Building Research Establishment before heading up strategic forecasting for Rank Xerox. He co-founded a software development company, Quantec plc, and ever since this was sold he has pursued interests in systems thinking, environment, and social justice. He was President of the UK Systems Society from 2005-2008. He is co-founder of a charity called The Converging World and is on the boards of Low Carbon South West CIC, the UK Systems Society, and Initiative Homes Ltd.

James C. Schopf received a PhD in political science from the University of California, San Diego and a BA from Oberlin College. His research primarily concerns the effects of democratization on corruption, and he specializes in the politics of Korea at Keimyung University in Daegu. He was among the first in 1998 to use the newly passed Korean Freedom of Information Act to gather previously classified material for his dissertation. His articles have appeared in numerous journals including Asian Survey, the Journal of East Asian Studies, the Asian Journal of Social Science, Journal of Comparative Asian Development and Pacific Review.

Olga Tikhomirova, PhD, University Professor and researcher with in-depth knowledge in sustainable development and system thinking. Accomplished researcher in system approach in management. Olga is author of more than 65 scientific articles and 10 books on management and entrepreneurship. The scientific fields of interest are the following: entrepreneurship and innovation, sustainable development and system approach, evolution of social systems. From 2000 to current she is an Associate Professor of Management Department of Production Management and Technology Transfer, Saint-Petersburg National Research University ITMO, Russia. Olga has several state awards in teaching, she is a speaker at some conferences and seminars are dedicated to sustainable development and entreprenuership, works as a business coach. She lectures and motivates student through in-depth lecture and discussion for developing there independent critical thinking and creativity. Olga has developed her own training system based on releasing creative potential and imagination both in business sphere and engineering.

David Wainwright is an Emeritus Professor of Information Systems at the Newcastle Business School, Northumbria University. He is a former President of the UK Association for Information Systems and has over 30 years of experience at both Northumbria and Durham universities researching, teaching and consulting in IS strategy, implementation and adoption. His research projects and major journal publications focus on the development of innovative sociotechnical/material methods using ‘soft operations management’ and process modelling methods aimed at understanding systems complexity, developing new forms of organizational design and enabling digital technology implementation. The impact from this research has resulted in improved facilitation of information systems adoption and implementation both within the public (healthcare information systems in hospitals and primary care) and private (construction management, manufacturing and SMEs) sectors.

Teresa Waring is an Emeritus Professor of Business Systems and Information Management at the Newcastle Business School, Northumbria University. She has extensive experience, researching, teaching and consulting, at both Northumbria and Newcastle university business schools. Her main interests include developing and applying critical approaches to the implementation of information systems and systems integration. Teresa also focuses on the IS adoption challenges and problems faced by large Public Sector organizations and in particular the UK National Health Service. She has published in major academic journals and used action research methods and approaches to conduct qualitative and quantitative studies of Hospital Information Systems strategy which has impacted on the development of new and improved approaches to systems integration.

Daune West is a Senior Lecturer in the Information Systems Engineering Group in the School of Engineering and Computing, at the University of the West of Scotland. Her main teaching and research interests are in the use of systems concepts to support the development of computer-based information systems that 'fit' user requirements and in the use of Action Research to undertake Information Systems research and consultancy. She is a past Chairperson of the United Kingdom Systems Society and a recipient of the Sir Geoffrey Vickers Memorial Award from the International Society of Systems Science. She gained her PhD for research into subjective methods of knowledge elicitation for the development of expert systems. Outside of the University she and her husband run a small livery yard and Equestrian Education Centre.

Index

K

M

O

P

R

S

T

U

V

www.ingramcontent.com/pod-product-compliance
Lightning Source LLC
LaVergne TN
LVHW080608300926
849162LV00005B/1085